# SPATIAL AND TEMPORAL ANALYSIS IN ECOLOGY

STATISTICAL ECOLOGY
Volume 8

a publication from the
satellite program in statistical ecology
international statistical ecology program

# *Statistical Ecology Series*

**General Editor: G. P. Patil**

Volume 1* SPATIAL PATTERNS AND STATISTICAL DISTRIBUTIONS
edited by G. P. Patil, E. C. Pielou, and W. E. Waters (1971)

Volume 2* SAMPLING AND MODELING BIOLOGICAL POPULATIONS AND POPULATION DYNAMICS
edited by G. P. Patil, E. C. Pielou, and W. E. Waters (1971)

Volume 3* MANY SPECIES POPULATIONS, ECOSYSTEMS, AND SYSTEMS ANALYSIS
edited by G. P. Patil, E. C. Pielou, and W. E. Waters (1971)

Volume 4 STATISTICAL DISTRIBUTIONS IN ECOLOGICAL WORK
edited by J. K. Ord, G. P. Patil, and C. Taillie (1979)

Volume 5 SAMPLING BIOLOGICAL POPULATIONS
edited by R. M. Cormack, G. P. Patil, and D. S. Robson (1979)

Volume 6 ECOLOGICAL DIVERSITY IN THEORY AND PRACTICE
edited by J. F. Grassle, G. P. Patil, W. K. Smith, and C. Taillie (1979)

Volume 7 MULTIVARIATE METHODS IN ECOLOGICAL WORK
edited by L. Orloci, C. R. Rao, and W. M. Stiteler (1979)

Volume 8 SPATIAL AND TEMPORAL ANALYSIS IN ECOLOGY
edited by R. M. Cormack and J. K. Ord (1979)

Volume 9 SYSTEMS ANALYSIS OF ECOSYSTEMS
edited by G. S. Innis and R. V. O'Neill (1979)

Volume 10 COMPARTMENTAL ANALYSIS OF ECOSYSTEM MODELS
edited by J. H. Matis, B. C. Patten, and G. C. White (1979)

Volume 11 ENVIRONMENTAL BIOMONITORING, ASSESSMENT, PREDICTION, AND MANAGEMENT — CERTAIN CASE STUDIES AND RELATED QUANTITATIVE ISSUES
edited by J. Cairns, Jr., G. P. Patil, and W. E. Waters (1979)

Volume 12 CONTEMPORARY QUANTITATIVE ECOLOGY AND RELATED ECOMETRICS
edited by G. P. Patil and M. L. Rosenzweig (1979)

Volume 13 QUANTITATIVE POPULATION DYNAMICS
edited by D. G. Chapman and V. Gallucci (1979)

*For these first three volumes, contact:
The Pennsylvania State University Press
University Park, PA 16802 USA

For all of the remaining volumes, contact:
International Co-operative Publishing House
P.O. Box 245
Burtonsville, MD 20730 USA

# SPATIAL AND TEMPORAL ANALYSIS IN ECOLOGY

*edited by*

R. M. CORMACK
*Department of Statistics*
*University of St. Andrews*
*St. Andrews, Scotland*

J. K. ORD
*Department of Statistics*
*University of Warwick*
*Coventry, England*

**International Co-operative Publishing House**
**Fairland, Maryland USA**

P. O. Box 245, Burtonsville, Maryland 20730

Library of Congress Catalog Card Number: 79-89707
Volume ISBN: 0-89974-005-7
Series ISBN: 0-89974-000-6

Printed in the United States of America

Mathematical ecology is moving out of its classical phase carrying with it untold promise for the future, but, as H.A.L. Fisher, the historian, remarks, progress is not a law of nature. Without enlightenment and eternal vigilance on the part of both ecologists and mathematicians there always lurks the danger that mathematical ecology might enter a dark age of barren formalism, fostered by an excessive faith in the magic of mathematics, blind acceptance of methodological dogma and worship of the new electronic gods. It is up to all of us to ensure that this does not happen.

J. G. SKELLAM (1972)
in Mathematical Models in Ecology
edited by J. N. R. Jeffers
Blackwell Scientific Publications
Oxford, England

For top management and general public policy development, monitoring data must be shaped into easy-to-understand indices that aggregate data into understandable forms. I am convinced that much greater effort must be placed on the development of better monitoring systems and indices than we have in the past. Failure to do so will result in suboptimum achievement of goals at much greater expense.

R. E. TRAIN (1973)
National Conference on Managing the Environment
The United States Environmental Protection Agency
Washington, D.C., USA

# *International Statistical Ecology Program*




SPONSORS

NATO Advanced Study Institutes Program
NATO Ecosciences Program

The Pennsylvania State University
The Texas A&M University
The University of California at Berkeley

National Marine Fisheries Service, USA
Environmental Protection Agency, USA
Fish and Wildlife Service, USA
Army Research Office, USA

Comunita Economica Europea
Universita degli Studi, Parma, Italy

Consiglio Nazionale delle Ricerche, Italy
Ministero dei Lavori Pubblici,
Affari Esteri, e Pubblica Istruzione, Italy
Societa Italiana di Statistica, Italy
Societa Italiana di Ecologia, Italy

The Participants and
Their Home Institutions and Organizations

## PARTICIPANTS

### SATELLITE A: COLLEGE STATION AND BERKELEY July 18-August 13, 1977

Andrews, P. L., Montana
Anthony, R. G., Oregon
Artuz, M. I., Turkey
Bagiatis, K., Greece
Bajusz, B. A., Pennsylvania
Baumgaertner, J., California
Bell, E., Washington
Bellefleur, P., Canada
Berthet, P., Belgium
Beyer, J., Denmark
Bingham, R., Texas
Boswell, M. T., Pennsylvania
Braswell, J. H., Georgia
Braumann, C. A., Portugal
Brennan, J. A., Massachusetts
Bruhn, J. N., California
Cairns, J. Jr., Virginia
Callahan, C. A., Oregon
Cancela da Fonseca, J. P., France
Caraco, T. B., Arizona
Chapman, D. G., Washington
Cho, A., Pennsylvania
Colwell, R. California
Cormack, R., Scotland
Coulson, R. N., Texas
DeMars, C. J., California
Dennis, B., Pennsylvania
Derr, J., Iowa
deVries, P. G., Netherlands
Doucet, P. G., Netherlands
Elterman, A. L., California
Engen, S., Norway
Ernsting, G., Netherlands
Fiadeiro, P. M., Portugal
Flores, R. G., Brazil
Flynn, T. S., California
Folse, L. J., Texas
Ford, R. G., California
Gallucci, V. F., Washington
Gates, C. E., Texas
Gautier, C., France
Gerald, K. B., Texas
Giles, R. H., Virginia
Gokhale, D. V., California
Grant, W. G., Texas
Guardans, R. C., Spain
Hart, D., Virginia
Hazard, J. W., Oregon
Hendrickson, J. A., Pennsylvania
Hennemuth, R., Massachusetts
Hogg, D. B., Mississippi
Innis, G. S., Utah
Janardan, K. G., Illinois
Johnson, D., Texas
Johnson, D. H., North Dakota
Jolly, G. M., Scotland
Kester, T., Belgium
Kie, J. G., California
Kobayashi, S., Japan
Kubicek, F., Czechoslovakia
Labovitz, M. L., Pennsylvania
Lamberti, G. A., California
Lasebikan, B. A., Nigeria
Lindahl, K. Q., California
Laurence, G. C., Rhode Island
Livingston, G. P., Texas
Ludwig, J. A., New Mexico
Ma, J. C. W., Texas
Macken, C. A., Minnesota
Marsden, M. A., Montana
Mason, R., Oregon
Matis, J. H., Texas
Matthews, G. A., Texas
Minello, T. J., Texas
Mizell, R. F., Mississippi
Monserud, R. A., Idaho
Myers, C. C., Illinois
Myers, R. A., Canada
Naveh, Z., Israel
Nebeker, T. E., Mississippi
Neyman, J., California
Norick, N. X., California
O'Neill, R. V., Tennessee
Ord, J. K., England
Overton, S., Oregon
Patil, G. P., Pennsylvania
Pennington, M. R., Massachusetts
Poole, R. W., Rhode Island
Pulley, P. E., Texas
Quinn, T. J., Washington
Rawson, C. B., Washington
Reynolds, J. F., North Carolina
Riggs, L. A., California
Robson, D. S., New York
Roman, J. R., New York
Rosenzweig, M. L., Arizona
Roughgarden, J., California
Roux, J. J. J., South Africa
Sanders, F., Tennessee
Sen, A. R., Canada
Serchuk, F. M., Massachusetts
Shoemaker, C., New York
Singh, K. P., India
Smith, W. K., Massachusetts
Solomon, D. L., New York
Southward, G. M., New Mexico
Stafford, S. G., New York
Steinhorst, R. K., Texas
Stenseth, N. C., Norway
Stiteler, W. M., New York
Stout, M. L., California
Stromberg, L. P., California
Taillie, C., Pennsylvania
Tracy, D. S., Canada
Usher, M. B., England
vanBiezen, J. B., Netherlands
Walter, G. G., Wisconsin
Waters, W. E., California
Wensel, L. C., California
West, I. F., New Zealand
Wiegert, R. G., Georgia
Williams, F. M., Pennsylvania
Wright, J. R., Alabama
Wu, Y. C., California
Yandell, B. S., California
Zweifel, J. R., California

### SATELLITE B: PARMA July 31-September 5, 1978

Arditi, R., France
Azzarita F., Italy
Balchen, J. G., Norway
Bargmann, R. E., Georgia
Barlow, N. D., England
Baxter, M. B., Australia
Behrens, J., Denmark
Beran, H. G., Austria
Berman, M., Maryland
Berryman, A. A., Washington
Boswell, M. T., Pennsylvania
Brambilla, C., Italy
Braumann, C. A., New York
Breitenecker, M., Austria
Buerk, R., West Germany
Callahan, C. A., Oregon
Cancela da Fonseca, J. P., France
Chieppa, M., Italy
Clark, W. G., Italy
Cobelli, C., Italy
Cooper, C., California
Curry, G. L., Texas
DeMichele, D. W., Texas
Derr, J., Iowa
Diggle, P. J., England
Drakides, C., France
Ebenhöh, W., West Germany
Engen, S., Norway
Feoli, E., Italy
Fiadeiro, P. M., Portugal
Fischlin, A., Switzerland
Framstad, E. B., Norway
Frohberg, K., Austria
Gallucci, V. F., Washington
Garcia-Moya, E., Mexico
Gatto, M., Italy
Geri, C., France
Giavelli, G., Italy
Ginzburg, L. R., New York
Gokhale, D. V., California
Goldstein, R. A., California
Granero Porati, M. I., Italy
Greve, W., West Germany
Grosslein, M. D., Massachusetts
Grümm, H. R., Austria
Gulland, J. A., Italy
Gutierrez, A. P., California
Gydesen, H., Denmark
Hanski, I., England
Hanson, B. J., Utah
Hau, B., West Germany
Helgason, T., Iceland
Hendrickson, J. A., Pennsylvania
Hengeveld, R., Netherlands

Hennemuth, R. C., Massachusetts
Hoff, J. M., Norway
Holling, C. S., Canada
Hotz, M. C. B., Belgium
Jancey, R. C., Canada
Kooijman, S., Netherlands
Lamont, B. B., Australia
Levi, D., Italy
Liu, C. J., Kentucky
Marshall, W., Canada
Martin, F. W., Maryland
Matis, J. H., Texas
Menozzi, P., Italy
Meyer, J. A., France
Mohn, R. K., Canada
Mosimann, J., Maryland
Naveh, Z., Israel
Noy-Meir, I., Israel
Olivieri-Barra, S. T., Belgium
Ord, J. K., England
Orloci, L., Canada
Pacchetti, G., Italy
Pagani, L., Italy
Patil, G. P., Pennsylvania
Patten, B. C., Georgia
Pennington, M. R., Massachusetts
Policello, G. E., Ohio
Pospahala, R. S., Maryland
Purdue, P., Kentucky
Radler, K., West Germany
Ramsey, F. L., Oregon
Reyment, R. A., Sweden
Reyna Robles, R., Mexico
Rinaldi, S., Italy
Robson, D. S., New York
Rossi, O., Italy
Russek, E., Washington
Russo, A. R., Hawaii
Sadasivan, G., India
Schaefer, R., France
Shoemaker, C. A., New York
Show, I. T., California
Shuter, B. J., Canada
Simberloff, D., Florida
Slagstad, D., Norway
Smetacek, V. S., West Germany
Smith, W. K., Massachusetts
Sokal, R. R., New York
Soliani, L., Italy
Solomon, D. L., New York
Spremann, K., West Germany
Steinhorst, R. K., Idaho
Stenseth, N. C., Norway
Stiteler, W. M., New York
Subrahmanyam, C. B., Florida
Szöcs, Z., Hungary
Taillie, C., Pennsylvania
terBraak, C. J. F., Netherlands
Torrez, W. C., New Mexico
Tracy, D. S., Canada
Tursi, A., Italy
Vale, C., Portugal
van Biezen, J. B., Netherlands
Vazzana, C., Italy
Wahl, E., West Germany
Walker, B. H., England
Walter, G. G., Wisconsin
Walters, C. J., Canada
Warren, W. G., Canada
Waters, W. E., California
White, G. C., New Mexico
Wise, M. E., Netherlands
Zanni, R., Italy

## SATELLITE C: JERUSALEM September 7-September 15, 1978

Austin, M. P., Australia
Baxter, M. B., Australia
Berthet, P., Belgium
Carleton, T. J., Canada
Eisen, P. A., New York
Galluci, V., Washington
Goldstein, R. A., California
Goodman, D., California
Hanski, I., Finland
Hendrickson, J. A., Pennsylvania
Hengeveld, R., Netherlands
Hennemuth, R. C., Massachusetts
Hirsch, A., Washington, DC
Kempton, R. A., England
Naveh, Z., Israel
Noy-Meir, I., Israel
Odum, H. T., Florida
O'Neill, R. V., Tennessee
Patil, G. P., Pennsylvania
Quinn, T. J., Washington
Resh, V. H., California
Rossi, O., Italy
Rosenzweig, M. L., Arizona
Safriel, U., Israel
Sheshinski, R., Israel
Simberloff, D. S., Florida
Sokal, R. R., New York
Solem, J. O., Norway
Stenseth, N. C., Norway
Subrahmanyam, C. B., Florida
Torrez, W., New Mexico
Waters, W. E., California
Whittaker, R. H., New York

## AUTHORS NOT LISTED ABOVE

Adams, J. E., Texas
Barber, M. C., Georgia
Barreto, M., Brazil
Batcheler, C. L., New Zealand
Beuter, K. J., West Germany
Bitz, D. W., Rhode Island
Brown, B. E., Massachusetts
Brown, G. C., Kentucky
Brown-Leger, L. S., Massachusetts
Chaim, S., Israel
Chardy, P. France
Clark, G. M., Ohio
Condra, C., California
Connor, E. F., Florida
Costa, H., Brazil
Dale, M., Australia
De LaSalle, P., France
Duek, J. L., Arizona
Elwood, J. W., Tennessee
Ferris, J. M., Indiana
Ferris, V. R., Indiana
Flinn, J. T., Massachusetts
Foltz, J. L., Texas
Gibson, V. R., Massachusetts
Giddings, J. M., Tennessee
Gittins, R., Australia
Godron, M., France
Grassle, J. F., Massachusetts
Green, R., California
Halbach, U., West Germany
Halfon, E., Canada
Helthshe, J. F., Rhode Island
Hildebrand, S. G., Tennessee
Hogeweg, P., Netherlands
Iwao, S., Japan
Jacur, G. R., Italy
Kaesler, R. L., Kansas
Kravitz, D., Massachusetts
Kruczynski, W. L., Florida
Lackey, R., Virginia
Laurec, A., France
Lepschy, A., Italy
Malley, J. D., Maryland
Marcus, A. H., Washington
Matern, B., Sweden
McCune, E. D., Texas
Moncrieff, R., California
Moroni, A., Italy
Nichols, J. D., Maryland
O'Connor, J. S., New York
Paloheimo, J. E., Canada
Perrin, S., California
Pickford, S. G., Washington
Plowright, R. C., Canada
Podani, J., Hungary
Ratnaparkhi, M. V., Pennsylvania
Rescigno, A., Canada
Rickaert, M., France
Rohde, C., Maryland
Rosen, R., Canada
Scott, E., California
Scott, J. M., Hawaii
Seber, G. A. F., New Zealand
Siri, E., Italy
Skalski, J. R., Washington
Stehman, S., Pennsylvania
Steinberger, E. H., Israel
Swartzman, G., Washington
Taylor, C. E., California
Taylor, L. R., England
Tiwari, J. L., California
Watson, R. M., Scotland
Wehrly, T. E., Texas
Wigley, R. L., Massachusetts
Wissel, C., West Germany
Yang, M., Florida

# Foreword

The Second International Congress of Ecology was held in Jerusalem during September 1978. In this connection, a Satellite Program in Statistical Ecology was organized by the International Statistical Ecology Program (ISEP) during 1977 and 1978. The emphasis was on research, review, and exposition concerned with the interface between quantitative ecology and relevant quantitative methods. Both theory and application of ecology and ecometrics received attention. The program consisted of instructional coursework, seminar series, thematic research conferences, and collaborative research workshops.

The 1977 and 1978 Satellite Program consisted of NATO Advanced Study Institutes at College Station in Texas, Berkeley in California, and Parma in Italy; NATO Advanced Research Institute at Parma; ISEP Research Conferences, Seminars, and Workshops at College Station, Berkeley, Parma, and Jerusalem; and a Research Conference at Jerusalem.

The Satellite Program has been supported by NATO Advanced Study Institutes Program; NATO Ecosciences Program; National Marine Fisheries Service, USA; Environmental Protection Agency, USA; Fish and Wildlife Service, USA; Army Research Office, USA; The Pennsylvania State University; The Texas A&M University; The University of California at Berkeley; Universita degli Studi, Parma; Consiglio Nazionale delle Ricerche, Italy; Ministero dei Lavori Pubblici, Affari Esteri, e Pubblica Istruzione, Italy; Societa Italiana di Statistica, Italy; Societa Italiana di Ecologia, Italy; Communita Economica Europea; and the participants and their home institutions and organizations.

Research papers and research-review-expositions were specially prepared for the program by concerned experts and expositors. These materials have been refereed and revised, and are now available in a series of ten edited volumes.

## HISTORICAL BACKGROUND

The First International Symposium on Statistical Ecology was held in 1969 at Yale University with support from the Ford Foundation and the US Forest Service. The three symposium co-chairmen (G. P. Patil, E. C. Pielou, and W. E. Waters) represented the fields of statistics, theoretical ecology, and applied ecology. The program was well attended, and it provided a broad picture of where statistics and ecology stood relative to each other. While effort was apparent, communication between the two disciplines was inadequate.

It was clear that a focal forum was necessary to discuss and develop a constructive interface between quantifiable problems in ecology and relevant quantitative methods. As a partial solution to fill this need at professional organizations' level, the director of the symposium (G. P. Patil) made certain recommendations to the Presidents of the International Association for Ecology (A. D. Hasler), the International Statistical Institute (W. G. Cochran), and the International Biometric Society (P. Armitage). The International Association for Ecology (INTECOL) took a timely step in creating a section in the organization, namely, the statistical ecology section. The three societies together took a timely step in setting up a liaison committee on statistical ecology. The INTECOL Section and the Liaison Committee together developed the International Statistical Ecology Program. Since its inception in 1970, ISEP (as it has come to be known) has put emphasis on identifying the interdisciplinary needs of statistics and ecology at advanced instructional levels, and also at research conference and workshop levels.

The First Advanced Institute on Statistical Ecology in the United States was organized at Penn State for six weeks in 1972 with support from the US National Science Foundation, the US Forest Service, and the Mathematical Social Sciences Board. The participants of the Institute are all enjoying the benefits of their fruitful participation. With support from the UNESCO program of Man and Biosphere, a six month program was held in Venezuela for participants from Latin America in 1974 under the direction of Jorge Rabinovich, himself a participant in the 1972 Institute. With some initiatives from ISEP, special statistical ecology sessions have been held at the international conferences of the International Statistical Institute and the Biometric Society.

While plans were being made for the Second International Congress of Ecology, the then Secretary General and current President of INTECOL (G. A. Knox), and the ISEP chairman (G. P. Patil) discussed the need and the timeliness of a program in statistical ecology. The Satellite Program in Statistical Ecology took its final shape from this beginning under the care and concern of its director, advisors, coordinators, and sponsors.

## SCIENTIFIC BACKGROUND AND PURPOSE

The perceptions of Skellam and Train quoted on page v in this volume recapitulate the cautions, inspirations, and objectives responsible for the Satellite Program. The rigorous formulation of a quantitative scientific concept requires and in a sense creates empirically measurable quantities. Conversely, the scientific validity of the concept is totally dependent upon the measured values of those quantities. This is a capsule version of the feedback process known as the "scientific method." In crude modern terms, we might label the first procedure "modeling" and the second "curve-fitting." For reasons of complexity, historical accident, or whatever, these mutually dependent, complementary components have never become firmly integrated within ecology and its application to environmental studies.

Both procedures involve forms of mathematics: In the "modeling" process, the mathematics is used *relationally* — as a system of logic to ensure rigor and clarity of reasoning. This is in the historical tradition of Volterra and Lotka. Validation is most often based on *qualitative* agreement: the right trend, or the correct shape of the curve. This is as it should be, especially for broad general theories: it is the ideas and understanding that count, not so much the quantitative detail.

In the "curve-fitting" tradition, the mathematics is used *numerically* — as a system for precise measurement and prediction. Validation is most often based on *quantitative* agreement: *n*-place accuracy, or minimum uncertainty. This is also as it should be, especially for application and management: it is the forecast and ability to act confidently that count, not so much the underlying concept.

It need not be taken as a sign of "physics envy" to assert that as a science matures, these two processes must converge — more quantification must be used in concept validation and more concepts must be incorporated into the methodology of quantification.

The purpose of the Satellite Program is to encourage that convergence within the science of ecology and promote its application in the study of the environment and environmental stress. Monitoring and assessment activities to be meaningful and defensible need: (i) a conceptual and philosophical basis, (ii) a theoretical framework, (iii) methodological support, (iv) a technological toolbox, and (v) administrative management. The ultimate purpose of the program is to help identify and integrate the specifics of these important factors responsible for protecting the environment.

We take as our theme the better melding of fundamental ecological concepts with rigorous empirical quantification. The overall result should be progress toward a stronger body of general ecological theory and practice.

## PLANNING AND ORGANIZATION

The realization of any program of this nature and dimension often fails to fully meet the initial expectations and objectives of the organizers. Factors that are both logistic and psychological in nature tend to contribute to this general experience. Logistic difficulties include optimality problems for time and location. Other difficulties which must be attended to involve conflicting attitudes towards the importance of individual contributions to the proceedings.

We tried to cope with these problems by seeking active advice from a number of participants and special advisors. The advice we received was immensely helpful in guiding our selection of the best experts in the field to achieve as representative and balanced a coverage as possible. Simultaneously, the editors together with the referees took a rather critical and constructive attitude from initial to final stages of preparation of papers by offering specific suggestions concerning the suitability, and also the structure, content and size. These efforts of coordination and revision were intensified through editorial sessions at the program itself as a necessary step for the benefit of both the general readership and the participants. It is our pleasure to record with appreciation the spontaneous cooperation of the participants. Everyone went by scientific interests often at the expense of personal preferences. The program atmosphere became truly creative and friendly, and this remarkable development contributed to the maximal cohesion of the program and its proceedings within the limited time period available.

In retrospect, our goals were perhaps ambitious! We had close to 350 lectures and discussions during 50 days in the middle of the summer seasons of 1977 and 1978. For several reasons, we decided that an overworked program was to be preferred to a leisurely one. First of all, gatherings of such dimension are possible only every 5-10 years. Secondly, the previous meetings of this nature occurred some 5-10 years back, and the subject area of statistical ecology had witnessed substantial growth in this time. Thirdly, but most importantly, was the overwhelming response from potential participants, many of whom were to come across the continents!

Satellite A at College Station, Texas, and at Berkeley, California covered a four week period during July 18-August 13, 1977 and had 125 participants. Satellite B at

Parma, Italy had 130 participants spread over a six week period during July 31-September 5, 1978. Satellite C at Jerusalem, Israel had 35 participants. Approximately, one-third of the participants were graduate students, one-half were university faculty, and one-third were agency scientists. Approximately, one-third of the participants had an affiliation with one mathematical science or another, one-half an affiliation with one environmental science or the other, and one-quarter had an affiliation with one environmental management program or another. Thus the group was a good mix of great variety contributing to the effectiveness of the program. Not only what one heard was enlightening, but what one over-heard was equally enlightening!

Professors G. P. Patil, Paul Berthet, J. K. Ord, and Charles Taillie served as scientific directors of the program with Professor Patil assuming the responsibility of its direction from its conception to its conclusion. The inaugural speakers were: Professor H. O. Hartley at College Station, Professor J. Neyman at Berkeley, Professor C. S. Holling at Parma, and Professor G. P. Patil at Jerusalem.

## SCIENTIFIC CONTENT AND PUBLICATION

The following summary information on the subjects and corresponding coordinators of the satellite program may be of some interest. The details of each subject and its publication volume are reported elsewhere.

Statistical Distributions in Ecological Work: J. K. Ord, G. P. Patil, and C. Taillie.

Spatial and Temporal Analysis in Ecology: R. M. Cormack and J. K. Ord.

Quantitative Population Dynamics: D. G. Chapman and V. Gallucci.

Sampling Biological Populations: R. M. Cormack, G. P. Patil, and D. S. Robson.

Ecological Diversity in Theory and Practice: J. F. Grassle, G. P. Patil, W. K. Smith, and C. Taillie.

Multivariate Methods in Ecological Work: L. Orloci, C. R. Rao, and W. M. Stiteler.

Systems Analysis of Ecosystems: G. S. Innis and R. V. O'Neill.

Compartmental Analysis of Ecosystem Models: J. H. Matis, B. C. Patten, and G. C. White.

Environmental Biomonitoring, Assessment, Prediction, and Management-Certain Case Studies and Related Quantitative Issues: J. Cairns, Jr., G. P. Patil and W. E. Waters.

Contemporary Quantitative Ecology and Related Ecometrics: G. P. Patil and M. L. Rosenzweig.

Three more subjects were organized in the program.

Scientific Modeling and Quantitative Thinking with Examples in Ecology: G. P. Patil, D. S. Simberloff, and D. L. Solomon.

Conceptual Foundations of Ecological Theory and Applications: M. B. Usher and F. M. Williams.

Optimizations in Ecological Theory and Management: C. S. Holling and C. Shoemaker.

It would be fruitful to reorganize these subjects and add a few more for the next satellite program when it occurs.

It should be mentioned here that the close coordination and cooperation between the coordinators and the authors/speakers of potential contributions to the Proceedings (which are particularly intensive during the satellite program) paid themselves handsomely when the editors were confronted with the technical work related to the publications after the close of the program at Jerusalem. It is therefore very satisfying to report that the edited research papers and research-review-expositions prepared for the program are ready for distribution within 12 months of the conclusion of the program. For purposes of convenience, the contributions are organized in ten volumes in the Statistical Ecology Series published by the International Co-operative Publishing House. Altogether, they consist of an estimated 4,000 pages of research, review, and exposition, in addition to this common foreword in each followed by individual volume introductions. Subject and author indexes are also prepared at the end. Every effort has been made to keep the coverage of the volumes close to their individual titles. May this ten volume set in its own modest way provide an example of synergism.

## FUTURE DIRECTIONS

We wish there was no need for a program of this nature and dimension. It would be ideal if the needs of an interdisciplinary program were satisfactorily met in the existing institutions. Unfortunately, universities and governmental agencies have not been able to find effective ways to foster healthy interdisciplinary programs. The individuals attempting to do something in this direction tend to feel disheartened or disillusioned.

The satellite-like-programs help create and sustain enthusiasm, inward strength, and working efficiency of those who desire to meet a contemporary social need in the form of some interdisciplinary work. It should be only proper and rewarding for everyone involved that such programs are planned from time to time.

Plans are being made for a satellite program in conjunction with the next Biennial Conference of the International Statistical Institute and with the next International Congress of Ecology. Care should be exercised that the next program not become a mere replica of the present one, however successful it has been. Instead, the next program should be organized so that it helps further the evolution of statistical ecology as a productive field.

The next program is being discussed in terms of subject area groups. Each subject group is to have a coordinator assisted by small committees, such as a program committee, a research committee, an annual review committee, a journal committee, and an education committee. This approach is expected to respond to the need for a journal on statistical ecology, and also to the need of bringing out well planned annual review volumes. The education committee would formulate plans for timely manuals, modules, and monographs. Interested readers may feel free to communicate their ideas and interests to those involved in planning the next program. With mutual goodwill and support, we shall have met a timely need for today's science, technology, and society.

July 1979 G. P. Patil

# *Program Acknowledgments*

For any program to be successful, mutual understanding and support among all participants are essential in directions ranging from critical to constructive and from cautious to enthusiastic. The present program is grateful to the members of the ISEP Advisory Board, and to the referees, editors, coordinators, advisors, sponsors and the participants for their timely advice and support.

The success of the program was due, in no small measure, to the endeavors of the Local Arrangements Chairmen: J. H. Matis and C. E. Gates at College Station, W. E. Waters at Berkeley, O. Rossi and R. Zanni at Parma, and I. Noy-Meir at Jerusalem. We thank them for their hospitality and support.

And finally those who have assisted with the arduous task of preparing the materials for publication. Barbara Alles has been an ever cheerful and industrious secretary in the face of every adversity. Charles Taillie managed both scientific and non-scientific aspects. Bharat Kapur copyedited and proofread. Bonnie Burris, Bonnie Henninger, and Sandy Rothrock prepared the final versions of the manuscripts. Marllyn Boswell helped with the subject and author indexes. So did Bharat Kapur, Satish Patil, and Rani Venkataramani.

All of these nice people have done a fine job indeed. To all of them, our sincere thanks.

July 1979 G. P. Patil

# *Reviewers of Manuscripts*

With appreciation and gratitude, the program acknowledges the valuable services of the following referees who have served as reviewers of manuscripts submitted to the program for possible publication. The editors thank the reviewers for their critical and constructive reviews.

J. Balchen
University of Trondheim

R. E. Bargmann
University of Georgia

C. Chatfield
University of Bath

R. Colwell
University of California

A. Berryman
Washington State University

J. Beyer
Danish Institute of Fisheries and Marine Research

M. T. Boswell
Pennsylvania State University

C. Braumann
Instituto Universitario de Evora

M. A. Buzas
Smithsonian Institution

J. Cairns, Jr.
Virginia Polytechnic Institute and State University

C. Cobelli
Laboratorio per Ricerche di Dinamica dei Sistemi e di Bioingegneria

R. Cormack
University of St. Andrews

B. Coull
University of South Carolina

M. B. Dale
Commonwealth Scientific and Industrial Research Organization

A. P. Dawid
The City University, London

B. Dennis
Pennsylvania State University

F. Diemer
Office of Naval Research

P. Diggle
University of Newcastle upon Type

P. G. Doucet
Vrije University

J. E. Dunn
University of Arkansas

L. Eberhardt
Battelle Pacific Northwest Laboratories

S. Engen
University of Trondheim

R. G. Flores
Fundacao Instituto Brasileiro de Geografia e Estatistica

E. D. Ford
Institute of Terrestrial Ecology

C. E. Gates
Texas A&M University

K. B. Gerald
Rockwell International

L. Ginzburg
State University of New York

J. C. Gittins
University of Oxford

N. Glass
Environmental Protection Agency

D. V. Gokhale
University of California

R. Goldstein
Electric Power Research Institute

M. Granero-Porati
Istituto di Fisica, Parma

J. F. Grassle
Woods Hole Oceanographic Institution

J. C. Griffiths
Pennsylvania State University

I. Hanski
University of Oxford

W. Harkness
Pennsylvania State University

T. Helgason
University of Iceland

J. Hendrickson
Academy of Natural Sciences

R. C. Hennemuth
National Marine Fisheries Service

D. Hildebrand
University of Pennsylvania

A. Hirsch
Fish and Wildlife Service

P. Holgate
Birkbeck College

H. Horn
Princeton University

G. Innis
Utah State University

I. James
Commonwealth Scientific and Industrial Research Organization

K. G. Janardan
Sangamon State University

G. M. Jolly
University of Edinburgh

C. Jones
Woods Hole Oceanographic Institution

R. Kaesler
University of Kansas

R. A. Kempton
Plant Breeding Institute, Cambridge

J. Kirkley
National Marine Fisheries Service

G. Knott
National Institutes of Health

S. Kotz
Temple University

A. M. Kshirsagar
University of Michigan

S. Kullback
The George Washington University

R. C. Lewontin
Harvard University

B. F. J. Manly
University of Otago

A. H. Marcus
Washington State University

F. Martin
Patuxent Wildlife Research Center

J. H. Matis
Texas A&M University

J. R. McBride
University of California

D. Mollison
Heriot-Wett University

R. V. O'Neill
Oak Ridge National Laboratory

J. Newton
University of St. Andrews

J. K. Ord
University of Warwick

L. Orloci
University of Western Ontario

G. P. Patil
Pennsylvania State University

B. C. Patten
University of Georgia

M. Pennington
National Marine Fisheries Service

S. Pimm
Texas Tech University

K. H. Pollock
University of Reading

R. W. Poole
Brown University

R. S. Pospahala
Patuxent Wildlife Research Center

E. Preston
Environmental Protection Agency

F. Preston
Preston Laboratories

P. Purdue
University of Kentucky

F. L. Ramsey
Oregon State University

C. R. Rao
Indian Statistical Institute

P. A. Rauch
University of California

E. Renshaw
University of Edinburgh

R. Reyment
Uppsala University

D. S. Robson
Cornell University

M. L. Rosenzweig
University of Arizona

O. Rossi
University of Parma

W. E. Schaaf
National Marine Fisheries Service

T. Schopf
University of Chicago

H. T. Schreuder
Forest Service

G. A. F. Seber
University of Auckland

J. Sepkoski
University of Rochester

I. T. Show, Jr.
Science Applications, Inc.

D. Simberloff
Florida State University

D. B. Siniff
University of Minnesota

W. K. Smith
Woods Hole Oceanographic Institution

R. R. Sokal
State University of New York

G. M. Southward
New Mexico State University

S. Stehman
Pennsylvania State University

R. K. Steinhorst
University of Idaho

W. M. Stiteler
Syracuse University

P. Switzer
Stanford University

C. Taillie
International Statistical Ecology Program

C. E. Taylor
University of California

C. Tsokos
University of South Florida

E. Ursin
Danish Institute of Fisheries and Marine Research

D. Vaughan
Oak Ridge National Laboratory

G. G. Walter
University of Wisconsin

W. G. Warren
Western Forest Products Laboratory

W. E. Waters
University of California

S. D. Webb
University of Florida

G. C. White
Los Alamos Scientific Laboratory

R. H. Whittaker
Cornell University

M. E. Wise
Leiden University

S. Zahl
University of Connecticut

J. Zweifel
National Marine Fisheries Service

# Contents of Edited Volumes

**STATISTICAL DISTRIBUTIONS IN ECOLOGICAL WORK**
**J. K. Ord, G. P. Patil, and C. Taillie (editors)** **400 pp. approx.**
*Chance Mechanisms Underlying Statistical Distributions:* M. T. BOSWELL, J. K. ORD, and G. P. PATIL, Chance Mechanisms Underlying Univariate Distributions. C. TAILLIE, J. K. ORD, J. MOSIMANN, and G. P. PATIL, Chance Mechanisms Underlying Multivariate Distributions.

*Ecological Advances Involving Statistical Distributions:* M. T. BOSWELL and B. DENNIS, Ecological Contributions Involving Statistical Distributions. W. G. WARREN, Some Recent Developments Relating to Statistical Distributions in Forestry and Forest Products Research.

*Abundance Models:* J. A. HENDRICKSON, The Biological Motivation for Abundance Models. R.HENGEVELD, On the Use of Abundance and Species-Abundance Curves. S. ENGEN and C. TAILLIE, A Basic Development of Abundance Models: Community Description. S. ENGEN, Abundance Models: Sampling and Estimation. R. HENGEVELD, S. KOOIJMAN, and C. TAILLIE, A Spatial Model Explaining Species-Abundance Curves. S. KOBAYASHI, Species-Area Curves.

*Research Papers:* T. CARACO, Ecological Response of Animal Group Size Frequencies. J. DERR and J. K. ORD, Feeding Preferences of a Tropical Seedfeeding Insect: An Analysis of the Distribution of Insects and Seeds in the Field. R. A. KEMPTON, The Statistical Analysis of Frequency Data Obtained from Sampling an Insect Population Grouped by Stages. M. PENNINGTON, Fitting a Growth Curve to Field Data. M. YANG, Some Results on Size and Utilization Distribution of Home Range.

**SAMPLING BIOLOGICAL POPULATIONS**
**R. M. Cormack, G. P. Patil, and D. S. Robson (editors)** **425 pp. approx.**
P. G. DEVRIES, Line Intersect Sampling — Statistical Theory, Applications and Suggestions for Extended Use in Ecological Inventory. C. E. GATES, Line Transect and Related Issues. F. RAMSEY and J. M. SCOTT, Estimating Population Densities from Variable Circular Plot Surveys. G. A. F. SEBER, Transects of Random Length. G. M. JOLLY, Sampling of Large Objects. R. M. CORMACK, Models for Capture-Recapture. D. S. ROBSON, Approximations to Some Mark-Recapture Sampling Distributions. G. M. JOLLY, A Unified Approach to Mark-Recapture Stochastic Models, Exemplified by a Constant Survival Rate Model. J. R. SKALSKI and D. S. ROBSON, Tests of Homogeneity and Goodness-of-Fit to a Truncated Geometric Model for Removal Sampling. A. R. SEN, Sampling Theory on Repeated Occasions with Ecological Applications. W. G. WARREN, Trends in the Sampling of Forest Populations. G. M. JOLLY and R. M. WATSON, Aerial Sample Survey Methods in the Quantitative Assessment of Ecological Resources. W. K. SMITH, An Oil Spill Sampling Strategy. C. ROHDE, Batch, Bulk, and Composite Sampling.

**ECOLOGICAL DIVERSITY IN THEORY AND PRACTICE**
**J. F. Grassle, G. P. Patil, W. K. Smith, and C. Taillie (editors)** **350 pp. approx.**
*Species Diversity Measures: General Definitions and Theoretical Setting:* G. P. PATIL and C. TAILLIE, An Overview of Diversity. D. SOLOMON, A Comparative Approach to Species Diversity. S. ENGEN, Some Basic Concepts of Ecological Equitability. C. TAILLIE, Species Equitability: A Comparative Approach C. TAILLIE and G. P. PATIL, Diversity, Entropy, and Hardy-Weinberg Equilibrium.

*Ecological Theory And Species Diversity:* R. COLWELL, Toward an Unified Approach to the Study of Species Diversity. B. DENNIS and G. P. PATIL, Species Abundance, Diversity, and Environmental Predictability.

*Estimating Diversity From Field Samples:* J. E. ADAMS and E. D. MCCUNE, Application of the Generalized Jackknife to Shannon's Measure of Information as a Measure of Species Diversity. J. F. HELTHSHE and D. W. BITZ, Comparing and Contrasting Diversity Measures in Censused and Sampled Communities. J. HENDRICKSON, Estimates of Species Diversity from Multiple Samples. D. SIMBERLOFF, Rarefaction as a Distribution-Free Method of Expressing and Estimating Diversity. W. K. SMITH, J. F. GRASSLE, and D. KRAVITZ, Measures of Diversity with Unbiased Estimates.

*Application of Species Diversity Measures: A Critical Evaluation:* R. FLORES, M. BARRETO, and H. COSTA, On the Relationship Between Ecological and Physico-Chemical Water Quality Parameters: A Case Study at the Guanabara Bay, Rio de Janeiro. R. W. MONCREIFF and S. PERRIN, Effects of Increased Water Temperature on the Diversity of Periphyton. Z. NAVEH and R. H. WHITTAKER,

Determination of Plant Species Diversity in Mediterranean Shrub and Woodland Along Environmental Gradients. V. RESH, Biomonitoring, Species Diversity Indices, and Taxonomy. J. SOLEM, A Comparison of Species Diversity Indices in Trichoptera Communities. W. K. SMITH, V. R. GIBSON, L. S. BROWN-LEGER, and J. F. GRASSLE, Diversity as an Indicator of Pollution: Cautionary Results from Microcosm Experiments. C. B. SUBRAHMANYAM and W. L. KRUCZYNSKI, Colonization of Polychaetous Annelids in the Intertidal Zone of a Dredged Material Island in North Florida. C. E. TAYLOR and C. CONDRA, Competitor Diversity and Chromosomal Variation in Drosophia Pseudoobscura. B. DENNIS and O. ROSSI, Community Composition and Diversity Analysis in a Marine Zooplankton Survey.

*Bibliography:* B. DENNIS, G. P. PATIL, O. ROSSI, S. STEHMAN, and C. TAILLIE, A Bibliography of Literature on Ecological Diversity and Related Methodology.

**MULTIVARIATE METHODS IN ECOLOGICAL WORK**
**L. Orloci, C. R. Rao, and W. M. Stiteler (editors)** **400 pp. approx.**
R. BARGMANN, Structural Analysis of Singular Matrices Using Union Intersection Statistics with Applications in Ecology. M. DALE, On Linguistic Approaches to Ecosystems and Their Classification. D. V. GOKHALE, Analysis of Ecological Frequency Data: Certain Case Studies. J. HENDRICKSON, Examples of Discrete Multivariate Methods in Ecological Work. R. HENGEVELD and P. HOGEWEG, Cluster Analysis of the Distribution Patterns of Dutch Carabid Species. R. JANCEY, Species Weighting. A. LAUREC, P. CHARDY, P. DE LASALLE, and M. RICKAERT, Use of Dual Structures in Inertia Analysis: Ecological Implications. J. MOSIMANN and J. D. MALLEY, Size and Shape Analysis. L. ORLOCI, Non-Linear Data Structure and Their Description. J. PODANI, A Generalized Strategy of Homogeneity-Optimizing Hierarchical Classificatory Methods. R. REYMENT, Multivariate Analysis in Statistical Paleoecology. E. SCOTT, Spurious Correlation. W. K. SMITH, D. KRAVITZ, and J. F. GRASSLE, Confidence Intervals for Similarity Measures Using the Two Sample Jackknife. R. K. STEINHORST, Analysis of Niche Overlap. W. M. STITELER, Multivariate Statistics with Applications in Statistical Ecology. Z. SZOCS, New Computer Oriented Methods for Structural Investigation of Natural and Simulated Vegetation Patterns. B. LAMONT and K. J. GRANT, A Comparison of Twenty Measures of Site Dissimilarity. R. GITTINS, Ecological Applications of Canonical Analysis.

**SPATIAL AND TEMPORAL ANALYSIS IN ECOLOGY**
**R. M. Cormack and J. K. Ord (editors)** **400 pp. approx.**
J. K. ORD, Time-Series and Spatial Patterns in Ecology. P. J. DIGGLE, Statistical Methods for Spatial Point Patterns in Ecology. R. M. CORMACK, Spatial Aspects of Competition Between Individuals. R. W. POOLE, The Statistical Prediction of the Fluctuations in Abundance in Nicholson's Sheep Blowfly Experiments. W. G. WARREN and C. L. BATCHELER, The Density of Spatial Patterns: Robust Estimation Through Distance Methods. B. MATERN, The Analysis of Ecological Maps as Mosaics. J. A. LUDWIG, A Test of Different Quadrat Variance Methods for the Analysis of Spatial Pattern. S. A. L. M. KOOIJMAN, The Description of Point Patterns. R. HENGEVELD, The Analysis of Spatial Patterns of Some Ground Beetles (Col. Carabidae).

**SYSTEMS ANALYSIS OF ECOSYSTEMS**
**G. S. Innis and R. V. O'Neill** (editors) **425 pp. approx.**
R. K. STEINHORST, Stochastic Difference Equation Models of Biological Systems. R. V. O'NEILL, Natural Variability as a Source of Error in Model Predictions. R. K. STEINHORST, Parameter Identifiability, Validation and Sensitivity Analysis of Large System Models. R. V. O'NEILL, Transmutation Across Hierarchical Levels. R. V. O'NEILL, J. W. ELWOOD, and S. G. HILDEBRAND, Theoretical Implications of Spatial Heterogeneity in Stream Ecosystems. R. V. O'NEILL and J. M. GIDDINGS, Population Interactions and Ecosystem Function: Plankton Competition and Community Production. J. P. CANCELA DA FONSECA, Species Colonization Models of Temporary Ecosystems Habitats. E. HALFON, Computer-Based Development of Large Scale Ecological Models: Problems and Prospects. G. S. INNIS, A Spiral Approach to Ecosystem Simulation.

**COMPARTMENTAL ANALYSIS OF ECOSYSTEM MODELS**
**J. H. Matis, B. C. Patten, and G. C. White (editors)** **400 pp. approx.**
*Applications of Compartmental Analysis to Ecosystem Modeling:* R. V. O'NEILL, A Review of Linear Compartmental Analysis in Ecosystem Science. G. G. WALTER, A Compartmental Model of a Marine Ecosystem. M. C. BARBER, B. C. PATTEN, and J. T. FINN, Review and Evaluation of Input-Output Flow Analysis for Ecological Applications. I. T. SHOW, JR., An Application of Compartmental Models to Meso-scale Marine Ecosystems.

*Identifiability and Statistical Estimation of Parameters in Compartmental Models:* C. COBELLI, A. LEPSCHY, G. R. JACUR, Identification Experiments and Identifiability Criteria for Compartmental Systems. M. BERMAN, Simulation, Data Analysis, and Modeling with the SAAM Computer Program. G. C. WHITE and G. M. CLARK. Estimation of Parameters for Stochastic Compartment Models. R. E. BARGMANN, Statistical Estimation and Computational Algorithms in Compartmental Analysis for Incomplete Sets of Observations.

*Stochastic Approaches to the Compartmental Modeling of Ecosystems:* J. L. TIWARI, A Modeling Approach Based on Stochastic Differential Equations, the Principle of Maximum Entropy, and Bayesian Inference for Parameters. J. H. MATIS and T. E. WEHRLY, An Approach to a Compartmental Model with Multiple Sources of Stochasticity for Modeling Ecological Systems. P. PURDUE, Stochastic Compartmental Models: A Review of the Mathematical Theory with Ecological Applications. A. H. MARCUS, Semi-Markov Compartmental Models in Ecology and Environmental Health. M. E. WISE, The Need for Rethinking on Both Compartments and Modeling.

*Mathematical Analysis of Compartmental Structures:* G. G. WALTER, Compartmental Models, Digraphs, and Markov Chains. K. B. GERALD and J. H. MATIS, On the Cumulants of Some Stochastic Compartmental Models Applied to Ecological Systems. A. RESCIGNO, The Two-variable Operational Calculus in the Construction of Compartmental Ecological Models.

**ENVIRONMENTAL BIOMONITORING, ASSESSMENT, PREDICTION, AND MANAGEMENT — CERTAIN CASE STUDIES AND RELATED QUANTITATIVE ISSUES**
**J. Cairns, Jr., G. P. Patil, and W. E. Waters (editors) 450 pp. approx.**
*Biomonitoring:* J. CAIRNS, JR., Biological Monitoring — Concept and Scope. Z. NAVEH, E. H. STEINBERGER, and S. CHAIM, Use of Bio-Indicators for Monitoring of Air Pollution by Fluor, Ozone and Sulfur Dioxide.

*Environmental assessment and prediction:* G. P. PATIL, C. TAILLIE, and R. L. WIGLEY, Transect Sampling Methods and Their Application to the Deep-Sea Red Crab. W. E. WATERS, Biomonitoring, Assessment, and Prediction in Forest Pest Management Systems. C. A. CALLAHAN, V. R. FERRIS, and J. M. FERRIS, The Ordination of Aquatic Nematode Communities as Affected by Stream Water Quality. R. A. GOLDSTEIN, Development and Implementation of a Research Program on Ecological Assessment of the Impact of Thermal Power Plant Cooling Systems on Aquatic Environments.

*Environmental Management:* A. HIRSCH, Ecological Information and Technology Transfer, R. H. GILES, JR., Modeling Decisions or Ecological Systems. R. LACKEY, Appliction of Renewable Natural Resource Modeling in Public Decision-Making Process. F. MARTIN, R. S. POSPAHALA, and J. D. NICHOLS, Assessment and Population Management of North American Migratory Birds. J. E. PALOHEIMO and R. C. PLOWRIGHT, Bioenergetics, Population Growth and Fisheries Management. Z. NAVEH, A Model of Multiple-Use Management Strategies of Marginal and Untillable Mediterranean Upland Ecosystems.

*Case Studies and Quantitative Issues:* M. D. GROSSLEIN, R. C. HENNEMUTH, and B. E. BROWN, Research, Assessment, and Management of a Marine Ecosystem in the Northwest Atlantic — A Case Study. J. NEYMAN, Two Interesting Ecological Problems Demanding Statistical Treatment. B. DENNIS, G. P. PATIL, and O. ROSSI, The Sensitivity of Ecological Diversity Indices to the Presence of Pollutants in Aquatic Communities. D. SIMBERLOFF, Constraints on Community Structure During Colonization.

**CONTEMPORARY QUANTITATIVE ECOLOGY AND RELATED ECOMETRICS**
**G. P. Patil and M. L. Rosenzweig (editors) 725 pp. approx.**
*Community Structure and Diversity:* R. A. KEMPTON and L. R. TAYLOR, Some Observations on the Yearly Variability of Species Abundance at a Site and the Consistency of Measures of Diversity. G. P. PATIL and C. TAILLIE, A Study of Diversity Profiles and Orderings for a Bird Community in the Vicinity of Colstrip, Montana. M. L. ROSENZWEIG, Three Probable Evolutionary Causes for Habitat Selection. O. ROSSI, G. GIAVELLI, A. MORONI, and E. SIRI, Statistical Analysis of the Zooplankton Species Diversity of Lakes Placed Along a Gradient. S. KOBAYASHI, Another Model of the Species Rank-Abundance Relation for a Delimited Community. M. L. ROSENZWEIG and J. L. DUEK, Species Diversity and Turnover in an Ordovician Marine Invertebrate Assemblage.

*Patterns and Interpretations:* D. S. SIMBERLOFF and E. F. CONNOR, Q-Mode and R-Mode Analyses of Biogeographic Distributions: Null Hypotheses Based on Random Colonization. D. GOODMAN,

Applications of Eigenvector Analysis in the Resolution of Spectral Pattern in Spatial and Temporal Ecological Sequences. R. H. WHITTAKER and Z. NAVEH, Analysis of Two-Phase Patterns. R. R. SOKAL, Ecological Parameters Infered From Spatial Correlograms. M. P. AUSTIN, Current Approaches to the Non-Linearity Problem in Vegetation Analysis. M. GODRON, A Probabilistic Computation for the Research of "Optimal Cuts" in Vegetation Studies. S. IWAO, The m*-m Method for Analyzing Distribution Patterns of Single- and Mixed-Species Populations.

*Modeling and Ecosystems Modeling:* D. L. SOLOMON, On a Paradigm for Mathematical Modeling. R. W. POOLE, Ecological Models and Stochastic-Deterministic Question. R. ROSEN, On the Role of Time and Interaction in Ecosystem Modelling. E. HALFON, On the Parameter Structure of a Large Scale Ecological Model. G. SWARTZMAN, Evaluation of Ecological Simulation Models. R. WIEGERT, Modeling Coastal, Estuarine and Marsh Ecosystems: State-of-the-Art.

*Statistical Methodology and Sampling:* J. DERR and J. K. ORD, Field Estimates of Insect Colonization, II. J. A. HENDRICKSON, JR., Analyses of Species Occurrences in Community, Continuum and Biomonitoring Studies. R. SHESHINSKI, Interpolation in the Plane. The Robustness to Misspecified Correlation Models and Different Trend Functions. W. C. TORREZ, The Effect of Random Selective Intensities on Fixation Probabilities. R. GREEN, A Graph Theoretical Test to Detect Interference in Selecting Nest Sites. I. NOY-MEIR, Graphical Models and Methods in Ecology. T. J. QUINN, The Effects of School Structure on Line Transect Estimators of Abundance. J. W. HAZARD and S. G. PICKFORD, Line Intersect Sampling of Forest Residue.

*Applied Statistical Ecology:* R. C. HENNEMUTH, Man as Predator. V. F. GALLUCCI, On Assessing Population Characteristics of Migratory Marine Animals. P. A. EISEN and J. S. O'CONNOR, MESA Contributions to Sampling in Marine Environments. W. E. WATERS and V. H. RESH, Ecological and Statistical Features of Sampling Insect Populations in Forest and Aquatic Environments. R. L. KAESLER, Statistical Paleoecology: Problems and Perspectives. S. A. L. M. KOOIJMAN and R. HENGEVELD, The Description of Non-Linear Relationship Between Some Carabid Beetles and Environmental Factors. P. E. PULLEY, R. N. COULSON, and J. L. FOLTZ, Sampling Bark Beetle Populations for Abundance.

*A Bibliography:* B. DENNIS, G. P. PATIL, M. V. RATNAPARKHI, and S. STEHMAN, A Bibliography of Selected Books on Quantitative Ecology and Related Ecometrics.

**QUANTITATIVE POPULATION DYNAMICS**
**D. G. Chapman and V. F. Gallucci, editors** **300 pp. approx.**
D. G. CHAPMAN and V. F. GALLUCCI, Population Dynamics Models and Applications. J. G. BALCHEN, Mathematical and Numerical Modeling of Physical and Biological Processes in the Barents Sea. A. BERRYMAN and G. C. BROWN, The Habitat Equation: A Fundamental Concept in Population Dynamics. C. A. BRAUMANN, Population Adaptation to a "Noisy" Environment: Stochastic Analogs of Some Deterministic Models. L. GINZBERG, Genetic Adaptation and Models of Population Dynamics. M. I. GRANERO-PORATI, Stability of Model Systems Describing Prey-predator Communities. K. J. BEUTER, C. WISSEL, and U. HALBACH, Correlation and Spectral Analysis of Population Dynamics in the Rotifer *Brachionus Calyciflorus Pallas*. G. G. WALTER, Surplus Yield Models of Fisheries Management. SOME MORE PAPERS IN PREPARATION.

# *Contributors to this Volume*

Batcheler, C. L.
Forest and Range Experiment Station
New Zealand Forest Service

Cormack, R. M.
Department of Statistics
University of St. Andrews

Diggle, Peter J.
Department of Statistics
University of Newcastle upon Tyne

Hengeveld, R.
Department of Geobotany
Catholic University, Nijmegen

Kooijman, S. A. L. M.
Department of Biology
Division of Technology for Society TNO

Ludwig, John A.
Biology Department
New Mexico State University

Matern, B.
Department of Forest Biometry
Swedish University of Agricultural Sciences

Ord, J. K.
Department of Statistics
University of Warwick

Poole, Robert W.
Division of Biology and Medicine
Brown University

Warren, W. G.
Forestry Directorate
Western Forest Products Laboratory

# PREFACE TO THE VOLUME

The concept of a plant or animal *population* is central to ecology, yet it is not always the well-defined entity we would desire. Thus, Poole (1978) states that

> A population is not a concrete, observable entity of the real world but rather is an arbitrarily designated group, usually of those individuals of a species within a specified geographical area.

Clearly, as Poole indicates, the group *may* be well defined but, on other occasions, the populations may be defined by political or economic considerations rather than natural environmental factors. The trapping records from the Hudson's Bay Company provide a classic example. Further, we are interested in a population as it develops over time, so that a static description may not be adequate. We hope that this volume will enable the reader to become more acquainted with the tools of time-series and spatial analysis and to assess the contribution of these methods to current ecological research.

Like Gaul, the volume is divided into three parts. In the first part, the paper by Ord reviews the standard methods for analyzing temporal and spatial patterns. Spatial patterns are considered in the second part by Diggle and Cormack. The paper by Diggle reviews the recent statistical and ecological literature and presents several new studies of mapped data on patterns of trees. Cormack provides an extensive review of the spatial competition between individuals, which considers both the analysis of spatial patterns and the design of experiments to study competition.

The final part of the volume contains six research papers. Poole's study of the results of Nicholson's blow-fly experiments shows how the dependence between the time-series may be modelled by Box-Jenkins methods. Since most ecological processes involve feed-back loops, such methods are likely to play an increasing role in population studies.

Warren and Batcheler examine the theoretical basis for empirical estimates of the density of trees using distance methods. The dependence of such methods upon the underlying spatial pattern is well-known, but the 'Poisson forest' has remained all too often the bench-mark in previous studies.

Matern's study of mosaics links the ecological and statistical concepts of a mosaic and provides some methods for analyzing such patterns.

The Greig-Smith analysis of contiguous quadrats has proved very popular over the years but sometimes the various refinements have been overlooked. Ludwig gives a review of these and then gives an empirical comparison of the performance of the different methods.

Point processes are often described in terms of their first and second order properties (or mean and covariance structure). Kooijman provides new methods for estimating this structure from both point and areal data.

Hengeveld's paper highlights some of the deficiencies of Gradient Analysis and outlines an alternative approach using a new form of 'factor analysis.' The variations in the spatial pattern of several species of Carabid beetles are monitored using the technique.

In editing the book, we have tried to ensure that there is appropriate cross-referencing between papers; in particular, the papers in the third part use those in earlier parts as "source" descriptions of particular methods. However, we are determined to retain the original flavor of each author's contribution so that the reader should be prepared for the omission of some topics and the multiple coverage of others. Indeed, where controversy exists we have studiously avoided taking a 'party line', so the reader may find occasional disagreements about the 'right' view. From such a diversity of opinion is born the research of the future and we hope that this volume may stimulate further collaboration between ecologists and statisticians in exploring the dimensions of space and time.

Each paper in the volume was refereed by two independent referees, who followed our request to apply the highest critical standards. Both as editors and authors, we can testify that this request was granted! We are most grateful to them for their work.

Any volume of this kind has many debts to acknowledge. In addition to the many heroes and heroines cited in individual papers, we should like to thank Dr. Charles Taillie for his diligent work as Managing Editor and Mrs. Barbara Alles and other members of the team for their excellent secretarial work.

June, 1979

Richard Cormack
Keith Ord

*Reference*

Poole, R. W. (1978). Some relative characteristics of population fluctuations. In *Time Series and Ecological Processes*, H. H. Shugart, ed. SIAM Institute for Mathematics and Society, Philadelphia. 18-33.

# ACKNOWLEDGMENTS

For permission to reproduce materials in this volume, thanks are due to the Editors of the Journal of the Royal Statistical Society for Figure 1.5 on page 100.

## TABLE OF CONTENTS

# SPATIAL AND TEMPORAL ANALYSIS IN ECOLOGY

R. M. Cormack and J. K. Ord, (eds).,
*Spatial and Temporal Analysis in Ecology*, pp. 1-94. All rights reserved.


# TIME-SERIES AND SPATIAL PATTERNS IN ECOLOGY

J. K. ORD

Department of Statistics
University of Warwick
Coventry CV4 7AL, ENGLAND

SUMMARY. Early analysis of time-series proceeded by the identification of trend, cylical and seasonal components. Current practice favors use of the autocorrelation structure and the spectrum after the series has been rendered stationary. Both approaches are reviewed and the methods used to study several series of ecological interest. Temporal point processes are then described briefly. In the second part of the paper, measures of spatial dependence are described, and related hypothesis testing and estimation procedures are developed.

TABLE OF CONTENTS

CHAPTER 6: MODELS FOR SPATIAL PROCESSES

CHAPTER 1

# THE BEGINNINGS

## 1.1 INTRODUCTION

Everything has its time and place. When studying any organism a thorough understanding of that organism's biology is necessary but it is not enough. We must also seek to comprehend the organism's environment and the forces for change which exert themselves over time. Indeed, such processes as growth and evolution become meaningless when divorced from their temporal context. Again, the ability to survive may be as much a function of location as it is of the individual's innate attributes.

While "everything depends upon everything else," some influences are much more important than others and the basis of most statistical and scientific modelling is the judicious neglect of secondary issues. Thus, by taking observations at a particular time we obtain a 'snapshot' view into which time need not enter explicitly. Simiarly, spatial interactions may be safely ignored on occasion. Thus, statisticians may often assume that a sample consists of independent drawings from a population. However, the purpose of this volume is to look at those situations where time and space are of primary importance and to show how we may better understand the underlying biological processes through appropriate analyses of the data. We shall feel our way step by step. First of all we develop methods for the analysis of time-series and then examine spatial data recorded at a particular time. Later papers in the volume deal with recent developments in the study of spatial and temporal processes: this paper confines itself to the basic ideas.

## 1.2 WHAT IS A TIME-SERIES?

Any process which is monitored over time provides a record which we may describe as a time-series. The variable we record may be continuous (in the sense of changing 'smoothly') or discrete (changing only occasionally by finite jumps). Again, time itself may be 'continuous' in that changes may occur in any small period or 'discrete' when changes occur only at certain well defined times (such as a breeding season). Further, 'biological' time may differ from 'real' time, since we may wish to consider the state of successive generations or the sizes of successive cells (as for the wood cells data in Section 1.3).

Finally, we note that a recorded value may refer to a time (such as a reading on an electro-cardiogram) or to an average, or sum, over a period (such as the lynx trapping records mentioned in the next section). For reference, the main variations are described below and some examples are given in Table 1.1.

*Continuous time.* The process develops continuously in time although it may well be recorded at regular intervals rather than continuously.

*Discrete time.* The process develops from one generation to the next. These generations are well defined and may be recorded separately. Regular time intervals will be assumed even if these may, on occasion, be 'generation' time rather than 'real' time.

*Continuous process.* Values of the process (in continuous time) can be plotted on an unbroken graph of $Y(t)$ against $t$ .

*Point process.* The process develops by a series of irregularly spaced jumps over time. When the size of the jump is small in relation to the level of the process, a continuous process approximation may be used.

It is evident that the boundaries between the different categories are rather fuzzy, but the choice of methods of analysis depends upon our perception of the type of problem at hand.

In this chapter and the next we concentrate upon time-series recorded at regular intervals. Immediately this raises the question, how big a time interval? When looking at successive cells or generations, the 'time' interval is self-evident. On other occasions, the interval may be dictated by necessity (historical trapping records or the number of field workers available). When the choice of interval is open to the experimenter then it is important to consider the nature of the phenomena we wish to study. For example, we may be interested in a heartbeat every second, diurnal variations in growth, seasonal fluctuations or a seventeen year reproductive cycle. In general, the *maximum* allowable time between observations is half the length of the shortest cycle we wish to monitor. So, to examine daily variation, we need observa tions at least every twelve hours and so on. It should be stressed that this is a minimal requirement, and data should be collected

*TABLE 1.1: Examples of different kinds of temporal processes.*

| Process | Time: Continuous — Continuous record | Time: Continuous — Discrete record | Time: Discrete |
|---|---|---|---|
| Continuous | Electro-cardiogram | Tree size in forest surveys | Yield of corn per acre |
| Point | Size of a family | Size of a herd of deer (regular survey) | Number of individuals in a generation |

more frequently than this where possible. Unless due note is taken of the design question, the subsequent data record may totally fail to reflect the major variations in the process. It is worth noting that when we are interested in frequency distributions rather than in the development of the process over time, a long time interval may be necessary to ensure (near) independence of the counts.

The discussion so far reveals considerable scope for variation in any definition of a time-series. We prefer to be pragmatic and say that a time-series is a set of data which can be usefully analysed by time-series methods! The examples of the next section will serve to cast light upon this circular description.

## 1.3 SOME EXAMPLES OF TIME-SERIES

To illustrate the methods discussed in later chapters we shall make use of four different data sets. These are described below. The data and figures appear on separate pages at the end of this section.

*Phytoplankton* (due to K. E. Damann, 1960). Over the period 1926-1958, daily samples were drawn from Lake Michigan in a region near Chicago, Illinois, and the numbers per milliliter of phytoplankton assessed. From this vast record of some 40,000 observations we look only at the annual averages, although detailed analysis of the original records could be a fascinating endeavor. It should be noted that the figures for 1926 and 1927 have been adjusted to allow for a change in enumeration techniques introduced from 1928. Otherwise, the entire program was carried out under the supervision of one person and there is no reason to suspect that the fluctuations in the series could be due to changes in technique. The data are given in Table 1.2 and plotted in Figure 1.1. For the most part the counts fluctuate around a steady value of around 1000 per ml, with peaks in the series from 1944-1947 and 1956-1958. Undoubtedly, it is difficult to understand these fluctuations without information on climatic, seasonal, and environmental changes or without any information on the composition of the counts (too big an interval between observations!). However, the series is useful in illustrating different approaches in the analysis of trends.

*Zooplankton* (due to R. L. Burgner and D. E. Rogers, 1973). For periods varying from six to nine days during the months June - September in 1961 to 1971, the authors took hauls from Lake Aleknagik in Alaska. These were vertical hauls with a ½ meter net of number 6 mesh, from a depth of either 60 meters or the lake bottom, whichever was less. The average densities are given in Table 1.3 for *Cyclops*, and for *Calanoid*, *Bosmina*, *Daphnia*, and *Holopedium* combined, for the years 1961-1971. The number of hauls on which the averages are based is also given. The *Cyclops* and 'Others' series are plotted in Figure 1.2 and the proportion of *Cyclops* in Figure 1.3. The motivation behind looking at both absolute numbers and proportions is to see whether the *Cyclops* counts rise and fall with the others or follow a distinct seasonal pattern.

*Canadian lynx*. One of the most famous time-series in ecology is the annual record of the numbers of Canadian lynx trapped in the Mackenzie River district of north-west Canada from 1821 to 1934. These data first appeared in Elton and Nicholson (1942), who discussed fluctuations in the size of lynx populations in Canada over some 200 years. These data are given in Table 1.4.

Many methods in time-series depend upon the assumption that fluctuations are *additive* (or that an accumulation of different effects may be 'added together'). However, fluctuations in population size are usually *proportional* to numbers of fecund females, so that fluctuations are likely to be larger in absolute size when the population is large, suggesting a *multiplicative* effect. If this is so, then the fluctuations will be additive in the logarithms (see Section 3.2).

Whether to analyze the original data or the logarithms (or even some intermediate transformed version) is a matter for debate. In his study of a similar series for the colored fox, Williamson (1972, 1975) argues for the logarithmic scheme, for reasons similar to those given above. However, Anderson (1977), in a study of the same series, disagrees and sticks with the original data. This is one of many areas where the investigator's judgement is more important than formal tools and sight should not be lost of the biological process underlying the series. Another factor which should not be forgotten is that the trapping records will be influenced by both climatic and economic considerations and may not wholly reflect the population in any year.

Notwithstanding the comments made above, the most striking feature of the data is the regular ten year cycle which can be seen clearly in Figure 1.4 (plotted for both the original data and the logarithms). A major food source for the lynx is the snowshoe rabbit and the regular cycles could be a reflection of this predator-prey system (see Bulmer, 1974). Although the simple deterministic Lotka-Volterra model does not allow such regular cycles, migration from one area to another is enough to sustain regular cycles (Bartlett, 1960, p. 43). For recent discussions of this series, see Campbell and Walker (1977), Tong (1977), and Bulmer (1978).

*New wood cells.* (I am indebted to Dr. E. D. Ford for making these data available.) In a study on short term tree growth samples of the new wood were taken from a Sitka spruce (*Picea sitchensis*) every twelve hours for fifteen days. A wedge of tissue with base 6mm by 4mm and containing a small amount of last years wood was cut by scalpel (for details of the sampling procedure, see Ford and Robards, 1976; Ford *et al.*, 1978). Thus a typical radial file of tracheids runs from the mature xylem through a zone of cells whose walls are in process of thickening on to a zone of expanding cells and then the immature cambial zone. The cells forming the phloem are ignored.

In this example we choose to treat successive cells on the file as equi-spaced points in 'time.' Although the development is obviously sequential, the development time may not correspond to real time. Nevertheless, we feel that the interactions between neighboring cells may be properly reflected by this kind of analysis. The cell diameters are given in Table 1.5 and are plotted in Figure 1.5. The wall thickening and expanding cell zones are clearly seen in the diagram.

PHYTOPLANKTON DATA SET

*TABLE 1.2: Annual average total (number per ml) of algal plankton for Lake Michigan at Chicago, Illinois (from K. E. Damann, 1960).*

| Year | Total plankton | Year | Total plankton |
|---|---|---|---|
| 1926 | 872 | 1943 | 970 |
| 1927 | 878 | 1944 | 1298 |
| 1928 | 1116 | 1945 | 1778 |
| 1929 | 1135 | 1946 | 1603 |
| 1930 | 869 | 1947 | 1443 |
| 1931 | 760 | 1948 | 842 |
| 1932 | 1064 | 1949 | 1108 |
| 1933 | 942 | 1950 | 1109 |
| 1934 | 810 | 1951 | 1061 |
| 1935 | 763 | 1952 | 963 |
| 1936 | 1041 | 1953 | 1059 |
| 1937 | 889 | 1954 | 772 |
| 1938 | 815 | 1955 | 914 |
| 1939 | 840 | 1956 | 1310 |
| 1940 | 945 | 1957 | 1561 |
| 1941 | 1253 | 1958 | 1762 |
| 1942 | 1277 | 1943-58 Average | 1222 |
| 1926-42 Average | 957 | 1926-58 Average | 1086 |

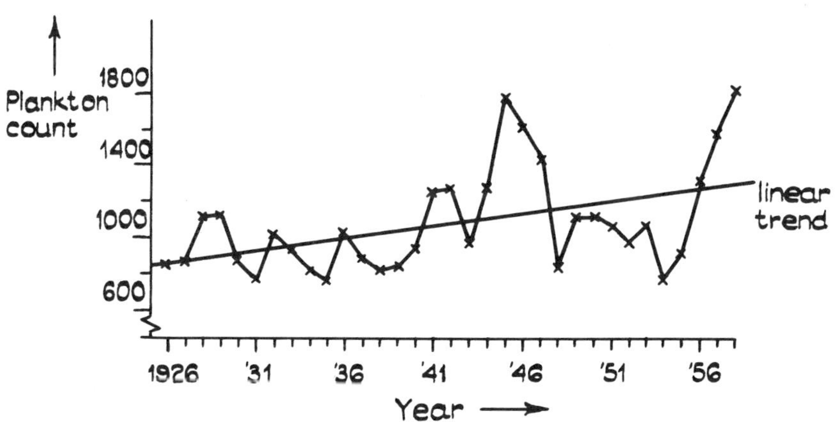


*FIG. 1.1: Plot of annual average totals.*

ZOOPLANKTON DATA SET

*TABLE 1.3: Average zooplankton densities (numbers per $m^3$) in Lake Aleknagik, Alaska (from Burgner and Rogers, 1973).*

| | June 21-26 | | July 11-20 | | August 5-13 | | Sept. 3-11 | |
|---|---|---|---|---|---|---|---|---|
| Year | *Cyclops* | Others | *Cyclops* | Others | *Cyclops* | Others | *Cyclops* | Others |
| 1961 | n.a. | | n.a. | | 2284 | 2819 | 1748 | 2750 |
| 1962 | 2058 | 629 | 2020 | 691 | n.a. | | 743* | 2496* |
| 1963 | 1864 | 1096 | 1632 | 1007 | 2621 | 5418 | 2309 | 3802 |
| 1964 | n.a. | | 1746 | 495 | 2216 | 1774 | 2395 | 4951 |
| 1965 | 2228 | 890 | 2212 | 2709 | 1922 | 2348 | 480 | 1331 |
| 1966 | 1324 | 390 | 2528 | 972 | 2540 | 1319 | 1616 | 4365 |
| 1967 | 1554 | 429 | 2028 | 983 | 999 | 528 | 1138 | 1495 |
| 1968 | 2242 | 447 | 3026 | 2049 | 2946 | 3564 | 3561 | 4471 |
| 1969 | 1540 | 564 | 2144 | 923 | 1888 | 3581 | 1604 | 5477 |
| 1970 | 1708 | 793 | 1657 | 1139 | 2106 | 3871 | 1768 | 4585 |
| 1971 | 1797 | 514 | 3160 | 1102 | 2508 | 1462 | 2165 | 3135 |

n.a. = not available. *Catches made on August 24*th*.

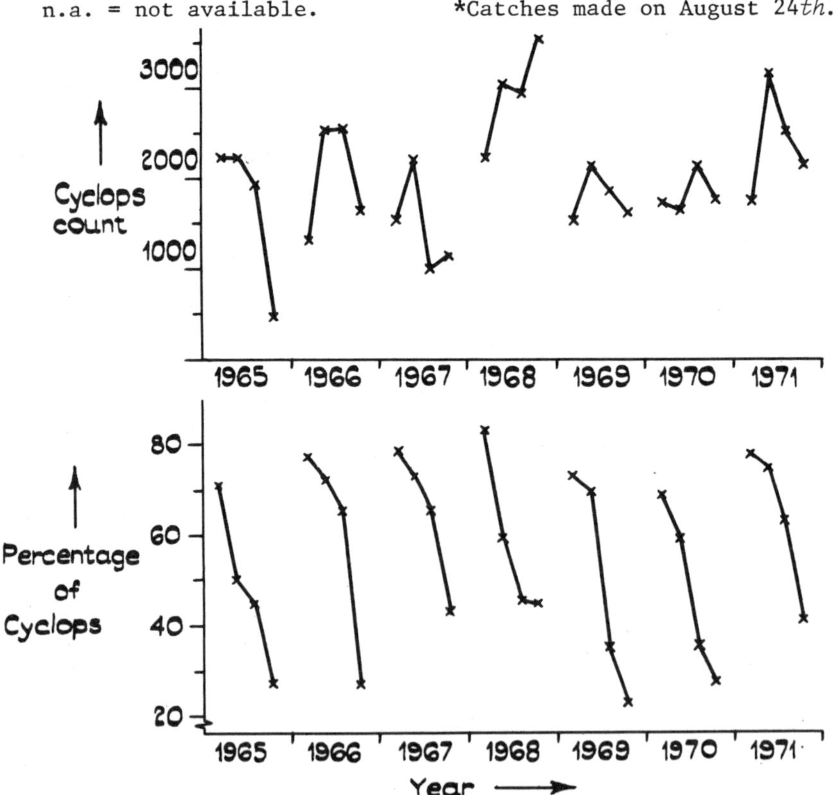

CANADIAN LYNX DATA SET

*TABLE 1.4: Numbers of lynx trapped in Mackenzie River district of NW Canada from* 1821 *to* 1934 *(from Elton and Nicholson,* 1942*).*

| 1821-40 | 1841-60 | 1861-80 | 1881-1900 | 1901-20 | 1921-34 |
|---|---|---|---|---|---|
| 269 | 151 | 236 | 469 | 758 | 229 |
| 321 | 45 | 245 | 736 | 1307 | 399 |
| 585 | 68 | 552 | 2042 | 3465 | 1132 |
| 871 | 213 | 1623 | 2811 | 6991 | 2432 |
| 1475 | 546 | 3311 | 4431 | 6313 | 3574 |
| 2821 | 1033 | 6721 | 2511 | 3794 | 2935 |
| 3928 | 2129 | 4254 | 389 | 1836 | 1537 |
| 5943 | 2536 | 687 | 73 | 345 | 529 |
| 4950 | 957 | 255 | 39 | 382 | 485 |
| 2577 | 361 | 473 | 49 | 808 | 662 |
| 523 | 377 | 358 | 59 | 1388 | 1000 |
| 98 | 225 | 784 | 188 | 2713 | 1590 |
| 184 | 360 | 1594 | 377 | 3800 | 2657 |
| 279 | 731 | 1676 | 1292 | 3091 | 3396 |
| 409 | 1638 | 2251 | 4031 | 2985 | |
| 2285 | 2725 | 1426 | 3495 | 3790 | |
| 2685 | 2871 | 756 | 587 | 674 | |
| 3409 | 2119 | 299 | 105 | 81 | |
| 1824 | 684 | 201 | 153 | 80 | |
| 409 | 299 | 229 | 387 | 108 | |

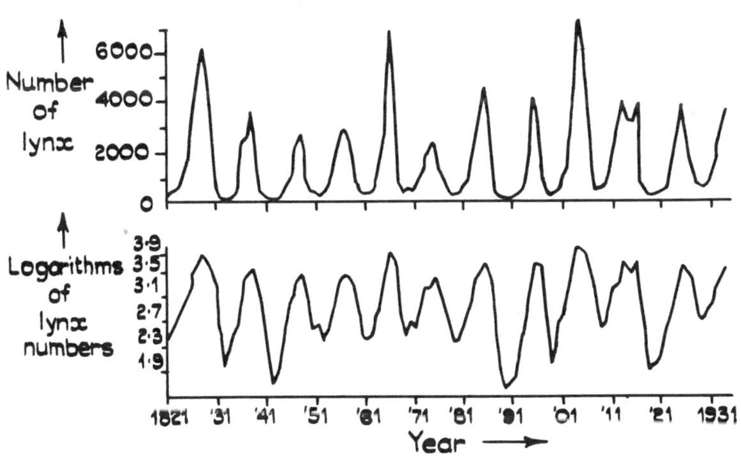

## WOOD CELLS DATA SET

*TABLE 1.5: Radial cell diameters (in μ) for cells along an entire file of tracheids (from E. D. Ford, unpublished).*

| Cell numbers | | | | | | | |
|---|---|---|---|---|---|---|---|
| 1-20 | 21-40 | 41-60 | 61-80 | 81-100 | 101-20 | 121-40 | 141-59 |
| 39.00 | 37.00 | 42.75 | 43.50 | 46.75 | 35.50 | 27.75 | 15.00 |
| 41.50 | 28.75 | 40.75 | 40.50 | 52.75 | 36.50 | 26.50 | 11.88 |
| 34.75 | 32.50 | 40.50 | 45.50 | 38.25 | 36.50 | 24.50 | 13.75 |
| 31.75 | 30.25 | 45.50 | 41.50 | 37.50 | 31.25 | 22.50 | 15.75 |
| 35.00 | 32.00 | 40.75 | 49.50 | 46.25 | 30.00 | 28.50 | 14.75 |
| 37.00 | 39.25 | 39.50 | 47.75 | 41.25 | 36.00 | 29.50 | 12.75 |
| 41.00 | 40.75 | 34.00 | 46.75 | 46.00 | 34.00 | 31.50 | 6.75 |
| 32.50 | 43.25 | 30.00 | 46.75 | 40.75 | 31.00 | 32.50 | 6.75 |
| 31.00 | 49.25 | 32.00 | 39.75 | 44.75 | 38.00 | 32.50 | 5.75 |
| 28.00 | 47.25 | 32.50 | 34.25 | 40.00 | 36.00 | 36.50 | 8.75 |
| 34.25 | 33.25 | 44.25 | 29.00 | 35.50 | 34.00 | 28.50 | 12.75 |
| 35.75 | 36.00 | 43.75 | 35.00 | 28.00 | 39.00 | 28.50 | 13.75 |
| 40.00 | 32.00 | 44.50 | 30.00 | 30.00 | 32.75 | 28.50 | 6.75 |
| 34.50 | 30.00 | 39.00 | 37.00 | 34.00 | 22.25 | 29.50 | 6.75 |
| 30.50 | 29.75 | 36.00 | 34.00 | 35.75 | 28.00 | 16.50 | 8.75 |
| 40.00 | 34.75 | 37.75 | 39.25 | 31.75 | 24.00 | 16.50 | 7.75 |
| 37.00 | 38.75 | 34.25 | 41.75 | 31.00 | 25.00 | 16.50 | 10.75 |
| 44.00 | 36.50 | 35.25 | 45.00 | 33.00 | 27.00 | 15.25 | 10.75 |
| 35.75 | 36.50 | 41.50 | 50.75 | 30.75 | 26.00 | 17.00 | 9.13 |
| 33.75 | 38.50 | 35.50 | 43.50 | 31.50 | 23.00 | 15.00 | |

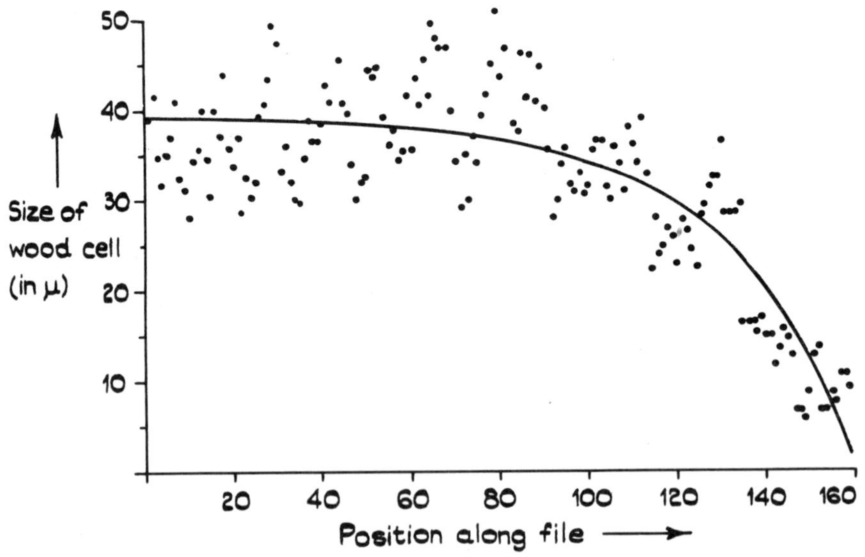

## 1.4 COMPONENTS OF A TIME-SERIES

In earlier sections, we have mentioned various features of a time-series, notably the long-term fluctuations (or trend), seasonal variations, and cyclic effects. Seasonal variations are those with known period, be it a day, a year or whatever, while cyclic effects are movements of unknown period; see Section 2.6. To this should be added a random (or irregular) component which encompasses the multitude of perturbations not included explicitly in a model of the series. Thus, the *additive* model for a time-series corresponds to a decomposition of the form

$$
\begin{aligned}
Y &= \text{systematic component} + \text{random component} \\
&= T + C + S + R \quad ,
\end{aligned}
$$

where $Y$ denotes the value of the time-series, with the time subscript supressed for convenience. The time-series has components: $T$, the trend or long-term movement of the series; $C$, the (regular) cyclic fluctuations about the trend; $S$, the seasonal component of the series; and $R$, the random (error) component.

The additive model is by no means the only model we could consider; the *multiplicative* model may be written as

$$Y = TCSR$$

or, upon taking logarithms,

$$\log Y = \log T + \log C + \log S + \log R \quad .$$

That is, after taking logs, we are back at an additive model for $\log Y$. This is the model underlying the use of logarithms for the lynx data.

Transformations are also used to stabilize variances, such as the Box-Cox family (Box and Cox, 1964),

$$Z = (Y + c)^{\lambda}/\lambda \quad ,$$

with $c$ a constant and $-1 \leq \lambda \leq 1$. This family includes the additive ($\lambda=1$) and multiplicative ($\lambda=0$) models as special cases. When an appropriate form for analysis is uncertain, a search over different $\lambda$ values may suggest a suitable scheme. The choice $\lambda = \frac{1}{2}$ corresponds to a variance stabilizing transformation for the Poisson distribution, which is often an appropriate assumption for count data.

Let us now suppose that an appropriate model has been identified; the next stage is to estimate the different components. In various forms this question will occupy our attention in the next two chapters. But why do this? Before plunging into analysis it is worth stating the possible objectives underlying time-series analysis (cf. Kendall, 1976, p. 12-16).

(a) In the spirit of the additive model introduced earlier, we may be concerned simply to *describe* the main features of the data series. This step is rarely an end in itself but may be a useful source of scientific hypotheses in the early stages of an investigation. However, separation of the components of the series is often a useful aid in understanding fluctuations, as in the deseasonalization of economic time-series.

(b) Leading on from the first stage we may develop a model to *explain* the fluctuations over time. Often this will require comparisons between actual and simulated series to examine the adequacy of the model.

(c) Either as a check on the validity of the model in (b) or out of direct interest, we may wish to *forecast* future values of the time-series. If the individual components of the series have been modelled adequately, then each of these may be projected forwards. However, there are many approaches to forecasting (see Section 3.7 and Poole, 1979).

(d) Given some understanding of the process, we may use forecast values to devise possible methods for controlling the system. The management of wildlife populations and the control of epidemics are obvious areas of potential application, although a single series is unlikely to give an adequate description of the system. One simple, but often overlooked, feature of attempts at control is that these attempts will affect the structure of the system, so that previously successful models may not be valid for future use.

(e) Once we recognize the *interactions* present in even the simplest ecosystem or socio-economic system, then multiple time-series methods become necessary. Unfortunately, greater realism brings with it greater mathematical complexity. Therefore for the most part, we shall concentrate upon single time-series and only give an outline of the extensions to several series. Further reading on these topics is given in Section 1.6.

Given sufficient knowledge it would be possible to develop a model which could be used for all the purposes listed above. In reality, the objective will often determine the mode of analysis.

For example, a simple *adaptive* model (see Section 3.7), may give good forecasts and adjust rapidly to changes in the mean level of the time-series, although a regression model may give a better description *within* the period of observation. Against this, the adaptive model may offer little by way of *explanation* for fluctuations in the time-series. There are few, if any, sure-fire methods in time-series analysis.

## 1.5 FUTURE CHAPTERS

In Chapter 2 we look at various methods for decomposing a time-series into its constituent parts and estimating these components. Such analyses may be described as data-analytic rather than model-building exercises. but will often provide interesting hypotheses for further investigation.

In Chapter 3 we examine autoregressive and moving average methods which are the main tools used in the statistical modelling of time-series. Then in Chapter 4 we look briefly at point processes in time where the intervals between successive events are of interest, rather than any particular measurement at a given time.

Chapters 5 and 6 relate to spatial processes. Chapter 5 is concerned with measures of spatial dependence while Chapter 6 considers various models for spatial processes. Spatial point patterns are considered by Diggle (1979) and Cormack (1979) in this volume.

## 1.6 BIBLIOGRAPHY

A brief introduction of this nature can do no more than scratch the surface of the vast literature of time-series analysis and spatial processes. Suitable texts at an intermediate level are those of Chatfield (1975) and Kendall (1976). which will be referenced repeatedly in the later chapters. On more specific topics, Jenkins and Watts (1968) is a useful reference on spectral analysis, while the work of Box and Jenkins (1970) on the modelling of series in the time domain has become a classic for the study of 'Box-Jenkins methods.' Anderson (1976) provides a simpler introduction to these methods. For those seeking a more rigorous treatment of the subject, Anderson (1971), Hannan (1970), or Kendall and Stuart (1976, chap. 44-50) are recommended.

The best source of time-series studies in ecology is the recent volume edited by Shugart (1978). This volume contains eighteen papers on a variety of topics ranging from general theoretical and methodological questions to several applications in ecology.

The literature for spatial processes is rather dispersed and relatively few texts exist. Two of the classics are those of Matérn (1960) and Matheron (1971); regrettably, these are available only by private circulation, as is the recent work of Kooijman (1977). Of books more readily available, Cliff and Ord (1973) discuss tests of spatial dependence in detail while Bartlett (1975) is a more theoretical work dealing with spatial models. The volume of papers edited by Davis and McCullagh (1975) includes several papers on the analysis of spatial data, although these are more oriented to geological problems.

CHAPTER 2

# TRENDS, SEASONS, AND CYCLES

## 2.1 INTRODUCTION

We begin this chapter by assuming that the additive model is appropriate, that is

$$Y = T + C + S + R \quad ,$$

so that we can operate on the various components (more or less) separately. To make life simpler, we shall describe the methods for one component at a time and choose to ignore the others. In practice, all components may be present and a combination of methods must be used.

The trend describes long-term movements in the series and, as such, may be expected to change rather slowly relative to the short-term oscillations induced by the random component (or random noise, to use the engineering term). Thus, the trend will show up more clearly if it is represented by a smooth function over time or by some averaging over adjacent terms in the time-series which allow the trend (or signal) to show through and reduce the effect of the noise. Conversely, we may take *differences,* such as

$$\nabla y_t = y_t - y_{t-1} \quad ,$$

where $y_t$ denotes the value of the series at time $t$ and $\nabla$ is known as the *backward difference operator*. Differencing has the effect of removing linear trends and is one of several methods of *trend-removal* which we examine in Sections 2.2, 2.3, and 2.4. The purpose of trend removal is to enable us to concentrate upon the other components of the series.

Looking at the de-trended series we may expect to see regular peaks and troughs corresponding to different times of the year (or other appropriate time interval). The data may be monthly or quarterly or relate only to certain periods of the year (as do the zooplankton data in Table 1.3). If observations are available for several years it becomes possible to estimate the seasonal component as we shall see in Section 2.5. Again, if data are recorded $k$ times per year then we may use differences to try to eliminate the seasonal effect, that is we take

$$\nabla_k y_t = y_t - y_{t-k} \quad ,$$

where $k=4$ for quarterly data, $k=12$ for monthly data, and so on.

The seasonal component is cyclical, but we know that the cycle is of one year's duration. The cycles described by the cyclical component will not be so regular and they may have almost any duration (or *wavelength*). Sometimes we might guess at likely cycle durations, such as the time from birth to maturity, but fluctuations induced by predator - prey interactions are often more difficult to predict. In Section 2.6, we develop a representation for a time-series in terms of cycles (sine waves) of different wavelengths, known as *spectral analysis*. This method can be likened to the use of a spectrometer to break a beam of light into its separate components each with a different wavelength.

## 2.2 ESTIMATING THE TREND

We now revert to a more standard statistical notation and let the random variable $Y_t$ describe the process at time $t$. Then we write the additive model as

$$Y_t = \alpha_t + \varepsilon_t \quad (t=1,\cdots,T) \quad , \tag{2.1}$$

where $\alpha_t$ denotes the systematic component (or trend), $\varepsilon_t$ denotes the random (error) component, and data are recorded at $T$ points in time. Since $\varepsilon_t$ describes fluctuations about the trend, we assume that it has zero mean; that is

$$E(\varepsilon_t) = 0 \quad .$$

Also, we shall assume that:

(a) the variance of the $\{\varepsilon_t\}$ is constant over time, or

$$\text{var}(\varepsilon_t) = \sigma_\varepsilon^2 \quad ; \text{ and}$$

(b) successive random errors are uncorrelated, or

$$\text{cov}(\varepsilon_t, \varepsilon_s) = 0 \quad , \text{ for all } s \neq t \quad .$$

Since the errors have zero means, (b) implies that

$$E(\varepsilon_t \cdot \varepsilon_s) = 0 \quad , \text{ for all } s \neq t \quad .$$

The assumption of uncorrelated errors is a natural starting point in a model for the trend component, but it is very restrictive. If a more general covariance structure is assumed then the least squares estimators described in the next section are inefficient; however, they remain unbiased. When the assumption of uncorrelated errors is made, it may be tested using the methods of Durbin and Watson (1950, 1951, 1971) for time-series and Cliff and Ord (1973, chap. 5, 6) for spatial data.

Given the assumptions regarding the structure of the random component, we now turn to $\alpha_t$ . We described the trend as relatively slow-moving so that a low order polynomial may give a reasonable description; that is, we might set

$$\alpha_t = \beta_o + \beta_1 t + \cdots + \beta_k t^k \quad , \tag{2.2}$$

where $\beta_0, \cdots, \beta_k$ are unknown parameters. Typically $k \leq 3$ should suffice. If a cubic does not describe the data adequately other methods should be tried since a high order of polynomial tends to be very erratic at the ends of the series. For the same reason, polynomial trends can be very misleading if used for forecasting. Nevertheless, the fitting of a low order polynomial can be a useful way of removing a trend so that attention may focus upon the estimated errors,

$$\hat{\varepsilon}_t = y_t - \hat{\alpha}_t \quad (t=1,\cdots,T) \quad , \tag{2.3}$$

in a search for seasonal and/or cyclical components. We now look at the problem of estimating the unknown parameters.

*2.2.1 Fitting Trend Curves.* The simplest and most widely used method is that of least squares; that is, we seek to minimize

$$S = \sum_{t=1}^{T} \varepsilon_t^2 = \sum_{t=1}^{T} (y_t - \alpha_t)^2$$

with respect to the $\{\beta_j\}$ . Whenever $\alpha_t$ is linear in the unknown parameters $\{\beta_j\}$ , as in equation (2.2), the resulting estimators are linear functions of the $\{y_t\}$ . The method is a special case of ordinary regression analysis (see Draper and Smith, 1966, p. 104-115). Thus, differentiating $S$ with respect to $\beta_j$ we obtain

$$\frac{dS}{d\beta_j} = -2 \sum_{t=1}^{T} (y_t - \alpha_t) \frac{d\alpha_t}{d\beta_j} .$$

But $d\alpha_t/d\beta_j = t^j$ so when we set $dS/d\beta_j = 0$ we obtain the estimating equations

$$\Sigma t^j y_t = \hat{\beta}_0 \Sigma t^{j+1} + \hat{\beta}_1 \Sigma t^{j+2} + \cdots + \hat{\beta}_k \Sigma t^{j+k} \quad (j=0,1,\cdots,k) ,$$

where all summations are taken over $t=1,\cdots,T$ . That is, we have a set of $(k+1)$ simultaneous linear equations in $\hat{\beta}_0,\cdots,\hat{\beta}_k$ which can be solved to give the least squares estimates. While solution of these equations poses no computational problems (provided $T \geq k$) there may be difficulties in interpreting the coefficients, since the addition of a quadratic term generally changes the coefficients of the constant and linear terms, with similar effects for every other term added. For this reason, we may use *orthogonal polynomials* (Kendall and Stuart, 1973, p. 372-377) in place of the straightforward polynomials in equation (2.2).

Given that the observations are recorded at times $t=1,\cdots,T$ let $u = t - (T+1)/2$ and define the orthogonal polynomials as

$$\phi_0 = 1, \qquad \phi_1 = u,$$
$$\phi_2 = u^2 - (T^2-1)/12, \qquad \phi_3 = u^3 - u(3T^2-7)/20,$$

and so on. These polynomials are constructed so that they are orthogonal; that is

$$\sum_{t=1}^{T} \phi_i(u)\phi_j(u) = 0 \text{ for all } i \neq j .$$

It must be noted that these functional forms apply only to equally-spaced data points; more specific functions must be constructed when the data are unequally spaced. Then, using the least squares method, we obtain the simplified estimators

$$\hat{\beta}_j = \sum_{t=1}^{T} y_t \cdot \phi_j(u) \Big/ \sum_{t=1}^{T} \{\phi_j(u)\}^2 .$$

This makes the addition of extra terms a simple procedure which does not disturb the earlier coefficients.

*Example 2.1.* The trend for the phytoplankton data is shown in Figure 1.1. Despite considerable variation, a slight upward trend is discernible.

The zooplankton and lynx series do not exhibit any noticeable trend, while the wood cell series describes a non-linear process of development to maturity. Since this is different in kind from anything encountered previously, we now examine this type of process separately.

*2.2.2 Growth Curves.* Cells develop over time and then approach maturity as growth ceases. Again, in capture, or sighting, experiments with a closed population the total number of individuals captured should level out over time approaching the population size $N$.

Polynomial schemes are inappropriate in such cases and expressions for the overall size or the cumulative total at time $t$ seem more appropriate. Also, the time origin has real meaning in such cases as the beginning of the growth or capture process. We first develop a model for the capture process and then turn to growth processes.

Suppose that the time to capture for any individual follows the exponential distribution with parameter $\gamma$, so that the distribution function, $F(t)$, is

$$F(t) = \Pr(T \leq t) = 1 - e^{-\gamma t} .$$

Let $Y_t$ denote the number of different individuals captured by time $t$ from a population of size $N$. Then the expected number caught is

$$E(Y_t) = N(1-e^{-\gamma t}) .$$

This expression might be used to describe the systematic component in equation (2.1), so that we might use the scheme

$$Y_t = N(1-e^{-\gamma t}) + \varepsilon_t \tag{2.3}$$

to reflect the levelling out over time. However, while there may be considerable variations in $Y_t$ at the beginning, these fluctuations will damp down as $t$ increases, since $Y_t$ approaches $N$

as $t$ becomes large. Thus, it may be better to look at the *rate* at which $Y_t$ approaches $N$, defined as

$$\frac{1}{1-F(t)} \cdot \frac{dF(t)}{dt} = \gamma \quad . \tag{2.4}$$

Provided that the interval between successive observations is not too large, we might estimate $F(t)$ by $Y_t/N$ and approximate the derivative by $(Y_{t+1} - Y_t)/N$, which yields, from equation (2.4).

$$Y_{t+1} - Y_t = \gamma N - \gamma Y_t + \varepsilon_t^* \tag{2.5}$$

where $\varepsilon_t^*$ is a revised error term. We may assume that N different individuals are sighted independently one of another (the usual assumptions in simple capture models for closed populations). It follows that, given $Y_t = y_t$, $Y_{t+1} - y_t$ has a binomial distribution with parameters $n_t = (N-y_t)$ and $p_t = 1-\exp(-\gamma)$. Provided that the means, $n_t p_t$, are not too small, least squares may be used to estimate the unknown parameters, $\gamma N$ and $\gamma$, in equation (2.5).

Although the model has been developed using the exponential distribution, it is valid under less restrictive conditions. Thus, for any individual not captured by time $t$, all we need to assume is that

$$\Pr(\text{capture in period } t+1 \mid \text{not captured by } t) = 1-e^{-\gamma} \; :$$

The risk of capture does not need to be uniform during the period; indeed, that is unlikely to be true in many applications.

*Example 2.2.* Ricker (1958) describes an experiment involving fishing a lake in Oregon for seven successive weeks. The method of fishing and the effort involved were the same at each stage, the fish were not replaced. Also, we assume that no fish born during this time were included in the catch. The catches in successive weeks were

| week (t+1) | 0 | 1 | 2 | 3 | 4 | 5 | 6 |
|---|---|---|---|---|---|---|---|
| catch | 25 | 26 | 15 | 13 | 12 | 13 | 5 |
| total ($Y_t$) | 0 | 25 | 51 | 66 | 79 | 91 | 104 |

The least squares estimators for scheme (2.5) are (Seber, 1973, p. 325-6) $\hat{\gamma} = 0.1895$ and $\widehat{(\gamma N)} = 26.11$, yielding an estimator for $N$,

$$\tilde{N} = \widehat{(\gamma N)}/\hat{\gamma} \doteq 142 \quad .$$

In the capture model, we observe the *same* system over time so that $Y_{t+1}$ must be equal to or greater than $Y_t$. In the case of a time-series like the wood cells data we observe individuals of different ages, so the difference between the sizes of adjacent cells may be negative. Also, cell size is a continuous random variable. Nevertheless, if we denote the size of a mature cell by $C$ then equation (2.4) can describe the growth to maturity where $F(t)$ now describes the size of an individual rather than a probability distribution. That is, $F(t)$ describes an *exponential growth curve* such that

$$F(t) = (C-\beta_0)(1-e^{-\gamma t}) + \beta_0 \tag{2.6}$$

denotes the *expected size* of an individual at time $t$ where $\beta_0$ is the size at time zero. Whether an individual ever *fully* matures is a moot point, but in practice the gap between $F(t)$ and $C$ is negligible for $\gamma t>5$, or thereabouts.

If we let $F(t)$ in equation (2.6) denote the systematic component $\alpha_t$, then the parameters $C$, $\beta_0$ and $\gamma$ may be estimated by nonlinear least squares; see Draper and Smith (1966, chap. 10) for details of the procedure. For the wood cell data with the newest cell relabeled as cell 1 and the oldest as 159, we find that

$$\hat{\beta}_o = 0.223 \quad , \quad \hat{C} = 38.968 \quad , \text{ and } \quad \hat{\gamma} = 0.0352 \quad .$$

The fitted growth curve is shown in Figure 1.5. The fit is not very good for the newest cells, but otherwise the curve fits quite well, accounting for 73% of the original variance.

Other capture models are described in Moran (1951). Another model for population growth is the logistic which has the form

$$F(t) = C/(1 + \alpha e^{-\beta t}) \qquad (0 \leq t < \infty) \quad ,$$

where the time origin is taken as zero without loss of generality. $F(t)$ increases from $C/(1+\alpha)$ to $C$ and allows a period of rapid growth (free from constraints) and a slowing down past $C/2$ as resource limitations restrict development. Iterative methods of fitting are needed unless C is known when we can write ($\ell n$ denotes natural logarithms)

$$\ell n[F(t)/\{C-F(t)\}] = -\ell n(\alpha) + \beta t \quad .$$

## 2.3 MOVING AVERAGES

The descriptions of trend used thus far are global. That is, a single functional form with fixed parameters is assumed to hold for all $t$ . This is non-adaptive and we might expect locally fitted trends to be more sensitive to changes in the level of the process. Clearly there is a trade-off to be made here. If the trend is defined in too local a way, then every minor fluctuation becomes part of the estimated trend.

A reasonable compromise, which smooths out local fluctuations and yet is responsive to changes in trend, is the method of *moving averages*. The trend value at time $t$ is defined by taking an average of the values of the time-series at time $t$ and neighboring times. One possibility is to average the $(2m+1)$ values $y_{t-m},\cdots,y_t,\cdots,y_{t+m}$ yielding the trend value at time $t$

$$\bar{y}_t(2m+1) = (y_{t-m} + \cdots + y_{t+m})/(2m+1)$$

$$(t=m+1,\cdots,T-m) \quad . \qquad (2.7)$$

A large value of $m$ gives a very smooth plot, but may lose essential features of the data whereas too small a value fails to remove irregular fluctuations. In practice, it is worth obtaining plots for several values of $m$ and selecting the best of these by eye. It should be noted that $m$ values at the beginning and end of the series are 'lost,' which is crucial when we wish to forecast, but need only be a minor inconvenience when we seek a general description of the data.

A simple average, as in equation (2.7) is not the only form we may select. In particular, we may wish to remove a cycle which is an even number of terms in length, such as a seasonal cycle for monthly or quarterly data. We could include an even number of terms in the moving average, but this would leave us half a time period 'out of phase.' That is, the moving average could not be used directly as the trend for any particular time period. To overcome this, we use *centered moving averages* (CMA), defined as follows: let

$$Z_t(1) = (y_{t-m} + \cdots + y_{t+m-1})/2m \quad ,$$

$$Z_t(2) = (y_{t-m+1} + \cdots + y_{t+m})/2m \quad ,$$

then the centered moving average is

$$\bar{y}_t(2m) = [Z_t(1) + Z_t(2)]/2 \tag{2.8}$$

$$= (y_{t-m} + 2y_{t-m+1} + \cdots + 2y_{t+m-1} + y_{t+m})/4m \quad .$$

Other averages may be designed to handle specific problems; for further details, see Kendall (1976, p. 29-38).

Plots of the moving average trend for the phytoplankton and wood cell series are given in Figures 2.1 and 2.2 for different values of $m$ using equation (2.7). The greater smoothing achieved by increasing $m$ is readily seen. In particular, for the wood cells series $m=2$ seems to give a much smoother trend than $m=1$. Ford and Robards (1976) note that cells formed at an average rate of 3.4 cells per radial file per day so that $m=2$ serves to smooth out much of the daily variation. As expected, the trend is broadly similar to the growth curve fitted in Section 2.2.2. The Canadian lynx series is more awkward to smooth because of the pronounced ten-yearly cycle. Any moving average other than a ten-yearly one will simply shift the peaks, damping them only slightly. To allow for this, we use a centered moving average with $m=5$ which has a dramatic effect on the lynx series (logarithms) as shown in Figure 2.3. There is little evidence of any trend over the period.

## 2.4 USE OF DIFFERENCES

In Section 2.2 and 2.3 we have concentrated upon methods of estimating the trend. Of course these methods could be used to eliminate the trend by looking at (observed value) minus (trend value). However, by using differences we go straight into trend removal. The most important differences are the first and second

$$\nabla y_t = y_t - y_{t-1} \quad , \text{ and}$$

$$\nabla^2 y_t = \nabla y_t - \nabla y_{t-1} = y_t - 2y_{t-1} + y_{t-2} \quad .$$

which remove linear and quadratic trends respectively. We shall see in Section 3.4.2 how to decide whether to make use of the original series or first or second differences when building adaptive models.

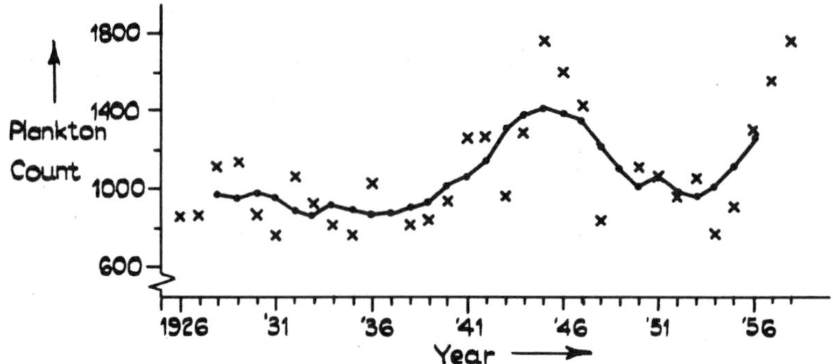


*FIG. 2.1: Moving averages for phytoplankton series* (m=2).

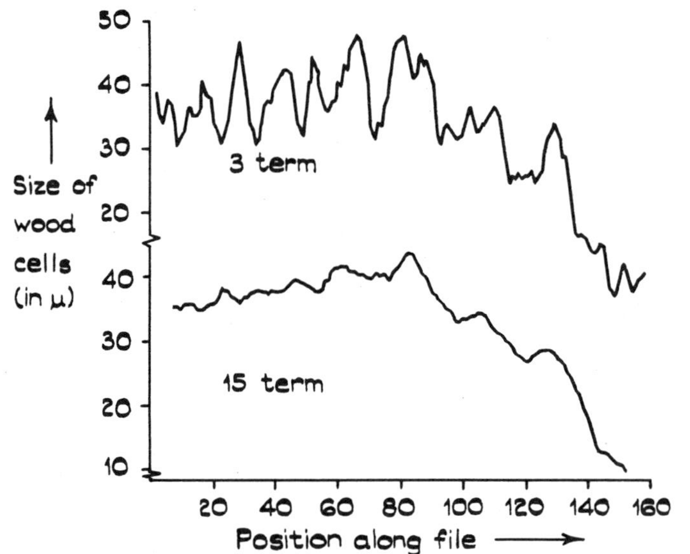


*FIG. 2.2: Moving averages for wood cells series* (m=1,7).

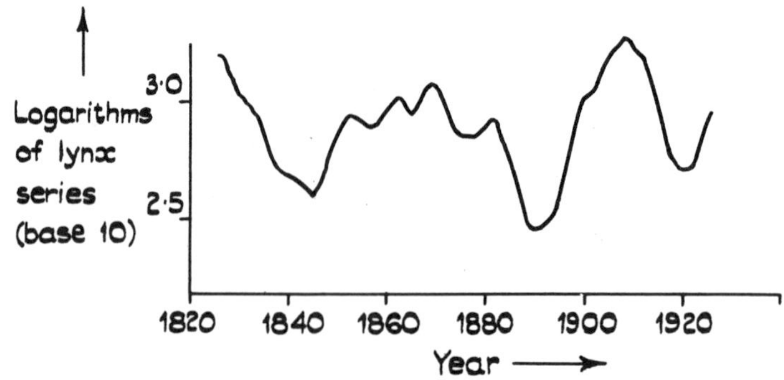


*FIG. 2.3: Centered moving averages for Canadian lynx series* (m=5).

## 2.5 THE SEASONAL COMPONENT

The seasonal effects represent deviations from the trend for each observation within a year (effects within a day or even between successive heartbeats could be considered in the same way). Such seasonal fluctuations may be due to the reproductive cycle, the weather, survival through the winter, and so on. For definiteness, we shall assume that we are dealing with four (quarterly) observations per year, but schemes for different numbers of observations are easily developed. One approach is to use the fourth order difference $\nabla_4 y_t = y_t - y_{t-4}$ . This is the approach used in the Box-Jenkins approach to modelling seasonal time-series (see Box and Jenkins, 1970, chap. 9; Poole, 1979). Alternatively, we may use a centered moving average, as in equation (2.8) with $m=2$. For this expression $\bar{y}_t(4)$ yields a deseasonalized trend since

$$\bar{y}_t(4) = (y_{t-2} + 2y_{t-1} + 2y_t + 2y_{t+1} + y_{t-2})/8$$

so that each of the four 'seasons' has equal representation. Using these trend values, we may compute the trend-free series $y_t - \bar{y}_t(4)$ for $t=3,\cdots,T-2$ . These deviations from the trend represent the seasonal-plus-random components (in the absence of cylical effects). An estimate of the seasonal effect for each quarter is then given by averaging the trend-free values for each quarter. That is, when $T = 4n$ , we estimate the seasonal effects by

$$\hat{S}_j = \sum_i [y_{j+4i} - \bar{y}_{j+4i}(4)]/(n-1)$$

with $i=1,2,\cdots,n-1$ for $j=1$ or 2 and $i=0,1,\cdots,n-2$ for $j=3$ or 4 . Similar expressions hold when $T$ is not a multiple of four. Conventionally, we adjust the seasonal effects so that $\Sigma_j S_j=0$ .

The *Cyclops* zooplankton (percentages) series given in Table 1.4 has been detrended and deseasonalized by this approach for the period 1965-71. The resulting trend line is given in Figure 2.4 and the seasonal effects are given in Table 2.1. It is evident that *Cyclops* are relatively more abundant earlier in the year and decline as the year progresses. Year-to-year fluctuations in this decline could well be due to variations in the weather, but we do not have the necessary data to explore this hypothesis.

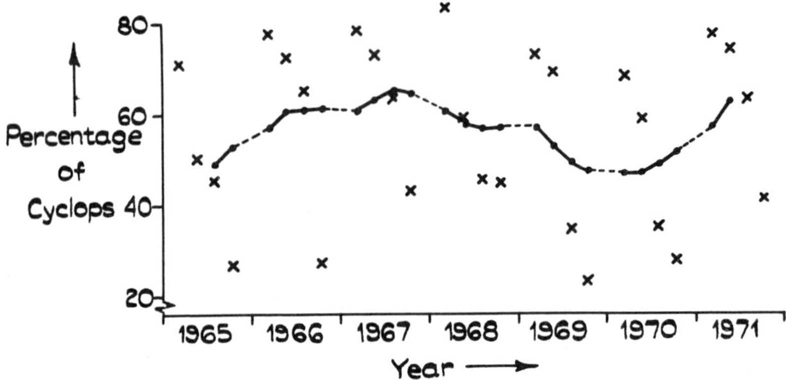


*FIG. 2.4: Centered moving averages for* Cyclops *zooplankton series (based on percentages;* m=2*).*

*TABLE 2.1: The seasonal effects* $\hat{S}_j$ *for the* Cyclops *data based upon (*a*) total numbers caught and (*b*) the percentage of* Cyclops *in the total catch.*

| Series | June 21-26 | July 11-20 | August 5-15 | Sept. 3-11 |
|---|---|---|---|---|
| | Time period | | | |
| (a) | 291.0 | -424.7 | -124.5 | 258.2 |
| (b) | 19.8 | 10.8 | -6.6 | -23.8 |

A third way of handling seasonal fluctuations is a regression method, particularly popular in econometrics. When the systematic part of equation (2.2) consists of seasonal and trend components we might write for $\alpha_{ji}$, in the $j$th quarter of the year $i$,

$$\alpha_{ji} = \beta_{j+4i} + \sum_{r=1}^{3} \lambda_r \delta_{jr} ,$$

where $\beta$ denotes the trend component, the $\{\gamma_r\}$ are seasonal coefficients, and the $\{\delta_{jr}\}$ are 'dummy' variables defined as

$$\delta_{jr} = \begin{cases} 1 & \text{if } r=j \\ 0 & \text{otherwise.} \end{cases}$$

Only three quarters have non-zero $\gamma$ coefficients to avoid a linear dependence among the variables in the regression equation. We shall not make use of this method, but it is particularly useful when multiple equation regression models are being considered.

## 2.6 CYCLICAL VARIATION

Suppose that we believe our process varies in a cyclical fashion over time. In particular, let us assume that the phenomenon follows a regular (co)sine wave with a cycle of period, or wavelength, C (see Figure 2.5). That is the cycle is repeated once every C time units. Then the systematic part of $Y_t$ may be written as

$$\begin{aligned}\alpha_t &= A\cos(2\pi t/C - \theta)\\ &= \beta\cos(2\pi t/C) + \gamma\sin(2\pi t/C) \quad , \qquad (2.9)\end{aligned}$$

where $\beta = A\cos\theta$ and $\gamma = A\sin\theta$ . The coefficient is known as the *amplitude* of the sine wave and $\theta$ is the *phase* shift relative to the time origin. When C is known, $\beta$ and $\gamma$ could be estimated by least squares in the usual way. For example, in his study of the lynx data, Bulmer (1974) used a sine wave with $C=10$ to describe the ten-year cycle.

More often, however, the period of a cycle (or the cycle *frequency* $\omega = 1/C$) is unknown and a primary aim of the study is to look for any meaningful cycles. For example, in a predator-prey situation the population size might exhibit cycles such as

| stage | 1 | 2 | 3 | 4 | 5 ... |
|---|---|---|---|---|---|
| prey | low | high | high | low | low ... |
| predator | low | low | high | high | low ... |

Whether or not these are of regular duration is debatable, but we might expect successive cycles to be of similar duration, provided that the system is not subject to any serious disturbances. In this data-analytic spirit, we might let the cyclical component include a wide range of frequencies and search for those waves with large amplitudes. Clearly, this is very much a first attempt at modelling the process and further progress is possible. For example, Bulmer (1975) has developed density dependent models of the predator-prey cycle using Fourier series.

If data are to be available at T equally spaced times, the cycle frequencies $\omega(j) = j/T$ $(1 \le j \le T/2)$ are known as the

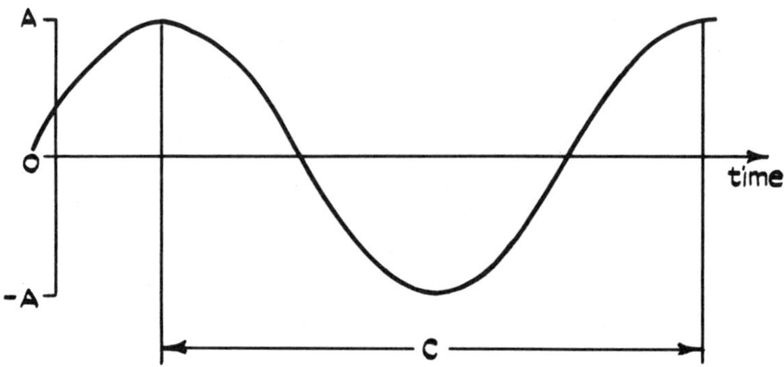


*FIG. 2.5: A sinusoidal wave with period* C.

*harmonics* (or Fourier frequencies) of the series. The Fourier representation theorem tells us that the data can always be represented by the set of cycles with these frequencies, while any higher frequency cycles cannot be distinguished using the data alone. The frequency $\omega(T/2) = 1/2$ for T even, known as the *Nyquist* frequency, is the highest frequency cycle that can be detected in discrete time series, (Kendall, 1976, p. 97). Referring back to Section 1.2, we see that the frequency of any cycles of scientific interest that we wish to study must be less than the Nyquist frequency.

*2.6.1 The Sample Spectrum.* The emphasis now shifts from the regression-type approaches considered so far, as our concern lies in breaking down the total variation into components corresponding to different frequencies (there is a direct parallel with the distinction between the fixed-effects and random-effects models for the analysis of variance (Kendall and Stuart, 1973, p. 58-60)). Thus, we specify the model, for $T=2M+1$ ,

$$y_t = \beta_o + \sum_{j=1}^{M} A_j \cos(2\pi\omega_j t - \theta_j) \quad , \qquad (2.10)$$

where $\omega_j = j/T$ and $A_j$ is the amplitude of the $j$*th* harmonic. We take T to be odd for convenience: the changes for T even are of technical interest only. The amplitude, $A_j$ , measures the contribution of the sine wave with frequency $\omega_j$ to the total variation in the time-series. If the sample variance of $Y_t$ is $s^2$ , this can be partitioned into

$$s^2 = (1/2) \sum_{j=1}^{M} A_j^2$$

so that $(1/2)A_j^2$ is the variance attributable to the *jth* harmonic. To demonstrate this, we first write

$$y_t = \beta_0 + \sum_{j=1}^{M} [\beta_j \cos(2\pi jt/T) + \gamma_j \sin(2\pi jt/T)] \quad , \qquad (2.11)$$

where $A_j^2 = \beta_j{}^2 + \gamma_j^2$ . Following Chatfield (1975, p. 130) we note that

$$\sum_{t=1}^{T} \cos(2\pi jt/T) = 0 \quad \text{for all } j \quad ,$$

$$\sum_{t=1}^{T} \cos(2\pi jt/T) \cdot \cos(2\pi kt/T) = \begin{cases} T/2 & \text{when } j=k \\ 0 & \text{otherwise.} \end{cases}$$

and the sine-sine terms have similar sums while

$$\sum_{t=1}^{T} \cos(2\pi jt/T) \cdot \sin(2\pi kt/T) = 0$$

for all $j$ and $k$ . Hence $\beta_0 = \bar{y}$ and the sample variance is

$$s^2 = \sum_{t=1}^{T} (y_t - \bar{y})^2/T = (1/2) \sum_{j=1}^{M} (\beta_j^2 + \gamma_j^2)$$

$$= (1/2) \sum_{j=1}^{M} A_j^2 \quad . \qquad (2.12)$$

Further,

$$\beta_j = (2/T) \sum_{t=1}^{T} y_t \cos(2\pi jt/T)$$

and

$$\gamma_j = (2/T) \sum_{t=1}^{T} y_t \sin(2\pi jt/T) \quad ,$$

for $j = 1,\cdots,M$ , which provides a way of evaluating the $A_j^2$ . Equation (2.12) gives a decomposition of the variance into the components attributable to each harmonic; this is known as the *spectral decomposition.*

The earliest graphical representation of this variance partition was the *periodogram,* wherein the $\{A_j^2\}$ were plotted against the wavelengths of the cycle. However, the *line spectrum,* showing the plot of the $\{A_j^2\}$ against the frequencies $\{\omega_j\}$ is now preferred.

*2.6.2 The Theoretical Spectrum.* The model in equation (2.10) is reasonable if we can be sure that only those $M$ different sine waves contribute to the variation in the time-series. However, the particular cycles included in (2.10) are a function of $T$ , the number of terms, which is scarcely an appropriate criterion for selecting frequencies. Also, it will rarely be true of ecological data that a particular frequency persists throughout the entire series. In these circumstances it is more appropriate to think of $Y_t$ comprising a continuous average of sine waves; that is

$$Y_t = \int_0^{\frac{1}{2}} A(\omega) \cos[2\pi\omega t - \theta(\omega)]d\omega \quad .$$

Then the plot of $A^2(\omega)$ against $\omega$ is the *theoretical spectrum.* Alternatively, when $\mathrm{Var}(Y_t) = \sigma^2$ , we refer to the plot of $A^2(\omega)/2\sigma^2$ against $\omega$ as the *spectral density function.* The term density function indicates that the function is non-negative and integrates to unity as does a probability density function.

Because of the Fourier representation theorem, the spectral density function gives a complete description of variation in the time-series. Spectral methods operate in the so-called *frequency domain* (our other methods are all defined in the *time domain*) and from a complete specification in one domain we can recover full information on the series in the other domain.

The reader should consult Kendall (1976, p. 95-99) or Chatfield (1975, p. 110-24) for the theoretical details.

*2.6.3 Estimation Problems.* This theoretical equivalence is encouraging but how far can we use it in practice? The first thing we find is that the sample line spectrum does not provide consistent estimators for the theoretical spectrum. Essentially, this happens because each $A_j^2$ is based upon only two degrees of freedom so that its sampling variance remains finite as $T \to \infty$. Therefore, to obtain consistent estimators of the spectral density we must smooth the sample line spectrum.

The first scheme which comes to mind is to use a moving average over adjacent frequencies such as

$$\overline{A_j^2}\,(2m+1) = \sum_{k=j-m}^{j+m} A_k^2/(2m+1) \quad . \tag{2.13}$$

This will produce biased estimators but they will be consistent provided that $m \to \infty$ as $T \to \infty$, but in such a way that $\lim(m/T) \to 0$. A reasonable ad-hoc rule for data analysis is to take $m$ near $\sqrt{T}$. Equation (2.13) defines the Daniell *spectral window*, which is one of several possible smoothing operations for the sample line spectrum. Others are discussed in Section 3.2.2.

*2.6.4 Analysis of the Lynx Data.* The form of equation (2.8) makes it clear that the model structure does not vary over time; this is known as the assumption of *stationarity* (cf. Section 3.1). Thus, in applications, it is necessary to suppose that the underlying conditions do not drastically change. Also, experience has shown that at least thirty observations are necessary before estimates are at all reliable and many time-series analysts would put the figure at fifty or more. For other comments on spectral analysis in practice, see Jenkins and Watts (1968, section 7.3).

The Canadian lynx series contains 114 terms and does not appear to exhibit any marked trend. We have evaluated the spectrum for the original series and for $Z_t = \ln Y_t$. The logarithmic transform allows for the multiplicative effect of births and deaths (steady *rates* rather than steady numbers). The spectra are plotted in Figure 2.6. There is a clear peak in each case at $\omega=0.1$ corresponding to the ten year cycle. The spectrum was smoothed using the Parzen window (see Section 3.2.2) with $m=25$. Various explanations for the ten year cycle have been put forward involving climatic and economic factors, but the predator-prey cycle seems the most logical biological explanation. This conclusion is reinforced by the results of Tanner (1975), who showed that the hare-lynx system is one of the few systems likely to have a stable limit cycle rather than a single equilibrium.

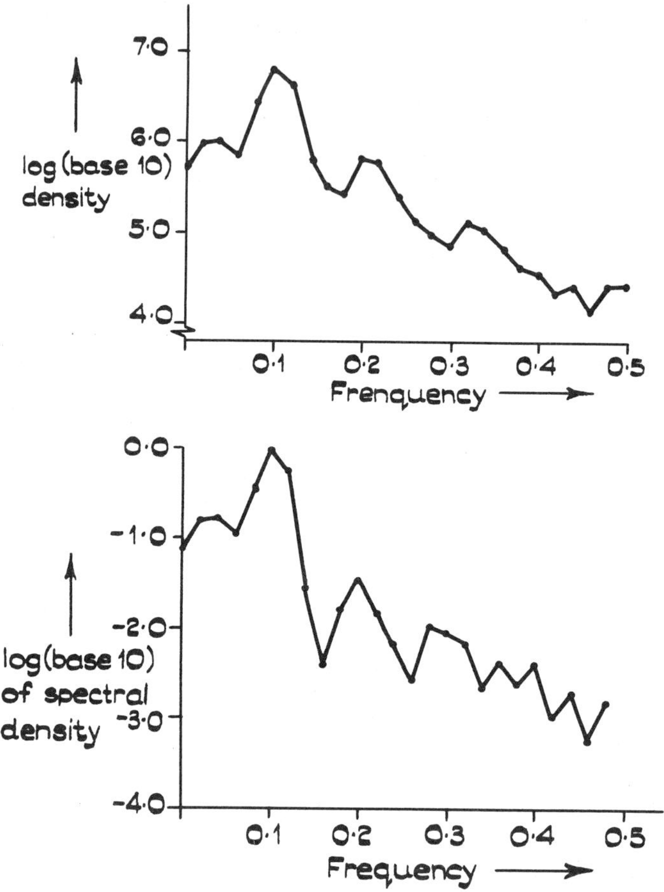


*2.6.5 Further Reading.* Our approach in this section has been deliberately 'data-analytic.' The reader wishing to delve deeper into estimation methods and the construction of confidence intervals should consult Chatfield (1975, p. 138-68) or Jenkins and Watts (1968, p. 209-318). Further, we have not considered cross-spectral methods for two or more series, for which appropriate references are Chatfield, chapters 8 and 9, or Jenkins and Watts, chapters 8, 9, and 10.

CHAPTER 3

# COVARIANCE MODELS FOR TIME-SERIES

## 3.1 INTRODUCTION

In this chapter we shall continue to look at continuous random variables that are observed at regular time intervals, while using the continuous time framework when convenient. In the previous chapter, we emphasised the 'signal plus noise' model where changes in the mean level of the process were of primary importance. However, in many biological systems, the idea of an adaptive process with feedback loops is more appealing. For example, consider the development of a biological population over time. The numbers in the population at any time clearly depend upon the offspring produced by earlier generations and thus upon the numbers in those earlier generations. Often, as in a predator-prey system, we may need to look at the interactions between two or more different species and such processes are more interesting biologically, see Poole (1979). However, at this stage, we shall restrict attention to a single time-series. To model an adaptive process through time, we shall consider the covariance between $Y(t)$ and the state of the system at earlier time periods, $Y(s)$ $(s<t)$. Thus we are led to consider the (auto)covariance structure of the process. This topic is considered in Section 3.2. The detailed specification and estimation of models is considered in Sections 3.3 - 3.6 and the chapter concludes with a brief discussion of forecasting problems.

## 3.2 AUTOCORRELATIONS

Let $Y(t)$ be the random variable describing the time-series at time $t$. We now adopt a *wide sense* specification in that we look only at the mean and covariance structure of the series, without requiring any distributional assumptions. Let the mean at time $t$ be

$$E[Y(t)] = \mu(t) \quad ;$$

the function $\mu(t)$ describes the *first-order* properties of the process. We denote the variance at time $t$ by

$$\operatorname{var}[Y(t)] = \sigma^2(t)$$

while the covariance between any pair of variates is

$$\operatorname{cov}[Y(s),Y(t)] = \sigma(s,t) \quad ;$$

the variance and covariance functions determine the *second-order* properties of the process. With such a general structure statistical analysis is difficult. The most common assumption made is that the series is *weakly stationary*; that is, we assume that the mean is constant and that the covariance structure is determined by the relative positions of points $s$ and $t$, rather than their absolute positions. This also implies that the variance is constant. Thus, for a series describing a biological population to be stationary, we would need to assume that there were no trends in the size of that population and that the 'time scale' of changes was constant; that is, an interval of so many days or months is not more critical at one place in the series than another. Thus, a fixed breeding season might produce non-stationarity within a year, but a regular series of annual observations could still be stationary.

Although the assumption of stationarity is a strong one, we can often avoid its full impact. For example, if $X(t)$ describes the size of a tree at the end of each season, the variance of $X(t)$ may increase over time and the mean certainly increases. However, the proportional increase in size, defined by

$$Y(t) = \log[\{X(t) - X(t-1)\}/X(t-1)]$$

or some similar function, may well conform to a stationary series. The choice of a suitable transformation is difficult, since transformations may be used for different reasons; e.g. in attempts to achieve both normality and stationarity. We shall use only transformations which seem justifiable on biological grounds except for the use of differences to remove polynomial trends (see Section 3.4). For further details on transformations, see Box and Cox (1964) and Poole (1979).

Henceforth it will be assumed that stationarity holds. This allows us to write

$$E[Y(t)] = \mu \quad , \text{and}$$

$$\operatorname{cov}[Y(s),Y(t)] = \sigma^2\rho(j) \quad , \quad j = |t-s| \quad ,$$

where $\rho(j)$ is the *autocorrelation function* with $\rho(0) = 1$. We note that $\rho(j)$ satisfies the usual properties of a correlation in that $|\rho(j)| \leq 1$, while $\rho(j) = 0$ when the random variables $Y(t)$ and $Y(t-j)$ are uncorrelated. We refer to $j$ as the *lag*.

Given $T$ equally spaced observed observations $y_1,\cdots,y_T$ we may estimate the unknown parameters as follows:

$$\hat{\mu} = \bar{y} = \sum_{t=1}^{T} y_t/T \quad , \tag{3.1}$$

$$\hat{\sigma}^2 = \sum_{t=1}^{T} (y_t-\bar{y})^2/T \quad , \text{ and} \tag{3.2}$$

$$\hat{\rho}(j) = r(j) = \sum_{t=j+1}^{T} (y_t-\bar{y})(y_{t-j}-\bar{y})/T\hat{\sigma}^2 \qquad (j=1,2,\cdots). \tag{3.3}$$

where $r(j)$ denotes the sample autocorrelation of order $j$. There is some debate about use of $(T-j)$ in place of $T$ in (3.3); this tends to reduce the bias but to increase the sampling fluctuations. The sample autocorrelation function enables us to look at the structure of the time-series and to search for possible cyclical variations through the covariance structure.

*Example 3.1.* For the lynx data introduced in Section 1.3, the first 25 sample autocorrelations for the transformed series (to logarithms) are given below. These results are calculated using $(T-j)$ rather than $T$ in equation (3.3). The sample autocorrelation function is plotted in Figure 3.1, this plot is commonly known at the *correlogram*.

| lag | r(j) | lag | r(j) | lag | r(j) | lag | r(j) | lag | r(j) |
|---|---|---|---|---|---|---|---|---|---|
| 1 | .792 | 6 | -.515 | 11 | .424 | 16 | -.473 | 21 | .267 |
| 2 | .346 | 7 | -.168 | 12 | -.014 | 17 | -.085 | 22 | -.143 |
| 3 | -.136 | 8 | .253 | 13 | -.434 | 18 | .301 | 23 | -.501 |
| 4 | -.512 | 9 | .589 | 14 | -.692 | 19 | .546 | 24 | -.643 |
| 5 | -.649 | 10 | .664 | 15 | -.703 | 20 | .541 | 25 | -.531 |

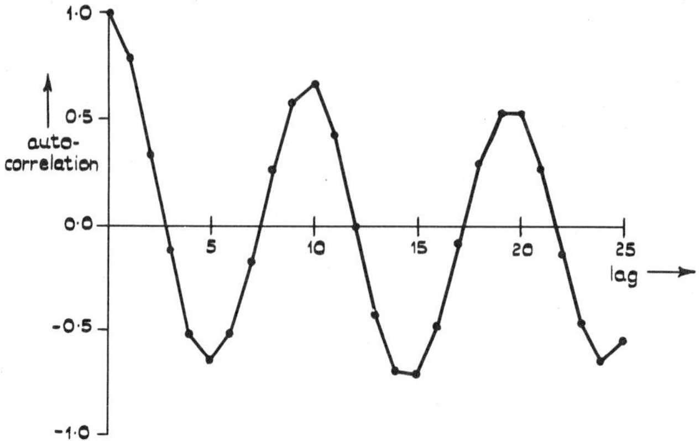


*FIG. 3.1: Correlogram for Canadian lynx series.*

In general, a high value of r(j) indicates that Y(t) is closely related to Y(t-j) , though typically the sample autocorrelations near these peaks will be quite high also. In this example, we see that there is a considerable short term effect (at lag 1) which is scarcely surprising given that the data are trapping returns. Further, we see that the empirical evidence for a ten year cycle is again unequivocal. We shall return to questions of an appropriate model in Section 3.6.

*3.2.1 A Test for Autocorrelation.* When successive terms in a time-series are independent it may be shown that the *jth* sample autocorrelation r(j) has approximate variance

$$\mathrm{var}[r(j)] \doteq n^{-1} + \text{terms of order } n^{-2} \quad .$$

Thus, a simple test of the hypothesis $H_0$: $\rho(j) = 0$ against the alternative $H_1$: $\rho(j) \neq 0$ may be carried out using a normal approximation to the distribution of r(j) provided that the sample size is not too small. For further discussion see Kendall (1976, chap. 7). Tests for successive j are *inter-related* and, in practice, the tool is used more informally. For example, we may plot limits $\pm 2/\sqrt{n}$ to check for 'interesting' autocorrelations. For the example just described n = 114 and the limits are $\pm$ 0.187, so that there is plenty of information to be gleaned from the series.

*3.2.2 The Spectral Representation.* The information conveyed by the correlogram appears rather similar to that demonstrated by our earlier analysis of the spectrum. It transpires that, given the autocorrelation function $\rho(j)$ $(j=0,1,\cdots)$, the spectral density function at frequency $\omega$ is given by

$$f(\omega) = 2[1 + 2 \sum_{j=1}^{\infty} \rho(j) \cos 2\pi j\omega] \quad . \tag{3.4}$$

That is, given the autocorrelation function for all $j$, we can evaluate the spectral density at every wavelength; the converse holds also. We say that the autocorrelation function and the spectral density function are a Fourier transform pair (cf. Chatfield, p. 110-19; Jenkins and Watts, p. 213-16). We recall from Section 2.6.2 that the spectral density function may be represented in terms of $A^2(\omega)$, where $A(\omega)$ is the amplitude at wavelength $\omega$. Thus the spectrum may be defined either directly as a partition of the variance or indirectly as the transform of the autocorrelation function. The practical consequence of these results is that the spectrum may be estimated directly (see Section 2.6.3) or indirectly using the autocorrelations. We now describe the second approach.

Given a data series containing $T$ values, we only have available the estimators $r(1),\cdots r(T-1)$ and, as we saw in Section 2.3, we cannot hope to construct consistent estimators for $f(\omega)$ without use of smoothing. Over the years, a considerable amount of effort has gone into the construction of *windows* to obtain satisfactory estimators. In Section 2.6.3 we used a *spectral window* to smooth the estimated amplitudes. Typically, when working from the autocorrelation function, we use estimators of the form

$$\hat{f}(\omega) = 2[\lambda_o + 2 \sum_{j=1}^{m} \lambda_j\, r(j) \cos 2\pi j\omega] \quad ,$$

where the $\{\lambda_j\}$ are a set of weights called the lag window and $m$ $(<T)$ is a suitable cut-off or truncation point. Two windows in common use (Chatfield, p. 138-42; Jenkins and Watts, p. 243-48) are the

*Tukey window:* $\lambda_j = (1 + \cos\pi a)/2 \quad (j=0,1,\cdots,m)$ ; and the

*Parzen window:* $\lambda_j = \begin{cases} 1 - 6a^2 + 6a^3 & (0\leq j\leq m/2) \\ 2(1-a)^3 & (m/2\leq j\leq m) \end{cases}$ ,

where $a = j/m$. The value of $m$ chosen must reflect a trade-off between increased bias (by smoothing over a lot of different frequencies) and increased variance (by smoothing over too few). The ad hoc choice of $m \doteq 2\sqrt{T}$ seems to provide a reasonable compromise between these conflicting criteria. If the estimators are to be consistent we must have $m\to\infty$ as $T\to\infty$, in such a way that $\lim(m/T)\to 0$. Our ad hoc choice conforms to this requirement, although the researcher is advised to try several $m$ values for any given $T$ to check that a reasonable balance between bias and variation has been achieved. When $m$ is too large, the spectrum will be very flat, while too small values of $m$ retain the excessive spikiness of the unsmoothed sample spectrum.

We used the Parzen window, with $m=25$, to construct the spectra for the lynx series (original and transformed) given in Figure 2.6. We now provide two estimates of the spectrum for the wood cells data of Section 1.4 using (a) all 159 observations, with $m=26$ and (b) 90 observations, corresponding to mature cells only, with $m=20$. These spectra are plotted in Figure 3.2. The first series is clearly non-stationary since the mean cell thickness is smaller for immature cells. This is reflected in the spectral plot where the components for the two smallest frequencies are about 10 times larger than the rest (remember the log plot!) Such results are typical of series with trends. Be warned!

The picture is altogether different for the 90 cell series where there is a clear peak in the spectrum at $\omega=0.075$ or period 13.3 cells and secondary peaks at $\omega=0.275$ (period 3.6) and $\omega=0.425$ (period 2.4). The main peak is consistent with Ford and Robard's (1976) suggestion of a growth surge every few days (the period observed corresponds to a four day cycle, since the average rate of production of new cells per day is 3.4). The minor peaks could correspond to the daily cycle, although their interpretation is less clear. The correlogram for the 90 cell series (Figure 3.3) shows a peak of lag 14, consistent with the spectral peak at $\omega=0.075$. Minor peaks at high frequencies often occur in sample spectra at frequencies which are multiples of the frequency at which a low frequency peak occurred. As a second example, we observe from Figure 2.6b that the peak at $\omega=0.1$ (ten year cycle) has secondary peaks at $\omega=0.2$ and $\omega=0.4$ (5 and $2\frac{1}{2}$ years respectively). In interpreting the spectrum it is necessary to be on the look-out for these echoes.

## 3.3 AUTOREGRESSIVE MODELS

The methods described so far have been primarily descriptive in nature. While the data analytic approach is an important first stage in any study, usually we wish to proceed to develop models of the process. For example, we may use a regression model of the form

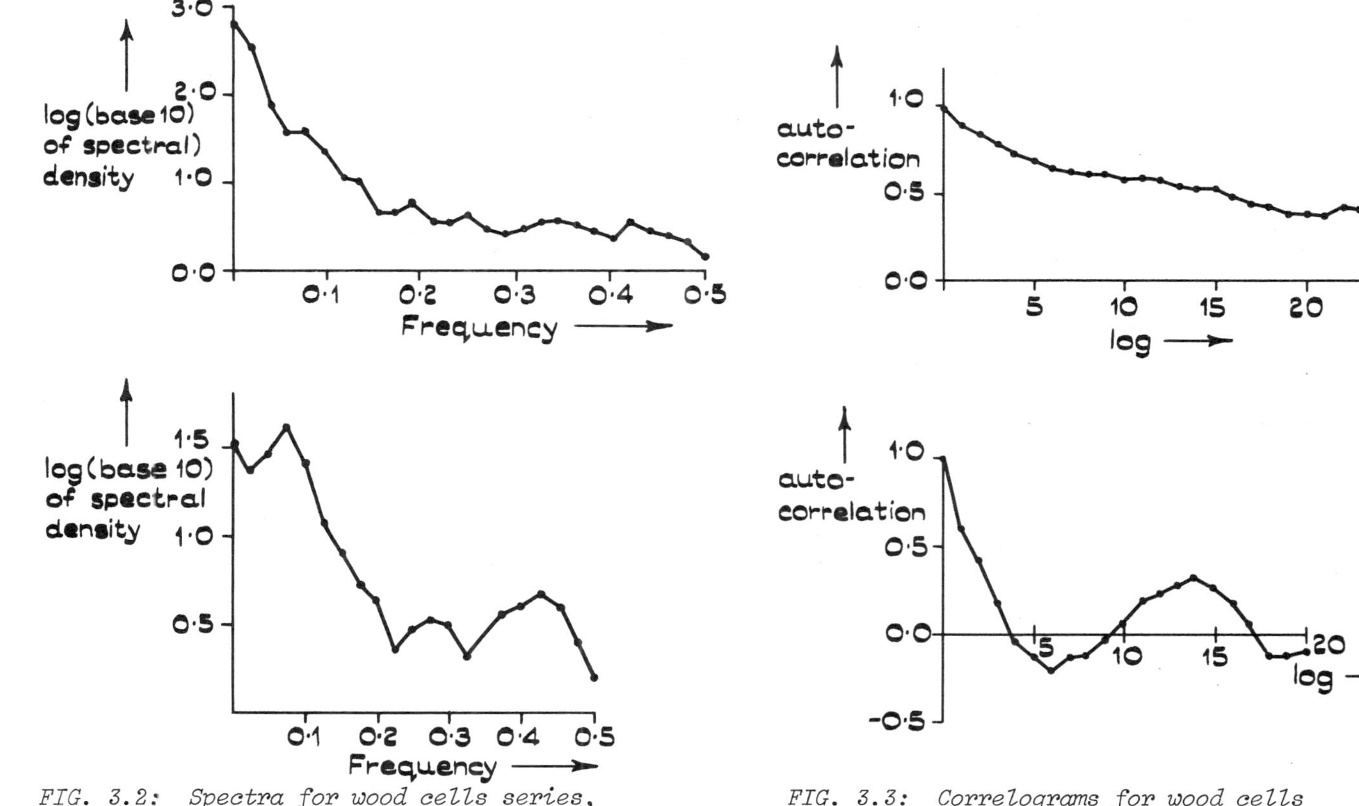


*FIG. 3.2: Spectra for wood cells series, smoothed by Parzen window, using (a) all* 159 *observations with* m=26 *and* (b) 90 *observations, corresponding to mature cells only, with* m=20.

*FIG. 3.3: Correlograms for wood cells series using* (a) 159 *observations and* (b) 90 *observations, as in Figure 3.2.*

$$Y_t = \beta_0 + \sum_{j=1}^{k} [\beta_j \cos 2\pi\omega_j t + \gamma_j \sin 2\rho\omega_j t] + \varepsilon_t$$

for selected cycles with frequencies $\omega_1,\cdots,\omega_k$ . The estimation of unknown parameters could proceed by ordinary least squares in the usual way or we might use a stepwise procedure to sift out the most important frequencies. When the set of possible frequencies is given by the harmonics of the series, this procedure is readily carried out by choosing the frequencies with the largest amplitudes.

Standard stepwise regression packages are not appropriate for this analysis, since the frequencies should be selected according to the magnitudes of the amplitudes $A_j$ , rather than the sample coefficients $\hat{\beta}_j$ and $\hat{\gamma}_j$ .

The inherent weakness in such an approach is that the model is not adaptive. For example, a sudden increase in the size of a natural population will be reflected in future generations, but this is not reflected in the cyclical model. Given the changing pressures on wildlife populations and the dependence of each generation upon its predecessors, this would seem a major flaw for population series, although other ecological data may be appropriately modelled by such means.

To introduce an adaptive element while retaining the simplicity of a linear model, we might assume that the current level of the process is determined entirely by earlier values; that is

$$Y_t = \sum_{j=1}^{p} \phi_j Y_{t-j} + \varepsilon_t \quad , \tag{3.5}$$

where we take the mean of $Y_t$ to be zero, without any loss of generality. We refer to equation (3.5) as an *autoregressive model of order* p , written as AR(p) to denote the inclusion of p lagged values of the series. There is no reason to suppose that $\phi_j \neq 0$ for all $j \leq p$ . Indeed, useful autoregressive schemes can be built up in a stepwise fashion using only a few scattered values of j .

To understand the nature of AR schemes, we look at the first order model:

$$Y_t = \phi Y_{t-1} + \varepsilon_t \quad . \tag{3.6}$$

Since $Y_{t-1} = \phi Y_{t-2} + \varepsilon_{t-1}$ and so on, successive substitution yields

$$Y_t = \varepsilon_t + \phi\varepsilon_{t-1} + \phi^2\varepsilon_{t-2} + \cdots \quad .$$

The standard assumptions concerning the random errors, $\varepsilon_t$, are that they have zero means, constant variances, and zero covariances. That is, we set $E(\varepsilon_t) = 0$, for all $t$, $\text{Var}(\varepsilon_t) = \sigma_\varepsilon^2$, for all $t$, and $\text{Cov}(\varepsilon_t, \varepsilon_s) = 0$, for all $s \neq t$. If $|\phi| \geq 1$, the variance of $Y_t$ will increase with $t$ and the process is non-stationary. Thus, provided that $|\phi| < 1$, the process will be *stationary* and we find that

$$\text{Var}(Y_t) = \sigma^2 = \sigma_\varepsilon^2/(1-\phi^2) \quad .$$

Likewise

$$\text{Cov}(Y_t, Y_s) = \sigma_\varepsilon^2\ \phi^j/(1-\phi^2) = \sigma^2\phi^j \quad , \text{ where } j = |s-t|$$

so that the autocorrelation function shows a steady (exponential) decay as $j$ increases (see Figure 3.4).

The spectral density is given by equation (3.4) and we find that, for the AR(1) process,

$$f(\omega) = \frac{2(1-\phi^2)}{(1 + \phi^2 - 2\ \phi\ \cos\ 2\pi\omega)} \qquad (0 \leq \omega \leq 1/2) \quad ;$$

this is shown in Figure 3.4.

A more direct derivation of the spectral density is possible, as we now show. Let $B$ be the *backward shift operator* so that

$$BY_t = Y_{t-1} \quad \text{and} \quad B^jY_t = Y_{t-j} \quad .$$

Then (3.6) may be written as

$$(1 - \phi B)Y_t = \varepsilon_t \quad .$$

In many respects, $B$ may be treated as a standard algebraic quantity (see Box and Jenkins, 1970, p. 8-9), so we may write this as

$$Y_t = H(B)\ \varepsilon_t \quad ,$$

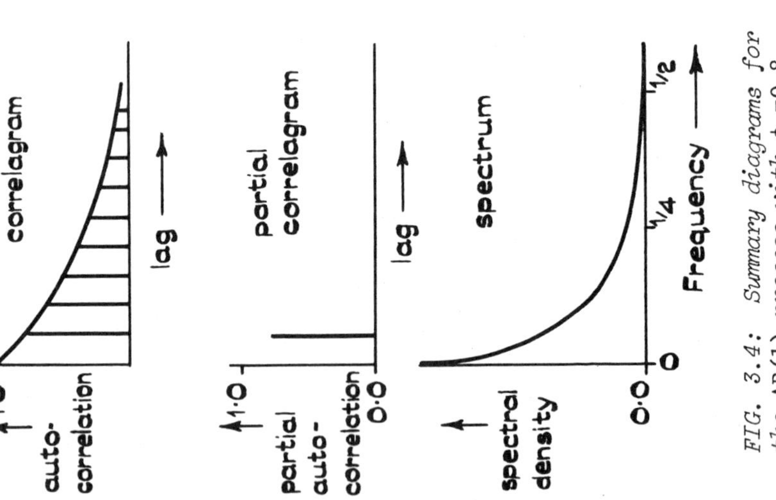


*FIG. 3.4: Summary diagrams for the* AR(1) *process with* $\phi$ =0.8.

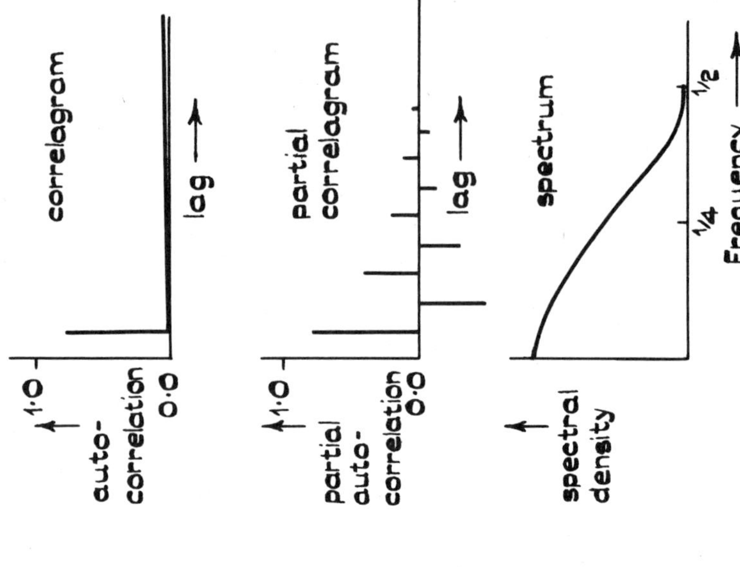


*FIG. 3.5: Summary diagrams for the* MA(1) *process with* $\theta$ =0.8.

where $H(B) = 1/(1-\phi B)$ in this case.

If we set $B = \exp(2\pi i\omega)$ , then the spectral density function for any AR process may be written as

$$f(\omega) = 2\sigma_\varepsilon^2 H(B)\, H(B^{-1})/\sigma^2 \quad (0 \leq \omega \leq 1/2) \quad . \tag{3.8}$$

Thus, the stationary AR(p) process may be written as

$$(1 - \phi_1 B - \cdots - \phi_p B^p)Y_t = \varepsilon_t \quad \text{or}$$

$$\left\{ \prod_{j=1}^{p} (1 - G_j B) \right\} Y_t = \varepsilon_t \quad ,$$

where the $G_j$ are either real roots or complex conjugate pairs. Again to ensure stationarty, we require that $|G_j| < 1$ (all the roots lie inside the unit circle). If any $|G_j| \geq 1$ , then the model is non-stationary. Non-stationarity is discussed further in Sections 3.4 and 3.6. Given stationarity, the spectral density function for the AR(p) process is of form (3.8) with

$$H(B) = \prod_{j=1}^{p} (1 - G_j B)^{-1} \quad .$$

To obtain the covariance function we multiply equation (3.5) by $Y_{t-j}$ and take expected values, so that

$$E(Y_{t-j} Y_t) = E[Y_{t-j} (\phi_1 Y_{t-1} + \cdots + \phi_p Y_{t-p} + \varepsilon_t)]$$

for $j = 1,\cdots,p$ . These expected values yield the set of $p$ simultaneous equations

$$\rho(j) = \phi_1 \rho(j-1) + \phi_2 \rho(j-2) + \cdots + \phi_p \rho(p-j)$$

$$(j=1,\cdots,p) \quad , \tag{3.9}$$

where $\rho(p-j) \equiv \rho(j-p)$ and $\rho(0) = 1$ . These are known as the *Yule-Walker equations* (Chatfield, 1975, p. 48).

Equations (3.9) may be solved to yield the $\{\rho(j)\}$ in terms of the coefficients $\phi_1,\cdots,\phi_p$ . Higher order $\rho(j)$ for $j > p$ then follow from (3.9) directly. When $\phi_1,\cdots,\phi_p$ are unknown, we may use the Yule-Walker equations as *estimating equations*. First, we use the sample autocorrelations to estimate the population values $\rho(j)$ $(j=1,\cdots,p)$ and then we solve the $p$ simultaneous equations to obtain the estimators $\hat{\phi}_1,\cdots,\hat{\phi}_p$ . Thus, the estimating equations are

$$r(j) = \hat{\phi}_1 r(j-1) + \cdots + \hat{\phi}_p\, r(p-j) \quad (j=1,\cdots,p) \quad . \tag{3.10}$$

*3.3.1 A Model for the Lynx Data.* In his pioneering study of the lynx data, Moran (1953) postulated the second order autoregressive scheme as an appropriate model:

$$Y_t = \phi_1 Y_{t-1} + \phi_2 Y_{t-2} + \varepsilon_t \quad .$$

Using (3.9) the Yule-Walker estimating equations become, from the table in Section 3.2,

$$\begin{pmatrix} 0.792 \\ 0.346 \end{pmatrix} = \begin{pmatrix} 1 & 0.792 \\ 0.792 & 1 \end{pmatrix} \begin{pmatrix} \hat{\phi}_1 \\ \hat{\phi}_2 \end{pmatrix}$$

yielding $\hat{\phi}_1 = 1.390$ and $\hat{\phi}_2 = -0.755$ . The estimates correspond to the pair of complex roots $0.695 \pm 0.521i$ so that the estimated model is stationary. Using the first two autocorrelations and the estimates $\hat{\phi}_1$ and $\hat{\phi}_2$ we may generate the higher order autocorrelations suggested by the model from (3.9). The results are as follows:

| Lag | 3 | 4 | 5 | 6 | 7 | 8 | 9 | 10 | 11 |
|---|---|---|---|---|---|---|---|---|---|
| *Autocorrelations* (×100) | | | | | | | | | |
| Observed | -14 | -51 | -65 | -52 | -17 | 25 | 59 | 66 | 42 |
| Generated by model | -12 | -42 | -50 | -38 | -14 | 8 | 23 | 25 | 18 |

The model fails to provide sufficient upturn in the autocorrelation function for lags 9-11, although it is noteworthy that the AR(2) model does capture the negative autocorrelation required over lags 3-7. The sample correlogram always tends to oscillate more than the theoretical correlogram but, even so, the model does seem to damp the oscillations too quickly. The analysis raises the important question - how is a suitable model to be chosen? We return to this topic after examining other linear processes.

## 3.4 THE MOVING AVERAGE PROCESS

Instead of introducing lags in the $Y_t$ terms, we may consider lagged $\varepsilon_t$ terms such as

$$Y_t = \varepsilon_t + \theta_1 \varepsilon_{t-1} + \cdots + \theta_q \varepsilon_{t-q} \quad ; \qquad (3.11)$$

this is known as a *moving average* process of order $q$, designated MA(q). This scheme postulates a series of random shocks to the system whose effects persist for $q$ subsequent time periods. For example, suppose that $Y_t$ designates the increase (or decrease) in the size of a population from one period to the next. An exceptionally large increase at time $t-1$ $(\varepsilon_{t-1} > 0)$ may well have an effect upon the increase during period $t$. The effect could be inhibitory (scarcity of resources) or contagious (population explosion) depending upon the life cycle of the animal or plant under consideration and upon the resources available. However, once that temporary effect has worked itself through, the system may return to its usual state.

Using the $B$ operator of the previous section we may rewrite equation (3.11) as

$$Y_t = (1 + \theta_1 B + \cdots + \theta_q B^q)\, \varepsilon_t \ . \qquad (3.12)$$

Then, by the methods of the previous section, we obtain the autocovariance function as

$$\mathrm{Var}(Y_t) = \sigma_\varepsilon^2 (1 + \theta_1^2 + \cdots + \theta_q^2) \ ,$$

$$\mathrm{Cov}(Y_t, Y_{t-j}) = \begin{cases} \sigma_\varepsilon^2 (\theta_j + \theta_1 \theta_{j+1} + \cdots + \theta_{q-j} \theta_q) & (1 \le j \le q) \\ 0 & (q < j) . \end{cases}$$

We note that the MA(q) process has only q non-zero autocovariances, while the AR(p) processes has all autocovariances non-zero. This distinction will be useful when we try to select an appropriate model (see Section 3.6).

As before we take the operator part of equation (3.12) and form the function $H(B) = \prod_{j=1}^{q} (1+F_j B)$, where the $F_j$ are the roots of the polynomial and we require $|F_j|<1$ for all $j$ to ensure stationarity. Then the spectral density function is given by equation (3.8) as before. For example when $q = 1$, the MA(1) process has

$$\mathrm{Var}(Y_t) = \sigma_\varepsilon^2 (1 + \theta_1^2) \quad ,$$

$$\mathrm{Cov}(Y_t, Y_{t-j}) = \begin{cases} \sigma_\varepsilon^2 \theta_1 & (j=1) \\ 0 & (j>1) \end{cases} \quad ,$$

$$f(\omega) = 2(1 + \theta_1^2 + 2\theta_1 \cos 2\pi\omega)/(1 + \theta_1^2) \qquad (0 \leq \omega \leq 1/2).$$

The typical shapes of these functions for $q = 1$ are given in Figure 3.5. The unknown parameters may be estimated using the autocorrelations, but this procedure is not very satisfactory and maximum likelihood estimators are to be preferred.

*3.4.1 Autoregressive Moving Average Processes.* An immediate extension of the last two sections is to produce a combined scheme with $p$ autoregressive lags and $q$ moving average components, designated ARMA(p,q) . The ARMA process is at the heart of the Box-Jenkins approach to time-series modelling and forecasting (Box and Jenkins, 1970; Kendall, 1973; Chatfield, 1975; Anderson, 1976). The full model may be written as

$$Y_t - \phi_1 Y_{t-1} - \cdots - \phi_p Y_{t-p}$$

$$= \varepsilon_t + \theta_1 \varepsilon_{t-1} + \cdots + \theta_q \varepsilon_{t-q} \quad , \qquad (3.13)$$

where $Y_t$ is taken to be stationary. It follows directly that the spectral density function for $Y_t$ is of the form given in equation (3.8) with

$$H(B) = \prod_{j=1}^{q} (1 + F_j B) \Big/ \prod_{j=1}^{p} (1 - G_j B) \quad , \qquad (3.14)$$

where the $F_j$ and $G_j$ are the roots of the polynomials and we require $|F_j| < 1$ and $|G_j| < 1$ in all cases to ensure stationarity.

The ratio form of the spectral density function provides an insight into the success of the Box-Jenkins approach in supplying a parsimonious representation of a time-series. Whenever two or three terms of either type are included in (3.14), a very flexible model is obtained for short term variations and additional terms add little to the model's descriptive power. For example, a series generated by the (2,2) model

$$(1 - 0.6B)\ (1 + 0.5B)Y_t = (1 - 0.8B)\ (1 - 0.4B)\varepsilon_t$$

may be closely approximated by the (1,1) scheme

$$(1 + 0.5B)Y_t = (1 - 0.6B)\varepsilon_t \quad ,$$

since the operator component for the (2,2) scheme may be written as

$$\frac{(1 - 0.8B)\cdot(1 - 0.4B)}{(1 - 0.6B)\cdot(1 + 0.5B)} = \frac{(1 - 0.6B)^2 - 0.04B^2}{(1 - 0.6B)\cdot(1 + 0.5B)}$$

$$\doteq \frac{(1 - 0.6B)}{(1 + 0.5B)} \quad .$$

The similarity of these two schemes should sound a cautionary note to users of these models. When fitting ARMA models it is very difficult to discriminate between models, although their structure may be quite different, see Section 3.7. Also, such models build upon the previous history of the single time-series so that while they may produce good short-term forecasts, their usefulness for making decisions about biological processes is limited.

Most studies using the Box-Jenkins framework have tended to work with models in which $p \leq 4$ and $q \leq 3$ . Such low order models may provide useful short term forecasting models for certain economic and physical processes, but longer term movements are clearly missed by such schemes. We consider the choice of model in Section 3.6.

*3.4.2 Differencing.* So far we have assumed that the time-series is stationary, but we have not suggested what is to be done when this requirement is not met. After any transformation considered necessary to ensure normality the time-series may still exhibit trends. In Section 2.4 we noted that the first and second order differences will remove linear and quadratic trends respectively. Thus, the final ingredient of the Box-Jenkins procedure is to consider differencing to remove trends from the data. In general, we define the new series

$$Z_t = \nabla^m Y_t \quad ;$$

so that $Z_t$ may be considered as a stationary time-series; $m = 0,1$ or $2$ are the cases of practical interest. Thus, the full Box-Jenkins scheme involves p autoregressive terms, m differencing operations, and q moving average terms in the AutoRegressive Integrated Moving Average process known as an ARIMA(p,m,q) scheme.

Differencing to remove trends is not a method to be used lightly in modelling time-series. For example, the two schemes

$$p = 1,\ m = q = 0 \quad \text{or } (1,0,0)$$

$$(1 - 0.9B)Y_t = \varepsilon_t$$

and $p = 0$, $m = q = 1$ or $(0,1,1)$

$$(1 - B)Y_t = (1 - 0.1B)\varepsilon_t$$

look similar, but the (1,0,0) process has $\mathrm{Var}(Y_t) = 5.26\ \sigma_\varepsilon^2$ whereas, given $Y_0$, the (0,1,1) process has unconditional variance $\mathrm{Var}(Y_t) = (0.1 + 0.9t)\ \sigma_\varepsilon^2$. The autoregressive process stays in the neighborhood of the mean of $Y_t$, since the process is stationary, but the IMA (1) process is non-stationary as its time-dependent variance indicates. This 'willingness to wander' makes the IMA(1) process responsive to changes in the mean level of the process and for this reason it is often preferred as a model for short-term forecasting (see Section 3.7).

To model time-series with seasonal components, the ARMA scheme again resorts to differencing. For example, when monthly data are available we would use $Z_t = Y_t - Y_{t-12}$, written as $Z_t = \nabla_{12} Y_t$. Likewise, we would use the operator $\nabla_4$ for quarterly data, and so on. When both trends and seasonal fluctuations are present, we may combine the difference operators and set

$$Z_t = \nabla\nabla_{12} Y_t = Y_t - Y_{t-1} - Y_{t-12} + Y_{t-13} .$$

This operator also removes quadratic trends so it is unlikely that any further differencing would be necessary. In general, if first or second order differencing does not work, a transformation should be considered in preference to higher order differences.

## 3.5 ESTIMATION FOR ARIMA MODELS

So far we have avoided any explicit distributional assumptions although this generality is rather illusory since least squares methods are identical to maximum likelihood procedures only when the error terms are independent and identically *normally* distributed. Such an assumption may be violated because the error terms (a) are correlated (or generally dependent); (b) have different variances (or different distributions); and (c) are not normally distributed.

As we saw in Chapter 1, biological populations vary in a multiplicative rather than an additive fashion. This feature suggests that the fluctuations in the time-series will be more marked when the population is large. Further the error term has a lower bound since population numbers cannot become negative. For these reasons we would suggest that the transformation of logarithms will yield a time-series whose random errors are more nearly normal and with constant variances.

If we treat $Y_t$ as normally distributed when it is not we arrive at a model imputing *linear* structure when this is not justified. While this may not have marked effects in short-term studies, long-term projections may be wildly inaccurate. The Box-Cox procedure for transforming to normality involves use of a suitable

$$Z = (Y + c)^{\lambda}/\lambda \quad (-1 \leq \lambda \leq 1) \quad ,$$

where $c$ and $\lambda$ are constants chosen to make $Z$ as near normal as possible. This procedure is described and applied to a specific problem in Poole (1979).

Once the time-series is in a form which may be viewed as consistent with the assumptions of normality, the parameters may be estimated by maximum likelihood. For details, see Box and Jenkins, (1970, p. 208-242, 267-284).

## 3.6 MODEL BUILDING

*3.6.1 Model Identification.* Once a particular model for a process has been selected, parameter estimation can proceed. But how do we select a model in the first place? The Box-Jenkins approach spells out three steps in modelling: (a) model identification; (b) parameter estimation; and (c) model diagnosis. If we decide that some form of ARIMA model is appropriate we must then select a particular scheme from that class. Clearly, any available scientific information such as the length of the reproductive cycle will be important in the selection of particular lags that ought to be included. Beyond that, we must turn to statistical methods of greater or lesser objectivity. We now look at these in turn.

I1: *Correlogram.* If we plot $r(j)$ against $j$ for $j=1,2,\cdots$ the shape of the correlogram gives an indication of the type of process. Thus, an MA(q) model would be suggested by a correlogram which died away sharply after $q$ periods, whereas the damping effect is more gradual for AR processes; see Figures 3.4 and 3.5. When the autocorrelations die away slowly (or not at all) this may be indicative of non-stationarity in the mean and differencing should be tried. If the correlograms for $m = 0,1$ and 2 are compared this will help indicate the order of differencing required.

I2: *Partial Correlogram.* Using the sample Yule-Walker equations (3.10) we may obtain the estimate $\hat{\phi}_j$ from the *jth* equation for $j=1,2,\cdots$ . Alternatively, these may be computed more efficiently using the procedure due to Durbin (1960). Given $b_1(1) = r(1)$ we evaluate successively

$$b_{j+1}(j+1) = [r(j+1) - \sum_{i=1}^{j} r(j+1-i)\cdot b_i(j)]/[1 - \sum_{i=1}^{j} r(i)b_i(j)]$$

$$b_i(j+1) = b_i(j) - b_{j+1-i}(j)\cdot b_{j+1}(j+1) \qquad (j=1,\cdots,p) \quad .$$

Then the partial autocorrelations are $b_j(j)$ $(j=1,2,\cdots)$ . The *jth partial* autocorrelation is the autocorrelation between $Y_t$ and $Y_{t-j}$ after taking due account of the relationship between $Y_t$ and the intervening variates $Y_{t-1},\cdots Y_{t-j+1}$ . Alternatively, if an AR(j) scheme is fitted, the *jth* partial autocorrelation is the coefficient of $Y_{t-j}$ in the fitted model. The plot of the partial autocorrelations against $j$ represents the partial correlograms. For an AR(p) process, these coefficients are zero for $j > p$ , while there is only a gradual decay for MA processes. Figures 3.4 and 3.5 illustrate these tendencies. Again, partial correlograms should be evaluated for $m=0,1$, and 2 to check for non-stationarity.

The identification of an appropriate model is *difficult* and this brief summary can do no more than illustrate some of the methods available. The analysis of the lynx data in section 3.6.3 and the studies by Poole (1979) help to show how these methods can be applied.

*3.6.2 The Diagnostic Phase.* From these diagrams, a short list of possible ARIMA models may be drawn up and each fitted in turn. Given the fitted schemes we need to select the most appropriate. To do this we must examine the performance of the model from several viewpoints and the final choice may not be clearcut. The following methods of analysis are useful in this context.

D1: *Residual variance.* For each scheme we compute the estimate, $\hat{\sigma}_{\varepsilon}^{2}$, which gives a guide to the overall fit of the model. Approximate standard errors may be computed for the coefficients enabling the researcher to identify terms that do not make a significant contribution. A stepwise search for important terms might be carried out.

D2: *Analysis of residuals.* For $t = p+1, \cdots, T$ the residuals $\hat{\varepsilon}_t = z_t - \hat{z}_t$ may be computed, where $\hat{z}_t$ is the value given by (3.13) using the estimated parameter values and lagged values of $z_t$ and $\hat{\varepsilon}_t$. Visual inspection of the residuals plot can often uncover discrepancies in the model or sudden changes in the character of the series. For details, see Box and Jenkins (1970, p. 287-98).

D3: *Reconstructed correlograms.* Using the estimated parameter values, we may construct the correlogram and partial correlogram for the model and contrast these with the observed correlograms. This was done in Section 3.3.1, and served to highlight the need for longer term lags, showing the AR(2) model to be inadequate.

D4: *Reconstructed spectra.* In the same way, the 'predicted' spectrum may be computed using the estimated parameter values in (3.13) and the plot compared with that given by the original series. Particular cycles that have been neglected may thus be identified. Indeed, the whole estimation procedure may be developed from the spectral viewpoint, using the Wiener-Kolmogorov filter (Bhansali, 1974) but we shall not pursue that topic here.

D5: *Information criteria.* Akaike (1974) has suggested that an information criterion be used, defined as

$$\begin{aligned} IC = & -2(\text{maximum log-likelihood}) \\ & +2(\text{number of parameters in model}) \end{aligned} \quad (3.14)$$

The model which minimizes IC is regarded as the most parsimonious representation of the series. In practice, approximate log-likelihood functions may be used when fitting

ARIMA models, but this should not seriously disturb the ordering of the competing models.

D6: *Hypothesis tests.* Quenouille (1959) developed a useful portmanteau statistic for autoregressive models. Let $b_j(p)$ denote the *jth* sample *partial* autocorrelation for the residuals when a model of order p has been fitted. Under the null hypothesis that a *pth* order scheme is correct,

$$Q_t = (n-p) \sum_{j=p+1}^{p+s} b_j^2(p)$$

has, approximately, a $\chi^2$ distribution with s degrees of freedom, for any choice of s . Unfortunately, these tests tend to be over-optimistic in that they rarely show significant results unless the original model was a poor choice. Since Quenouille's result develops from the basic result that the $b_j(p)$ are (asymptotically) independent and identically normally distributed a better bet would seem to be to plot these values on normal probability paper and to look for peculiar behavior directly.

*3.6.3 The Lynx Data Revisited.* Having developed the Box-Jenkins framework, we are now in a position to review competing models for the lynx series more effectively. It was evident from our earlier work that the Moran model damped the larger-term fluctuations too heavily, while the pure sine wave model lacks the adaptive flexibility of the ARMA models. The first eleven partial autocorrelations (×100) are

| j | 1 | 2 | 3 | 4 | 5 | 6 | 7 | 8 | 9 | 10 | 11 |
|---|---|---|---|---|---|---|---|---|---|---|---|
| $b_j(j)$ | 79 | −75 | −12 | −23 | 17 | 08 | 27 | 12 | 17 | −38 | −30 |

which suggests including terms beyond j=1 and 2 , possibly j=4, j=10, and j=11 . Thus, our conclusions are similar to those of Tong (1977) who, using the Akaike criterion, arrived at the model

$$Y_t = 1.094\, Y_{t-1} - 0.357Y_{t-2} - 0.127Y_{t-4} + 0.324Y_{t-10} - 0.362Y_{t-11} , \quad (3.15)$$

which closely reproduces the sample spectrum and correlogram. Inspection of (3.15) reveals that the model gives due weight to both short term and longer term variations. A four year cycle is not very marked in these data, although it does feature in other series from the Hudson Bay Company's records.

Similar studies have been carried out for several other series, notably by Bulmer (1974) and Anderson (1977). Bulmer used a mixed cosine wave/autoregressive model for several species, trying to link the cycles in the predator populations to those of the snowshoe hare as prey. Anderson identifies an ARMA model for the series on the colored fox in Nain, Labrador from 1834-1925. This series displays two and four year, rather than ten year, cycles and the lower order scheme

$$Y_t + \phi_2 Y_{t-2} = \varepsilon_t + \theta_2 \varepsilon_{t-2}$$

seems to give an adequate representation. Chan and Wallis (1978) develop a bivariate Box-Jenkins scheme for mink-muskrat series from the Hudson Bay records. Their model gives a good representation of the interactions between the two populations.

*3.6.4 Non-stationary Models.* Apart from differencing or the removal of simple trends, we have done very little by way of anlysis for non-stationary series. Regrettably, the methodology available for non-stationary series is much more limited than that available in the stationary case. Priestley (1965) and his co-workers have made considerable progress in developing 'evolutionary' spectra for processes which change gradually over time, although practical experience with these models is very limited. In the time domain, the Kalman filter has recently attracted considerable attention. This model was developed by engineers to allow for processes which undergo random shocks. These may be of a purely transitory nature, as in our earlier models, or may change the level of the process, or may affect the growth rate of the process. For details, the reader should consult Harrison and Stevens (1976). Although such models have not been used thus far in ecology, the increased interest in predator-prey models with density-dependent coefficients (e.g. Bulmer, 1975) suggests that these methods may be of value.

## 3.7 FORECASTING

Forecasting procedures are valuable for two main reasons. They may be used both as a diagnostic device and, in their own right, for future policy decisions such as allowed kills of game, annual catches of whales, and so on.

The methods of model identification considered earlier are by no means fool-proof. For example, the two schemes

$$\text{AR}(1) : Y_t = \phi Y_{t-1} + \varepsilon_t$$

$$\text{IMA}(1) : Y_t - Y_{t-1} = \varepsilon_t + \theta\varepsilon_{t-1}$$

are often identified as alternatives. As we saw in Section 3.4.1 these models seem quite similar when $\phi \doteq 1-\theta$, $\theta$ relatively small, but the AR(1) model is stationary while the IMA(1) is not and the two produce quite different forecasts. Thus, if the AR(1) model is used to forecast the series, the L step ahead forecast is

$$\hat{y}_{t+L} = \alpha^L y_t \quad ;$$

that is, the forecast settles down and approaches the mean value; here taken as zero. However the IMA(1) scheme yields

$$\hat{y}_{t+1} = y_t - \theta^L e_t \quad ,$$

where $e_t = y_t - \hat{y}_t$, the observed error in the latest forecast. The IMA(1) forecast tends to stay at the present level of the process rather than revert to any 'mean,' reflecting its non-stationary nature. When $L=1$, we obtain

$$\hat{y}_{t+1} = (1-\theta)y_t + \theta\hat{y}_t \quad ;$$

this is known as the *exponetially weighted moving average* (EWMA). When $\theta$ is close to 1, the EWMA adapts slowly, while for $\theta$ near zero the EWMA adapts rapidly to changes in the level of the process.

From the statistical viewpoint, use of the model to forecast ahead is a useful diagnostic check. Indeed, it is good practice to split the data into two parts and retain about a quarter of the series for testing the model generated by the first three quarters of the time-series. In this way, the ability of the model to respond to changes in the process can be examined more objectively. For example, ten-step ahead projections using the Moran model fail to capture the regular ten year cycle in the lynx data.

In general, forecasts from AR processes presuppose that the system will settle back towards an equilibrium, whereas those from IMA schemes assume that the future is determined by current level of the process. The success of the ARIMA schemes is due to their ability to trade-off tendencies in a short-term forecasting model.

Forecasting is a hazardous business in any area and it must be evident that single-series methods can only hope to work when the system under study is free from outside disturbances. For example, the main prey for the lynx is the snowshoe hare and some authors (Bulmer, 1978) have argued that it is inherent fluctuations in the hare population which lead to the ten-year cycle. In the absence of outside disturbances, the single series approach can describe the fluctuations satisfactorily, but any changes which affect the hare population directly will be ignored by the model.

Thus, it would be more satisfactory to develop transfer function models which link the predator and prey populations. Such a development is beyond the scope of our work here and the reader is referred to Chan and Wallis (1978) and to Poole (1979) for studies using this approach.

In conclusion, the single series approach has a useful role to play in ecological modelling and forecasting, but we hope that the reader will bear in mind both the advantages and the limitations of the method.

CHAPTER 4

# TEMPORAL POINT PROCESSES

## 4.1 INTRODUCTION

In the two previous chapters, we supposed that the process was recorded at regular time intervals. Thus, time was a coordinate, serving to index the random variable of interest, population size or whatever. We now direct our attention to situations where the random variable is the time between successive events. Often, these events will involve the arrival and/or departure of 'individuals' and the random mechanisms are known as birth and/or death processes. First of all, we look at the Poisson process and in Section 2 we look at tests and estimation procedures based upon the Poisson process. The work in this chapter finds a natural extension to two or more dimensions in the papers of Diggle (1979) and Cormack (1979) on spatial point processes.

*4.1.1 The Poisson Process.* This is known as a 'pure birth' process, since individuals accumulate over time. However, this is a rather confusing terminology and we shall refer to *events* occurring within certain time intervals. Thus, we let $Y(t)$ denote the number of individuals captured or recaptured by time $t$, the number of distinct species of a certain genus identified in a study area, and so on. Our theoretical development will follow along the lines of Cox and Lewis (1966).

Consider the time interval $[0,t+h)$ broken into two parts $[0,t)$ and $[t,t+h)$, as in Figure 4.1. We suppose that the time interval $h$ is so small that the probability of more than one event occurring in the interval $[t,t+h)$ is negligible. (Technically, we assume that the probability is *of smaller order than* $h$, written as $o(h)$, which means that these terms will disappear when we consider limits as $h$ approaches zero.)

Henceforth we exclude the possibility of multiple events in a period of duration $h$ so that the only possibilities that

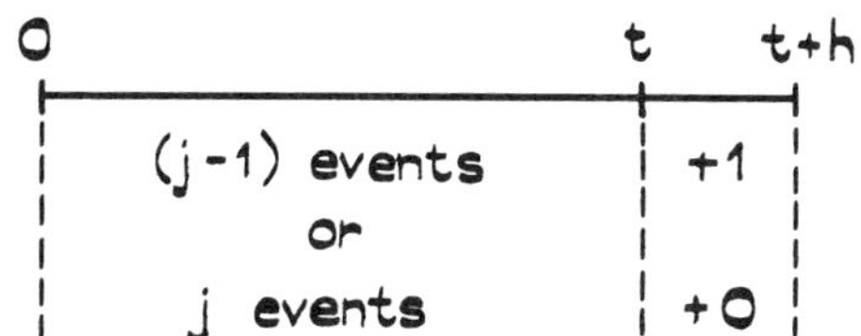


*FIG. 4.1: Possible ways of achieving events in* [0,t+h).

can occur (for some j) are as shown in Figure 4.1. These outcomes are mutually exclusive and exhaustive (only one can occur and one must occur) so we may write

$$\begin{aligned}&\Pr[j \text{ events by } (t+h)]\\ &\quad = \Pr[j \text{ events by } t]\,\Pr[\text{no event in } h, \text{ given } j \text{ by } t]\\ &\quad + \Pr[(j-1) \text{ events by } t]\,\Pr[1 \text{ event in } h, \text{ given } (j-1) \text{ by } t]. \qquad (4.1)\end{aligned}$$

We now write "$j$ events occur by time $t$" as "$Y(t) = j$" and we denote the probability of its occurrence by

$$\Pr[Y(t) = j] = p_j(t) \quad .$$

Further, we suppose that

$$\begin{aligned}&\Pr[\text{no event in } h, \text{ given } j \text{ by } t] = 1 - \lambda h + o(h)\\ &\Pr[1 \text{ event in } h, \text{ given } (j-1) \text{ by } t] = \lambda h + o(h)\end{aligned} \qquad (4.2)$$

so that the probability of multiple events in $h$ is of $o(h)$ . As we intend to let $h \to 0$ , these higher order terms will disappear. After these assumptions, equation (4.1) becomes

$$p_j(t+h) = p_j(t)\cdot(1-\lambda h) + p_{j-1}(t)\,\lambda h + o(h)$$

or

$$\frac{1}{h}[p_j(t+h) - p_j(t)] = \lambda p_{j-1}(t) - \lambda p_j(t) + o(1) \quad .$$

Letting $h \to 0$ yields the differential equations

$$\frac{d}{dt} p_j(t) = \lambda p_{j-1}(t) - \lambda p_j(t) \quad (j=1,2,\cdots) \quad ,$$

while for $j = 0$

$$\frac{d}{dt} p_0(t) = - \lambda p_0(t) \quad .$$

Recalling that $Y(0) \equiv 0$ so that $p_0(0) = 1$ , we may verify by direct substitution that

$$p_j(t) = e^{-\lambda t}(\lambda t)^j/j! \quad (j=0,1,\cdots) \quad . \tag{4.3}$$

These are the terms of the Poisson distribution, justifying the name Poisson process. We note that the key assumptions leading to the Poisson process are (i) no multiple events and (ii) the probability of an event is independent of the past history of the process (the Markov assumption). These are strong conditions to impose and are rarely if ever satisfied completely. Nevertheless, the Poisson process provides a useful benchmark in the study of point processes and it is in this spirit that we shall continue to use it.

The process obtained in (4.3) is known as the *homogeneous* Poisson process, since it can be shown that

$$\Pr[j \text{ events in } [t,t+u)] = \Pr[j \text{ events in } [t',t'+u)]$$

for any choices of $t$ and $t'$ , whether or not the intervals are partially overlapping. That is, the time origin has no significance beyond its role as the 'official' starting point of the record. Suppose now that the constant $\lambda$ in (4.2) is replaced by the time dependent rate $\lambda(t)$ so that

$$\Pr[1 \text{ event in } h, \text{ given } (j-1) \text{ by } t] = h\lambda(t) + o(h) \quad ,$$

and so on. If $\Lambda(t) = \int_0^t \lambda(u)du$ , it follows that

$$p_j(t) = \{\Lambda(t)\}^j \, e^{-\Lambda(t)}/j! \quad (j=0,1,\cdots) \quad . \tag{4.4}$$

This result may be checked by substitution. Equation (4.4) denotes the *non-homogeneous* Poisson process, as $\lambda(t)$ may be used to describe trends or seasonal variations in the rate of occurrence of events.

*4.1.2 Other Processes.* We may consider an even more general version of (4.2) by setting

$$\Pr[1 \text{ event in } h, \text{ given } j \text{ by } t] = h\omega(j,t) \quad .$$

where $\omega(j,t)$ denotes the rate of occurrence of new events. Some restrictions are necessary to make analytic progress and the most common assumption is to set

$$\omega(j,t) = a(j)\ \lambda(t) \quad .$$

Boswell and Patil (1972) describe the variety of distributions which arise from this class of processes. In Section 4.1.1 we took $a(j) = 1$ to obtain the Poisson distribution (with a non-homogeneous process when $\lambda(t)$ is not constant). With $a(j) = n-j$ we get the binomial and with $a(j) = k+j$ , the negative binomial. The process of most direct interest is the simple linear birth process (Cox and Miller, 1965, p. 156) with

$$\omega(j,t) = (k+j)/(c+t) \quad . \tag{4.5}$$

This yields the negative binomial form

$$p_j(t) = \binom{k+j-1}{j} t^j c^k/(c+t)^{j+k} \qquad (j=0,1,\cdots) \quad .$$

## 4.2 SPACINGS

Given a simple basic process like the Poisson, a natural starting point in many investigations is to see how far the Poisson assumptions are satisfied. Many procedures exist for goodness-of-fit tests for the Poisson distribution, but we shall not pursue that line of thought. Instead, we exploit the *duality* between 'events per unit time' and 'time to n*th* event,' as follows:

> let $Y(t)$ denote the number of events in $[0,t)$ and let $T(n)$ denote the elapsed time until the n*th* event occurs. Then, whatever the process,
>
> $$\Pr[Y(t) < n] = \Pr[T(n) > t] \quad .$$

In particular, for the Poisson process and $n = 1$

$$\Pr[Y(t) < 1] = e^{-\lambda t} = P[T(1) > t]$$

so that the length of the time interval up to the occurrence of the first event has an *exponential distribution* with density function

$$f(t) = \lambda e^{-\lambda t} \quad (t > 0) \quad . \tag{4.6}$$

By similar arguments, the length of the time interval up to the occurrence of the *nth* event has density function

$$f_n(t) = \lambda^n t^{n-1} e^{-\lambda t}/(n-1)! \quad ;$$

that is, a *gamma distribution* with index $n$ . For further details of these distributions, see Boswell, Ord, and Patil (1979, Chap. 4).

Further it follows from the Markov assumption quoted in the last section that past and future are independent, so that the time intervals between successive events are independent exponential random variables, each with density function given by equation (4.6). These intervals are known as spacings (Pyke, 1965) and a variety of test criteria have been developed from them. Suppose that the process has been observed for time $t$ and let

$$d_i = (t_i - t_{i-1})/t \quad (i=1,2,\cdots,n+1)$$

denote the *ith* spacing, where $t_0 = 0$ and $t_{n+1} = t$ . The spacings may be ordered in increasing size as

$$d_{(1)} \leq d_{(2)} \leq \cdots \leq d_{(n+1)} \quad .$$

Durbin (in the discussion on Pyke's paper) suggested the statistic

$$S = 2(n+1) - 2 \sum_{i=1}^{n+1} id_{(i)}$$

which gives large (negative) weight to the larger spacings so that the Poisson assumption should be rejected for extreme values of S. Durbin shows that, under the Poisson null hypothesis, the distribution of $S^* = (S - n/2)/\sqrt{(n/12)}$ rapidly approaches the normal, with zero mean and unit variance. Thus tests of the null hypothesis are readily performed.

To illustrate the use of this method, we use data on the intervals between successive pulses along a nerve fibre, measured in units of 1/50*th* of a second. These data were collected by Fatt and Katz and only the first 8 seconds of the record are used here. The full record of over 800 intervals is recorded in Appendix I of Cox and Lewis (1966). The intervals were (read down columns)

| | | | | | | | | | |
|---|---|---|---|---|---|---|---|---|---|
| 10.5 | 3.0 | 7.0 | 2.5 | 12.0 | 4.5 | 12.0 | 2.0 | 7.5 | 14.0 |
| 1.5 | 9.0 | 9.5 | 7.5 | 8.0 | 1.5 | 14.5 | 1.0 | 16.5 | 18.0 |
| 2.5 | 27.5 | 1.0 | 11.5 | 3.0 | 10.5 | 8.0 | 7.5 | 3.0 | 7.0 |
| 5.5 | 18.5 | 7.0 | 7.5 | 5.5 | 1.0 | 3.5 | 6.0 | 25.5 | (5.0) |
| 29.5 | 4.5 | 4.5 | 4.0 | 7.5 | 7.0 | 3.5 | 13.0 | 5.5 | |

The last interval denotes the gap from the last pulse up to the end of the recording period. We find that $n = 48$ and $S = 28.63$ whence $S^* = 2.42$, which is just significant at the 1% level, for a two tailed test. The positive value of $S^*$ suggests that the distribution of spacings may be slightly more regular than exponential. This may be due to the data recording procedure, since intervals are recorded to the nearest 0.5, with an apparent minimum of 1.0. Interestingly, Cox and Lewis (1966) do not detect any departure from the null hypothesis in their examination of the full data set.

As with any hypothesis test, departures from the null hypothesis may be due to the failure of one or more of several assumptions. Possible causes of clustering and time-heterogeneity should always be considered, along with the less constructive possibilities such as the limitations of the data-recording devices.

*4.2.1 The Broken Stick Model.* The standardized spacings, $d_i$, may be regarded as random partitions of the unit interval $[0,1]$. Thus, if the unit interval is used to describe total available 'resources,' its random partition is a simple model of resource allocation. This process corresponds to the 'broken-stick' model suggested (but later rejected) by MacArthur (cf. Pielou, 1975, p. 19-31) as a simple model for the relative abundances of species.

Cliff *et al.* (1975, chap. 3) found that the model was of some descriptive value for human populations (in different geographical areas), provided that a threshold was introduced. That is, we set

$$d_i = d_i^* + c \quad (c > 0) \quad ,$$

where $c$ represents a minimum allocation of resources to each area (or species). Then the $\{d_i^*\}$ may be considered as spacings on an interval of length $1 - (n+1)c$.

The parameter $c$ might describe the minimum level of resources needed to sustain a viable population of a species. In their study, Cliff *et al.* used the $S$ statistic to test the random allocation plus threshold model and found that it worked reasonably well when the geographical areas considered were homogeneous in character (e.g. all rural or all urban, but not mixed).

Returning to our example, if we consider a threshold of 1/50 second, corresponding to the minimum recorded interval between events, then $S$ is reduced to 26.16 and $S^* = 1.08$, consistent with the null hypothesis.

*4.2.2 Estimation for the Poisson Process.* If we are prepared to postulate a particular form of process, then interest focuses upon the estimation of unknown parameters, rather than hypothesis testing. Cox and Lewis (1966, p. 45-6) show that the likelihood function reduces to

$$[\prod_{i=1}^{n} \lambda(t_i)]\cdot\exp[\Lambda(T)] \quad . \tag{4.6}$$

where $\Lambda(T) = \int_0^T \lambda(u)du$. If the homogeneous Poisson is proposed, then $\lambda(t) = \lambda$, for all $t$ and the usual estimator $\hat{\lambda} = n/T$ follows directly. If a trend is suspected, Cox and Lewis recommend use of the log-linear form

$$\lambda(t) = e^{\alpha+\beta t}$$

which keeps $\lambda(t) \geq 0$ for all values of $\alpha,\beta$, and $t$. In this case the log-likelihood function becomes

$$n\alpha + \beta \sum_{i=1}^{n} t_i - e^{\alpha}(e^{\beta T}-1)/\beta \quad .$$

The estimates $\hat{\alpha}$ and $\hat{\beta}$ can be found by Newton-Raphson or other numerical methods. It is evident that $n$ and $\Sigma t_i$ are the sufficient statistics in this case.

An interesting extension of the non-homogeneous Poisson in the spatial case is discussed by Kooijman (1979) and Hengeveld (1979).

CHAPTER 5

# MEASURES OF SPATIAL DEPENDENCE

## 5.1 INTRODUCTION

We began by considering random variables $Y(\underset{\sim}{x},t)$ that are indexed both by their location $\underset{\sim}{x}$ and their position in time $t$. We then concentrated upon the time dimension only, which is often appropriate at an aggregate level when we deal with the behavior of entire populations. However, when we wish to study interrelations between members of the same population, their locations in space become important. Again, a variety of different situations arises as the examples in Table 5.1 show. The main variations are summarized below:

*Space as a continuum.* Many processes, such as the input of solar energy, may be defined for all points in space and time. For measurement purposes, it may be necessary to aggregate such variables in one or both 'dimensions' to consider solar energy input per unit area or time. Alternatively, we may have a finite network of measurement sites at which to monitor the process.

*Discrete space.* Other processes may occur only at a set of physically separate sites, and the random variable is defined only at those sites. Such sites may be regularly spaced (as on a lattice) or irregularly spaced.

*Type of random variable.* As for time-series, the random variable may be continuous or discrete or may relate to a point process describing the location pattern of individuals.

It will be evident from the examples in Table 5.1 that at the scale of activity of interest, the time dimension may or may not be very important. Clearly, time is important in the day to day competition between crowded plants struggling for survival; [see Cormack (1979)]. Equally, when we come to study the spatial pattern of trees in an established forest, the rate of ecological succession is so slow that time may reasonably be ignored. In

*TABLE 5.1: Examples of different kinds of spatial processes.*

| Nature of the random variable | Continuous space | Discrete space |
|---|---|---|
| Continuous | Solar energy, crop yield | Size of trees |
| Discrete | Number of plants in a quadrat | Egg masses on a tree or plant |
| Point process | Location of trees or plants | Location of diseased plants within an initial population |

this chapter we concentrate upon purely spatial patterns. Further, we restrict attention to variates measured at particular locations, rather than point patterns which are discussed fully in Cormack (1979) and Diggle (1979).

*5.1.1 Types of Analysis.* As in the study of time-series we may look initially at tests of spatial independence, to determine whether or not there is any spatial structure in the data. Such tests play a more important role in spatial analysis than in the study of time-series. In part this reflects the relative shortage of analytic tools for spatial processes, although the test statistics often provide useful summary measures. But there is more to it than that. The one dimensional development of a process through time allows considerable simplification in the identification of temporal patterns. In space, no such unilateral dependence exists and we mush search for regularities of scale, direction, and orientation as well as distance. Especially for point patterns, the comparison of data with the norms of some 'standard' processes seems a useful point of departure in tackling the overall analysis (see Diggle, 1979). Again, for processes defined at given locations, the diversity of possible patterns of dependence makes initial tests a useful data-monitoring technique.

When spatial dependence is suspected or established, we wish to develop models of spatial processes, which is the topic reserved for Chapter 6.

## 5.2 TESTS FOR SPATIAL DEPENDENCE

Suppose that we are given a set of sites, which we may label $1,2,\cdots,n$. For experimental data such sites often lie on a regular grid, either as individuals located on the points of the grid or as spatial aggregates (counts or averages over a quadrat).

```
- 1 1 - 1 1 - 1 1 1
1 - - - - - 2 2 1 1
- - 1 1 - 3 2 3 - 2
- 2 - 2 - - - 1 - 3
- - - 1 - 2 3 1 - 3
- - 1 - - 1 5 3 1 -
- - 1 - - 1 - 1 3 3
1 - - - 1 1 - 4 1 2
- - 1 - 1 3 1 1 2 2
- - - 1 - - 4 - - 4
```

*FIG. 5.1: Per cell abundances of* Cossura sp. *(from Jumars et al., 1977). Dashes indicate that no specimens were found.*

For example, Jumars *et al.* (1977) report the abundances for several polychaete species on a 10 × 10 array of contiguous one inch square (6.45 $cm^2$) cores at six meters depth in St. Andrews Bay, Panama City, Florida. The data were collected using a 'centuple-cell sampler' and the results for *Cossura sp.* are given in Figure 5.1.

The general question of interest is whether such data display spatial dependence, be it clustering or regular spacing, at the given scale of measurement. A glance at Figure 5.1 immediately suggests clustering in the right hand part of the grid. Our purpose is to develop formal methods capable of detecting this and other, less obvious, patterns of spatial dependence.

Let $Y_i$ be the random variable describing the *ith* site or cell; $i = 1, \cdots, n$ . We assume that the process is *spatially stationary* (cf. Section 3.1), so that

$$E(Y_i) = \mu, \quad Var(Y_i) = \sigma^2 \quad ; \text{ and}$$

$$Cov(Y_i, Y_j) = \sigma^2 \rho_{ij}$$

where the autocorrelation, $\rho_{ij}$ , depends upon the *relative* positions of sites $i$ and $j$ rather than their absolute positions. This assumption may also be represented by setting

$$\rho_{ij} = \rho(d_{ij}) \quad ,$$

where $d_{ij}$ is the 'distance' between sites $i$ and $j$ according to some suitable measure of distance. Then $\rho(d_{ij})$ is the spatial autocorrelation function.

The null hypothesis of interest is that the process does not exhibit any spatial dependence; that is,

$$\rho_{ij} = 0 \text{ for all } i \neq j \quad .$$

If we let $\underset{\sim}{V}$ denote the $n \times n$ covariance matrix for $\underset{\sim}{Y}$ then the null hypothesis may be expressed as

$$H_0 : \underset{\sim}{V} = \sigma^2 \underset{\sim}{I}$$

against the general alternative

$$H_1 : \underset{\sim}{V} \neq \sigma^2 \underset{\sim}{I} \quad .$$

for which $\rho_{ij} \neq 0$ in general. Without further assumptions, it is not possible to specify an appropriate test. However, Cliff and Ord (1973, p. 134-35) show that when the joint distribution of the $\underset{\sim}{Y}$ under $H_1$ is multivariate normal, with common mean $\mu$ and covariance matrix $\sigma^2 \underset{\sim}{V}$, then the test statistic takes the form

$$\underset{\sim}{z}' \underset{\sim}{V}^{-1} \underset{\sim}{z} \;/\; \underset{\sim}{z}'\underset{\sim}{z}$$

where the vector $\underset{\sim}{z}$ has elements $z_i = y_i - \bar{y} \quad (i=1,\cdots,n)$ .

A useful single parameter model for spatial dependence has the covariance matrix

$$\underset{\sim}{V} = \sigma^2 (\underset{\sim}{I} - \beta \underset{\sim}{W})^{-1} \quad ; \tag{5.1}$$

see Section 6.3 for details. Here $\beta$ is a parameter which measures the strength of the spatial dependence (the null hypothesis corresponds to $\beta = 0$) and $\underset{\sim}{W} = \{w_{ij}\}$ is a symmetric matrix of known *weights*, with $w_{ii} = 0$ for all $i$ , by convention.

The weights are selected to reflect the directions in which spatial dependence is suspected *a priori*. For example, if seedlings are planted close together on a square lattice, then a given seedling will compete for light and nutrients with its four immediate neighbors (to the north, east, west, and south). If the random variable describes the size of the seedling, the acute competition would be reflected in negative correlation between the size of neighboring plants. Therefore, to detect any such negative correlation, we might set

*TABLE 5.2: Common choices of weights for spatial autocorrelation coefficients (with obvious modifications at edges and corners).*

| Pattern of sites | Graphical form | Algebraic form |
|---|---|---|
| Square lattice (rook's case) | 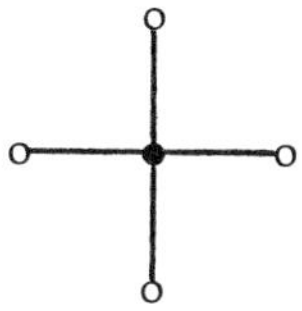 | $w_{ij}$ = 1 if j is a vertical or horizontal neighbor of i |
| Square lattice (queen's case) | 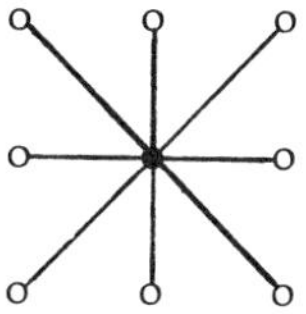 | $w_{ij}$ = 1 if j is a vertical, horizontal, or diagonal neighbor of i |
| Triangular lattice | 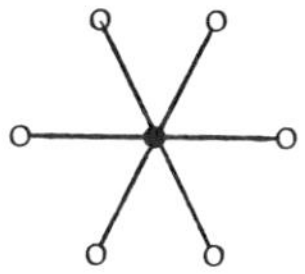 | $w_{ij}$ = 1 if j is one of six equidistant neighbors of i |
| (Possibly) irregular | 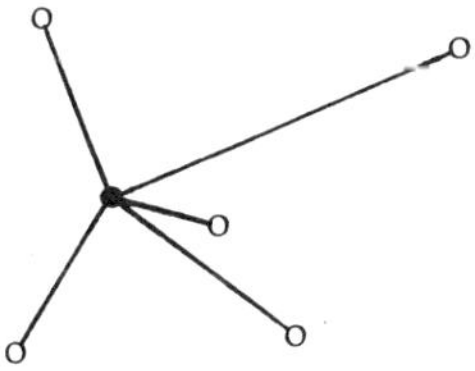 | $w_{ij} = \exp(-\alpha d_{ij})$ |
| (Possibly) irregular | 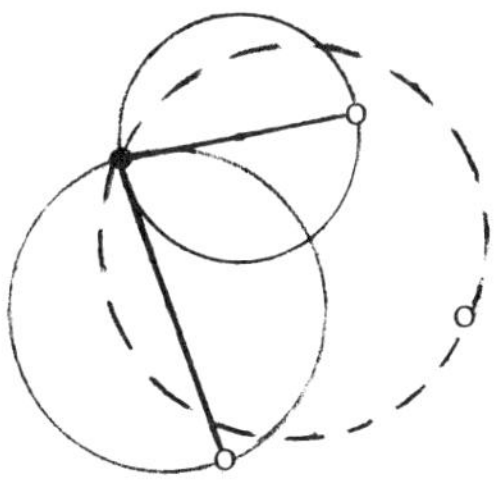 | $w_{ij}$ = 1 if circle of diameter $d_{ij}$ passing through i and j contains no other points (Gabriel graph) |

*Note:* $w_{ij}$ = 0, except as specified; $d_{ij}$ is the distance between points i and j.

$$w_{ij} = 1, \text{ if } i \text{ and } j \text{ are immediate neighbors}$$
$$w_{ij} = 0, \quad \text{otherwise.}$$

The weights are determined only up to an arbitrary constant, so that nothing is lost by setting $w_{ij} = 1$ rather than some other single number. Several other possible patterns of weights are summarized in Table 5.2; also, see Tobler (1975).

If we take $\underset{\sim}{V}$ as in equation (5.1) then

$$\underset{\sim}{z}'\underset{\sim}{V}^{-1}\underset{\sim}{z} = \underset{\sim}{z}'\underset{\sim}{z} - \beta\underset{\sim}{z}'\underset{\sim}{W}\underset{\sim}{z}$$

and the test statistic reduces to

$$I = \alpha\underset{\sim}{z}'\underset{\sim}{W}\underset{\sim}{z}/\underset{\sim}{z}'\underset{\sim}{z} = \alpha\Sigma_{(2)} w_{ij}z_iz_j/\Sigma_{(1)}z_i^2$$

where $z_i = y_i - \bar{y}$, $n\bar{y} = \Sigma_{(1)} y_i$ , $\alpha$ is a constant, and the summation notation is

$$\Sigma_{(1)} \equiv \Sigma_{i=1}^{n} \quad \text{and} \quad \Sigma_{(2)} \equiv \Sigma_{i=1}^{n} \underset{i\neq j}{\Sigma_{j=1}^{n}} .$$

We refer to $I$ as a *spatial autocorrelation* coefficient and take the constant $\alpha = n/S_o$ where $S_o = \Sigma_{(2)} w_{ij}$ .

*5.2.1 Distribution Theory for* I. Cliff and Ord (1973, p. 8) consider the distribution of $I$ under $H_0$ under two different regimes:

*Assumption* N. $Y_1,\cdots,Y_n$ are independent and identically distributed normal random variables;

*Assumption* R. (randomization) whatever the underlying distribution of the random variables, we consider the distribution of $I$ based upon the set of all possible random permutations of the data values $\{y_1,\cdots,y_n\}$ around the $n$ sites; there are $n!$ such permutations, which are regarded as equally likely.

Using the subscripts N and R to denote these assumptions, it follows that, under $H_0$ ,

$$E_N(I) = E_R(I) = -(n-1)^{-1} ,$$

$$E_N(I^2) = (n^2S_1 - nS_2 + 3S_0)/[(n^2-1)S_0^2] \quad ,$$

$$E_R(I^2) = \frac{n[(n^2-3n+3)S_1 - nS_2 + 3S_0^2] - b_2[(n^2-n)S_1 - 2nS_2 + 6S_0^2]}{(n-1)(n-2)(n-3)S_0^2} \quad ,$$

from which the variance of $I$ is just

$$Var(I) = E(I^2) - [E(I)]^2 \quad \text{for each of } N,R,$$

where

$$S_1 = 2\Sigma_{(2)}\, w_{ij}^2, \quad S_2 = 4\Sigma_{(1)}\, w_i^2, \quad w_i = \Sigma_{j=1}^{n}\, w_{ij} \text{ and } b_2 = \Sigma Z_i^4/(\Sigma Z_i^2)^2.$$

The detailed derivations of these results are given in Cliff and Ord (1973, p. 29-33). These authors go on to show (p. 34-37) that under fairly mild conditions the distribution of $I$ approaches normality as $n$ increases, under either assumption $N$ or assumption $R$. Recent work by Sen (1976) has extended this analysis and shown that the distributions under $H_0$ are asymptotically normal provided that the weights $(w_{ij})$ are such that

$$w_{ij} \geq 0, \quad w_i \leq C \text{ for all } i \quad , \quad \text{and}$$

$$0 < \lim_{n\to\infty}[\Sigma_{(2)}\, w_{ij}^2/nS_0] < \infty \quad ,$$

where $C$ is finite. These conditions are certainly satisfied whenever each site is linked to a finite set of 'neighbors,' and hold for all the sets of weights described in Table 5.2. Exceptions arise when one or a few sites dominate the lattice, as illustrated in Cliff and Ord (1973, p. 34-37). Further, these authors show how the normal approximation may be improved for smaller lattices. As a rule of thumb, approximate normality may be assumed for $n \geq 25$ provided that: (a) the distribution of the random variable is not markedly non-normal and (b) no small subset of sites dominates the lattice structure. These general guidelines are given more substance in Cliff and Ord (pp. 37-52) on the basis of Monte Carlo studies for different patterns of weights and various distributional assumptions.

*5.2.2 Choice of Weights.* For the user, the immediate question is how to specify the weights $(w_{ij})$. The diversity of possible spatial patterns makes weight selection difficult, although scientific interest will usually dictate declining weights as the distance between sites increases. There is no unique set of correct weights unless we are willing to specify a particular alternative hypothesis (form of $\underset{\sim}{V}$ in (5.1)), which begs the question anyway. In practice, similar sets of weights produce qualitatively similar results and excessive concern about the exact weights structure is unwarranted.

The problem has been inverted in an interesting manner by Kooijman (1976). Suppose that we partition the possible distances into $K$ distinct intervals $(0,d_1], (d_1,d_2],\cdots,(d_{K-1},d_K]$ where $d_K$ is the largest distance between any two sites in the study area. Then set

$$w_{ij} = u_k \quad \text{whenever} \quad d_{k-1} < d_{ij} \leq d_k \ .$$

That is, we collapse the possible weights down to $K$ distinct values, $u_1,\cdots,u_K$. Writing $a_k = \Sigma z_i z_j$ where the sum is taken over the $n_k$ distinct pairs for which $d_{k-1} < d_{ij} \leq d_k$, Kooijman maximizes $I$ with respect to the $\{u_k\}$, subject to the conditions $S_0 = \underset{\sim}{1}'\underset{\sim}{W}\underset{\sim}{1} = 1$, $\underset{\sim}{1}'\underset{\sim}{W}^2\underset{\sim}{1} = 1$ where $\underset{\sim}{1}' = (1,\cdots,1)$. He finds that

$$I_{max} + (n-1)^{-1} \doteq \sqrt{[\Sigma(a_k^2/n_k) - \{n/(n-1)\}]}$$

and

$$nu_k = \{\frac{a_k}{n_k}(n^2-n-1) + (n+I_{max})\}/\{1+(n-1)I_{max}\} \ .$$

These results enable us to construct distance related weights that are 'best' according to the $I_{max}$ criterion. Thus, if a certain separation distance is important, this will be reflected in a high $u_k$ value for that distance. Essentially, the results are the same when we set $\underset{\sim}{1}'\underset{\sim}{W}^2\underset{\sim}{1} = b$, where $b$ is some positive constant and $b \neq 1$. Further, Kooijman gives the moments of $I_{max}$, which enables a test of the null hypothesis of spatial independence to be carried out using a normal approximation. Extensions of this approach are given in Kooijman (1979).

*5.2.3 Some Examples.* Returning to the data given in Figure 5.1, Jumars *et al.* (1977) used

$w_{ij} = 1$ if i and j are horizontal or vertical neighbors,

$w_{ij} = 1/2$ if i and j are diagonal neighbors,

$w_{ij} = 0$ otherwise

and obtained $I = 0.142$ against $E_R(I) = -0.01$ . For a two tailed test of the null hypothesis, the tail area probability (using the normal approximation) is less than 0.01, confirming the visual impression that there is a strong spatial pattern.

```
4 1 - 1 - 1 - 2 1 1
4 1 - - 1 - 2 4 - -
5 5 - 1 1 4 7 3 1 1
2 - 1 - - 3 8 2 1 1
1 - - 1 - - 1 2 1 -
- 1 1 1 - 1 - 1 - -
- 2 - - - - - 1 4 -
3 3 1 1 1 - 1 - 9 6
6 7 - 1 1 1 - 4 2 3
5 2 2 - 1 - 1 4 7 7
```

*FIG. 5.2: Grid count on acorn barnacles (from Kooijman,* 1976*). Dashes indicate that no specimens were found.*

The data shown in Figure 5.2 are given by Kooijman (1976) and refer to a grid count for acorn barnacles settled on a Dutch cutter docked at Scheveningen, Holland. Using the weight selection procedure outlined in the previous section with the distance between horizontally or vertically adjacent cell centers equal to unity and the distances split into the intervals (0,1], (1,2],•••, Kooijman obtains the $u_k$ values shown in Figure 5.3. The $u_k$ values suggest a local effect for immediate neighbors and a larger scale effect close to the overall grid size, which would need a larger sampled area to be properly analyzed. Kooijman also reports the analysis with other structures for $u_k$ .

Sokal and Oden (1978a,b) give an extensive review of spatial autocorrelation methods and include many examples of genetic and ecological interest.

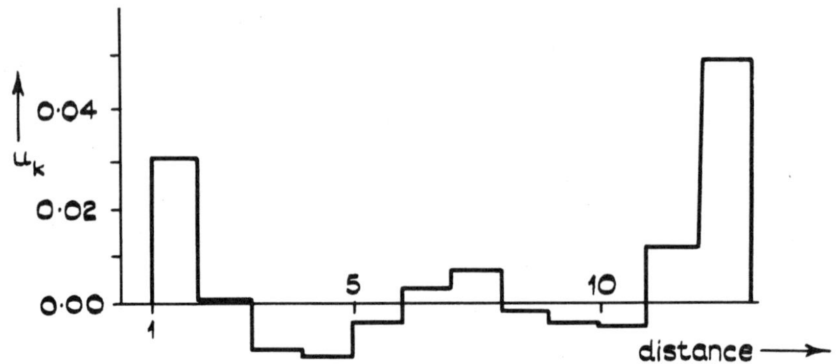


*FIG. 5.3: Fitted 'weights' function using the data on acorn barnacles (from Kooijman,* 1976*).*

## 5.3 SPATIAL DIVERSITY

Many of the early measures of spatial association concentrated upon binary random variables (sites coded black, B or white, W for example). When two neighboring sites are both coded black, we say that they form a BB join. Two possible statistics are

weighted number of BB joins: $BB = (1/2)\Sigma_{(2)} w_{ij} x_i x_j$

weighted number of BW joins: $BW = (1/2)\Sigma_{(2)} w_{ij}(x_i - x_j)^2$

$$\text{where } x_i = \begin{cases} 1 & \text{if } i\text{th site is coded B} \\ 0 & \text{if } i\text{th site is coded W;} \end{cases}$$

the factor of 1/2 is included to avoid double counting. Typically, the weights are binary also (see the queen's and rook's weights in Table 5.2), although such a restriction is not necessary.

These statistics are two dimensional analogues of the runs tests used for time-series and for line transect data. When data form a natural dichotomy such as the presence or absence of a particular species in a sample quadrat or the conditions of a tree (diseased or healthy), the join count measures are an obvious choice. However, an artificial dichotomy forced from interval scaled data may lose a lot of information. Under the null hypothesis of no spatial dependence, the distributions of the BB and BW statistics may be considered under either:

*free sampling,* when each site is independently coded B with probability $p$ and coded W with probability $(1-p)$; or

*non-free sampling,* when the numbers of B and W sites, $n_1$ and $n_2(=n-n_1)$ say, are held fixed and the set of all possible random permutations, $n!/(n_1!n_2!)$ in all, is considered.

The mathematical arguments are easier to apply in the free sampling case, but the conditional form of the non-free sampling scheme is usually preferred in practice.

The means and variances of these statistics are as follows:

*free sampling*

$$E(BB) = (1/2)\, S_0 p^2 \quad ,$$

$$Var(BB) = (1/4)\{S_1 p^2 + (S_2 - 2S_1)p^3 + (S_1 - S_2)p^4\} \quad ,$$

$$E(BW) = S_0 pq \quad ,$$

$$Var(BW) = (1/4)\{2S_1 pq + (S_2 - 2S_1)pq + 4(S_1 - S_2)p^2 q^2\} \quad ,$$

*non-free sampling*

$$E(BB) = (1/2)\, S_0 n_1^{(2)}/n^{(2)} = (1/2)\, S_0\, \alpha_2 \quad ,$$

$$Var(BB) = (1/4)\{S_1 \alpha_2 + (S_2 - 2S_1)\alpha_3 + (S_0^2 + S_1 - S_2)\alpha_4 - S_0^2\, \alpha_2^2\} \quad ,$$

$$E(BW) = S_0\, n_1 n_2/n^{(2)} = S_0\, \alpha_1\, \beta_1 \quad ,$$

$$Var(BW) = (1/4)S_2\, \alpha_1\, \beta_1 + (S_0^2 + S_1 - S_2)\alpha_2\, \beta_2 - S_0^2\, \alpha_1\, \beta_1 \quad ,$$

where $\alpha_j = \binom{n_1}{j}/\binom{n}{j}$ and $\beta_j = \binom{n_2}{j}/\binom{n-j}{j}$, $j = 1,2$, and $S_0$, $S_1$ and $S_2$ are defined previously. Again, provided that $n$ is not too small and $p$ (or $n_1/n$) is not too far from $1/2$, say $\min[np,nq] \geq 10$, a normal approximation to the distribution under $H_0$ may be employed. Derivations and further details are given in Cliff and Ord (1973, chap. 2).

Rather than give examples for the two class situation, we turn directly to the problem of spatial diversity. Two different measures for spatial diversity have been proposed in the literature and we describe these in turn.

*5.3.1 David's Measure of Spatial Diversity.* Consider n contiguous cells (or sampling locations), defined upon a rectangular grid of size $R \times C$ say. We suppose that each cell may be occupied by just one individual drawn from one of k distinct species A, B, C, D,•••, K say; alternatively, we may classify each cell by the species which is dominant in that cell. Then the possible occupancy patterns for each set of four adjacent cells are as follows:

| (a) | | (b) | | (c) | | (d) | | (e) | |
|---|---|---|---|---|---|---|---|---|---|
| A | A | A | A | A | A | A | A | A | B |
| A | A | A | B | B | B | B | C | C | D |

The letters may stand for any species so that pattern (a) means "all four the same species," (c) means "two pairs," and so on. If we weight horizontal, vertical, and diagonal joins equally, the number of joins between cells containing members of the same species is

6 for (a), 3 for (b), 2 for (c), 1 for (d), and 0 for (e). These are the scores proposed for each pattern of four by David (1971a). David then suggests that the total score of the n cells be used as a measure of spatial diversity. However, from our discussions earlier in the section we see that this measure may be written as

$$D = (1/2)\,\Sigma_{(2)}\, w_{ij}\, a_{ij} \quad ,$$

where $w_{ij} = 1$, if cells i and j are queen's case neighbors and $w_{ij} = 0$, otherwise, while $a_{ij} = 1$, if cells i and j are occupied by members of the same species and $a_{ij} = 0$, otherwise.

The weights for the cells on the grid boundary would need to be slightly different to agree with David's definition, but our specification gives a more even weighting to all pairs of cells. The differences will be very slight unless the lattice is small.

If T denotes the total number of joins between neighboring cells containing different species, we see that $D = M - T$ where $M = S_0/2$. T is a k-species extension of the BW join count statistic considered earlier.

The statistical properties of D are described by David (1971a) or, equivalently, those of T by Cliff and Ord (1973,

p. 14), based upon the early work of Krishna Iyer (1949). Also, see the discussion in Cormack (1979, section 4.3). The measure is extended in David (1971b) to cover multiple occupancies per cell.

If $n_r$ denotes the number of cells occupied by members of the *rth* species then the expected number of joins between two members of that species under *non-free sampling* is $Mn_r(n_r-1)/n(n-1)$ so that E(D) is proportional to $\sum_r n_r(n_r-1)/n(n-1)$ . The statistic [M-E(D)]/M is just the *Simpson index of diversity* (Pielou, 1977, p. 309). Thus, the D statistic reflects spatial variations 'within' overall diversity as measured by the Simpson index.

*5.3.2 Pielou's Measure of Spatial Diversity.* Again we consider n contiguous cells (or sampling locations) laid out on a rectangular grid. Typically, we might think of different species in a plant community which form single species clusters. The overall pattern is known as a *mosaic*, and has the form shown in Figure 5.4. If we define joins between neighboring locations by either the queen's or rook's case (Table 5.2), then we arrive at a set of counts $\{m_{rs}\}$ , which denote the numbers of joins between the *rth* and *sth* species, such that $\sum_{r\leq s}\sum m_{rs} = M$ . The

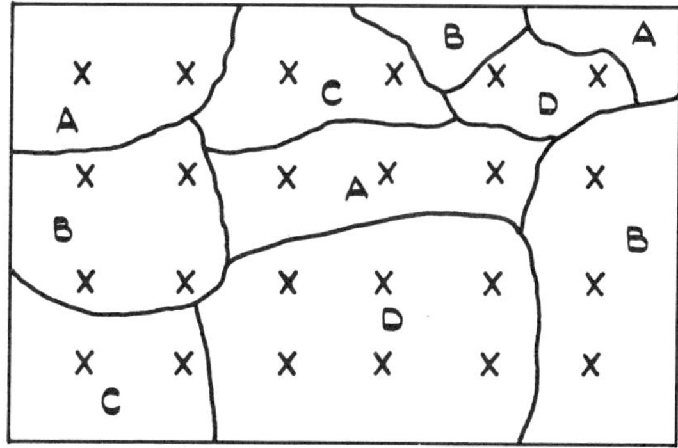


*FIG. 5.4: A four-species mosaic sampled at locations (marked ×) on a rectangular 4 × 6 grid.*

counts for Figure 5.4 (rook's case) are

| Species r | A | B | C | D |
|---|---|---|---|---|
| A | 3 | 4 | 3 | 4 |
| B | | 6 | 2 | 4 |
| C | | | 2 | 2 |
| D | | | | 8 |

Pielou's statistic (Pielou, 1975, p. 72-80) is of the entropy type and may be written as

$$P = -\{\sum_r \sum_s m_{rs} \log m_{rs} - \sum_r m_r \log m_r\}/M \quad , \qquad (5.2)$$

where $m_r = \sum_s m_{rs}$ . This may be linked to the Shannon entropy measure of diversity

$$S = -\{\sum_r m_r \log m_r - M \log M\}/M \quad ,$$

showing that $P$ measures the spatial diversity 'within' the overall species diversity. It should be noted Pielou developed her measure for line transect data, but the above extension to two dimensions is straightforward. However, the sampling properties of $P$ in the two dimensional case are unknown.

An alternative 'justification' for $P$ would be to consider $M$ independent neighboring pairs of cells and to let

$$\theta_{rs} = \Pr(\text{the cells form an r-s pair}) \quad (r,s=1,2,\cdots,k) \quad .$$

The likelihood ratio statistic for testing

$$H_0 : \theta_{rs} = \theta_r \theta_s \quad (\theta_r = \sum_s \theta_{rs})$$

against a general alternative, $H_1$ , takes the form of $P$ in (5.2) since under $H_0$ , $\hat{\theta}_{rs} = m_r m_s / M^2$ while, under $H_1$ , $\hat{\theta}_{rs} = m_{rs}/M$ .

*5.3.3 A Measure of Spatial Association.* To examine the extent to which particular species tend to occur together, we may develop a measure of spatial association (cf. Pielou, 1977, p. 208-15) as follows. Let $m_{rs}$ denote the number of joins between individuals of the *rth* and *sth* species as before and set

$$m_r^* = \sum_{s \neq r} m_{rs} \quad .$$

The null hypothesis of no spatial association may be represented by

$$H_0 : \theta_{rs}^* = \theta_r^* \theta_s^* \quad , \text{ all } r \neq s$$

where $\theta_r^* = \sum_{s \neq r} \theta_{rs}^*$ and $\theta_{rs}^*$ is the probability that cells form an r,s pair given that they contain different species. Finally, let $M^* = \sum_r m_r^*$ . Then, following the previous argument, the measure of spatial association becomes

$$P_A = -\{\sum_{r \neq s} \sum m_{rs} \log m_{rs} - \sum_r m_r^* \log m_r^*\}/M^* \quad ,$$

reflecting high values when certain species tend to occur in adjacent clumps.

## 5.4 FURTHER READING

Cliff and Ord (1973, chap. 5,6) have developed a procedure for testing regression residuals for spatial dependence. The effect of spatial dependence upon standard hypothesis tests has been examined by several authors; Cliff and Ord (1976) looked at the two sample t-test while Griffiths (1978) considers the k-sample analysis of variance.

A matter of continuing interest to ecologists is the scale of spatial pattern. The pioneering work of Greig-Smith (1952, 1964) introduced the hierarchical analysis of contiguous quadrats and this approach has been extended in several ways. For further details, the reader should consult three of the other papers in this volume: Cormack (1979, section 4.2), Diggle (1979, section 4.2), and Ludwig (1979).

CHAPTER 6

# MODELS FOR SPATIAL PROCESSES

## 6.1 INTRODUCTION

As we have seen already, tests for spatial pattern have a useful role in statistical ecology. However, the null hypothesis of no spatial dependence is often something of a 'straw man,' so we seek to go beyond the initial tests and to formulate spatial models. As in time-series analysis, the simplest models are the regression schemes which look at variations in the mean together with a simple covariance structure; these models are discussed in Section 6.2. In Section 6.3, we explore processes with an autogressive structure; estimation methods for such schemes are then considered in Section 6.4. The chapter concludes with a brief review of other approaches.

## 6.2 TREND SURFACE ANALYSIS

As before, we assume that it is possible to observe the random variables $\{Y_i : i = 1,\cdots,n\}$ at each of $n$ sites whose coordinates are $\{\underset{\sim}{x}_i\}$. We suppose that

$$E(Y_i) = g(\underset{\sim}{x}_i), \quad Var(Y_i) = \sigma^2 \quad , \text{ and}$$

$$Cov(Y_i, Y_j) = 0 \quad (i \neq j) \quad .$$

That is, we are using a *wide-sense* specification and supposing that

$$Y_i = g(\underset{\sim}{x}_i) + \varepsilon_i \quad ,$$

where the $\{\varepsilon_i\}$ denote random variation. The function $g(\underset{\sim}{x})$ is of known form though dependent upon unknown parameters. For example, when $\underset{\sim}{x} = (x_1, x_2)$ we might take

$$g(\underset{\sim}{x}) = \beta_0 + (\beta_{10}\, x_1 + \beta_{01}x_2) + (\beta_{20}\, x_1^2 + \beta_{11}\, x_1\, x_2 + \beta_{02}\, x_2^2) + \cdots, \tag{6.1}$$

where we have grouped linear, quadratic,··· terms *or* a similar version involving orthogonal polynomials for $x_1$ and $x_2$ (cf. Section 2.2). A third alternative would be to use cosine functions with a selection of different frequencies.

Since (6.1) is linear in the parameters, these unknowns may be estimated by least squares in the usual way. This type of model is known under the general heading of *trend surface analysis* and is widely used in geography and geology. For recent reviews of the subject see Watson (1972) and Whitten (1975).

Low order polynomial trend surfaces are a convenient way of removing trends from spatial data and may accurately represent trends *within* the boundaries of the study area. However, at the edges of the study area and beyond, the estimated $g(\underset{\sim}{x})$ functions become increasingly erratic as the order of the polynomial is increased. Kooijman (1979, Section 2) shows how trends may be estimated for spatial point processes.

## 6.3 SPATIAL AUTOREGRESSIVE MODELS

Since autoregressive models for irregularly spaced data pose some problems with regard to an appropriate parametrization, we shall stick to a regular lattice in the development of the model. Suppose that we are dealing with a rectangular lattice containing R rows and C columns and $Y_{ij}$ denotes the random variable corresponding to the (i,j)*th* cell. Further, we assume that $E(Y_{ij}) = \mu$ and set $Z_{ij} = Y_{ij} - \mu$, for all i and j.

To define an autoregressive scheme we must specify the pattern of spatial dependence just as we specified the pattern of dependence for a time-series in Section 3.3. For example, for time-series, we specified the first order or AR(1) scheme by linking the current $(Z_t)$ and immediate past $(Z_{t-1})$ values of the series so that

$$Z_t = \rho Z_{t-1} + \varepsilon_t \quad , \tag{6.3}$$

where the $\{\varepsilon_t\}$ are uncorrelated random errors with equal variances.

On a rectangular lattice, the simplest scheme which comes to mind is to link the (i,j)*th* cell with its four immediate neighbors, the rook's case, as shown in the diagram

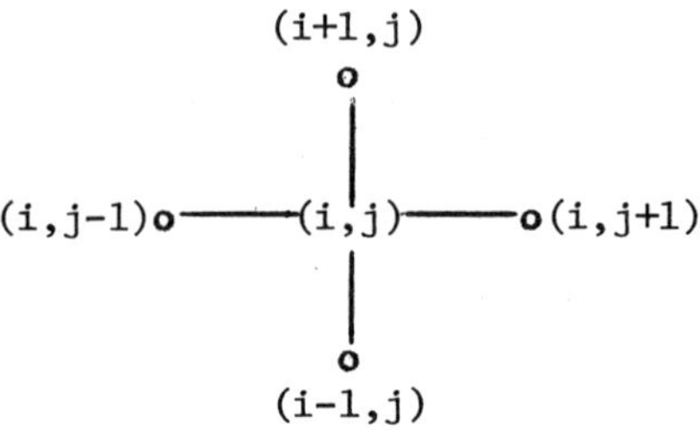


Thus, we are led to a first order spatial autoregressive scheme of the form

$$Z_{ij} = \beta(Z_{i-1,j} + Z_{i+1,j} + Z_{i,j-1} + Z_{i,j+1}) + \varepsilon_{ij} \quad , \qquad (6.4)$$

where the parameter $\beta$ measures the extent of the spatial interaction just as $\rho$ measures the temporal dependence in model (6.3). The specification of the model is completed by setting

$E(\varepsilon_{ij}) = 0$, $Var(\varepsilon_{ij}) = \sigma^2$ , for all i and j , and

$E(\varepsilon_{ij}\varepsilon_{rs}) = 0$ unless $i = r$ *and* $j = s$ .

The form of model (6.4) is not correct for cells on the boundary of the lattice and various approaches have been adopted to allow for this:

(a) when the study area forms a natural 'island,' we might replace $\beta$ by $4\beta^*/R_{ij}$ where $R_{ij}$ denotes the number of immediate neighbors of cell (i,j) ;

(b) simply include the actual neighbors without changing the parameter;

(c) when the study area is part of a larger area, we may surround the lattice by 'guard rows.' If a k*th* order model is specified, there should be at least k guard rows. The model may then be specified for the interior cells only, though the analysis is more effective if we condition upon the observed values in the guard rows.

It should be noted that the variances of the $Y_{ij}$ for boundary cells will differ from those for interior cells whenever (a) or (b) is adopted.

We note that the complete specification of model (6.4) corresponds to an assumption of *spatial stationarity*, so that the correlation structure between cells depends upon their *relative* position. Indeed, we have made a rather stronger assumption since, for model (6.4), direction does not affect the extent of the dependence between cells. When a process is direction-invariant, we say that it is *isotropic*. Direction invariance is not a necessary assumption and is relaxed in Section 6.4.1.

Model (6.4) was introduced by Whittle in a seminal paper on spatial processes (Whittle, 1954). While there is an obvious analogy with the first order temporal scheme given in (6.3) there is also a fundamental difference. For the time-series model the unidirectional nature of time means that we can assume that $Z_{t-1}$ and $\varepsilon_t$ are uncorrelated. Hence, expression (6.3) implies that

$$E[Z_t | Z_s = z_s \ , \text{ for all } \ s < t] = \rho z_{t-1} \ . \qquad (6.5)$$

However, for the spatial scheme there is *multilateral* dependence and equation (6.4) does *not* imply that

$$E[Z_{ij} | Z_{rs} = z_{rs} \ , \text{ for all } \ (r,s) \neq (i,j)]$$

$$= \beta(z_{i-1,j} + z_{i+1,j} + z_{i,j-1} + z_{i,j+1}) . \quad (6.6)$$

If we assume that model (6.4) holds, written in general matrix terms as $\underset{\sim}{Z} = \beta \underset{\sim}{W} \underset{\sim}{Z} + \underset{\sim}{\varepsilon}$ then, given the further assumption that the $\underset{\sim}{\varepsilon}$ are normally distributed we find that $\underset{\sim}{Z}$ is normal with zero mean and variance $\underset{\sim}{V} = \sigma^2[(\underset{\sim}{I} - \beta\underset{\sim}{W})(\underset{\sim}{I} - \beta\underset{\sim}{W})']^{-1}$ . This is known as the *joint model*, since the $\underset{\sim}{Z}$ are jointly specified in expression (6.4). Alternatively, we may wish to specify that the conditional expectations (6.6) are valid, together with the assumption that, for all $i$ and $j$ ,

$$\text{Var}[Z_{ij} | Z_{rs} = z_{rs}, \text{ for all } (r,s) \neq (i,j)] = \sigma^2 \ .$$

Then, given normality we find that the $\underset{\sim}{Z}$ are normally distributed with zero mean and variance $\underset{\sim}{V} = \sigma^2(\underset{\sim}{I} - \beta\underset{\sim}{W})^{-1}$ . This is known as the *conditional model*, since it is based upon the conditional expectations in (6.4); this model was introduced by Bartlett (1971) and further developed by Besag (1974). We note that $\underset{\sim}{W}$ must be symmetric and $\underset{\sim}{I} - \beta\underset{\sim}{W}$ must be positive definite for the conditional scheme.

The arguments for and against each model are given in Cormack (1979, section 8.2); the principal attraction of the conditional scheme is that it may be derived from a spatio-temporal process. For either model, we note that a similar specification may be adopted for any pattern of sites, regular or irregular. When the sites are irregularly spaced, or represent irregular spatial aggregates, it may be difficult to justify the use of a simple parametric structure such as (6.3) or (6.4), with $\underset{\sim}{W}$ known. It would seem more appropriate to specify the weights as a parametric function of distance and other positional variables, although such an approach leads to acute numerical problems when we attempt to estimate the unknown parameters.

Spatial moving average models have also been developed, notably by Haining (1977, 1978), but we shall not pursue this topic.

## 6.4 ESTIMATION FOR SPATIAL MODELS

The search for a suitable estimation procedure starts, naturally, with the method of maximum likelihood. We shall restrict attention to the special case of the conditional model where $\underset{\sim}{V} = \sigma^2(\underset{\sim}{I}-\beta\underset{\sim}{W})^{-1}$ and $\underset{\sim}{W}$ is a matrix of known weights. If we assume that the joint distribution of $\underset{\sim}{Z}$ is normal, then the log-likelihood function has the form

$$\ell n\ L = \ell = \text{const} + \frac{1}{2}\ \ell n\ |\underset{\sim}{I} - \beta\underset{\sim}{W}| - \frac{n}{2}\ \ell n\ \sigma^2 - \frac{1}{2\sigma^2}\underset{\sim}{z}'(\underset{\sim}{I}-\beta\underset{\sim}{W})\underset{\sim}{z} \quad , \tag{6.7}$$

where $\ell n$ denotes logarithms to base $e$ .

In autoregressive time-series models $\underset{\sim}{W}$ is triangular so that the term involving the determinant vanishes. This is not so for the spatial scheme which leads to numerical difficulties. Various proposals have been put forward to avoid this problem. Mead (1971) suggested taking observations in pairs, but the method has undesirable statistical properties. When the weights in $\underset{\sim}{W}$ are positive only for immediate neighbors, it is possible to use a coding method developed by Besag (1974). For the rectangular lattice (rook's case), the sites are coded as

```
x o x o x o x
o x o x o x o
x o x o x o x
o x o x o x o
```

and the analysis may then be carried out for the x sites given the observed values on the o sites, or vice-versa; see also Besag and Moran (1975). For one set of sites, given the values observed at the other set, the conditional maximum likelihood estimators reduce to ordinary least squares, eliminating the computational problem. However, the least squares estimator

$$\hat{\beta} = \underset{\sim}{Z}'\underset{\sim}{W}\underset{\sim}{Z}/\underset{\sim}{Z}'\underset{\sim}{W}'\underset{\sim}{W}\ \underset{\sim}{Z} \tag{6.8}$$

is also consistent and generally more efficient than the coding estimators. Even so, the least squares estimator is not fully efficient except for very restrictive forms of $\underset{\sim}{W}$ . Nevertheless, the consistency of the least squares estimator (for the conditional model) is a useful advantage, not shared by the joint scheme. In situations where data are fairly abundant, the loss of efficiency may be unimportant compared to the computational savings when n is large.

For small or moderate values of n , the least squares estimator can serve as a starting value in an iterative search for the maximum of the log-likelihood function. Following Ord (1975), if the eigenvalues of $\underset{\sim}{W}$ are $\lambda_1,\cdots,\lambda_n$ then $\ell$ becomes

$$\ell = \text{const} + \frac{1}{2}\sum_{i=1}^{n} \ell n(1-\beta\lambda_i) - \frac{n}{2}\,\ell n\, \sigma^2 - \frac{1}{2\sigma^2}\sum_{i=1}^{n} (1-\beta\lambda_i)a_i^2 \ , \tag{6.9}$$

where $a' = (a_1,\cdots,a_n)$ and $\underset{\sim}{a} = \underset{\sim}{U}'\underset{\sim}{z}$ ; $\underset{\sim}{U}$ being the matrix of eigenvectors corresponding to $\underset{\sim}{W}$ . Numerical evaluation of $\hat{\rho}$ from (6.9) is then a straightforward matter using the Newton-Raphson procedure. A similar procedure holds for the joint scheme, when $\underset{\sim}{W}$ is symmetric. All that is required is to replace $(1-\beta\lambda_i)$ by $(1-\beta\lambda_i)^2$ throughout (6.9).

*6.4.1 Spatial Variation in Wheat Yields.* Whittle (1954) investigated data due originally to Mercer and Hall on wheat yields over 500 quadrats (roughly 11 feet square) arranged in a 20 × 25 rectangle. Whittle gave the estimated autocorrelations as

| j | i: 0 | 1 | 2 | 3 | 4 |
|---|---|---|---|---|---|
| -3 | 0.1880 | 0.1602 | 0.1509 | 0.1276 | 0.1352 |
| -2 | 0.1510 | 0.0234 | 0.0020 | -0.0137 | -0.1039 |
| -1 | 0.2923 | 0.1853 | 0.1349 | 0.0788 | 0.0878 |
| 0 | 1.0000 | 0.5252 | 0.4055 | 0.3639 | 0.3561 |
| 1 | 0.2923 | 0.2354 | 0.1799 | 0.1205 | 0.1399 |
| 2 | 0.1510 | 0.1285 | 0.0999 | 0.0749 | 0.0859 |
| 3 | 0.1880 | 0.1935 | 0.2483 | 0.2415 | 0.2284 |

Here $i$ denotes the north-south direction while $j$ denotes east-west direction.

Besag (1974) proposed several conditional autoregressive models for these data. One simple scheme is of the form

$$E[Z_{ij} | Z_{rs} = z_{rs}, (r,s) \neq (i,j)]$$

$$= \beta_1(z_{i-1,j} + z_{i+1,j}) + \beta_2(z_{i,j-1} + z_{i,j+1}) \quad .$$

Using the coding procedure, the two codings yield the estimates

| | (1) | (2) | average |
|---|---|---|---|
| $\hat{\beta}_1$ | 0.332 | 0.354 | 0.343 |
| $\hat{\beta}_2$ | 0.128 | 0.166 | 0.147 |

while least squares yields $\hat{\beta}_1 = 0.334$ and $\hat{\beta}_2 = 0.132$ , with approximate standard errors 0.05 in each case. The results show the greater strength of the spatial dependence in the east-west direction, which can be seen also from the sample auto-correlations. Several other models are fitted by Besag (1974) and the reader should consult this paper for further details.

Spatial dependence in agricultural field trials has often been ignored; a recent approach which gives due weight to spatial factors is that of Bartlett (1978).

## 6.5 FURTHER READING

In this chapter we have introduced only the single parameter autoregressive model for a single random variable. In place of a linear model in the 'space' domain we could work with the spectrum; see Granger (1969). Ford's (1976) study of the canopy of a pine forest and Robinson's (1975) study of oil reserves show the potential effectiveness of this approach, although a major difficulty is the requirement for equally spaced data. Hsu (1975) suggested interpolation of the observations onto a regular grid, but this form of averaging is open to the objection that it may induce cycles not present in the original data. For the spectral analysis of point processes, see Bartlett (1964, 1975). An alternative approach to spatial modelling is through specification of the autocorrelation function. Over the years, Matheron (1971) and his colleagues have developed the theory of *regionalized* random variables. If $Y(\underset{\sim}{x})$ describes the random variable at location $\underset{\sim}{x}$, then the process is specified through the mean $E[Y(\underset{\sim}{x})] = \mu(\underset{\sim}{x})$ and the variogram

$$E\{[Y(\underset{\sim}{x})-Y(\underset{\sim}{x}+\underset{\sim}{h})]^2\} = \sigma^2(\underset{\sim}{h}) \qquad (\underset{\sim}{h} \neq \underset{\sim}{0}) \quad ,$$

while $\mathrm{Var}[Y(\underset{\sim}{x})] = \sigma^2 + \sigma_\varepsilon^2$. This specification allows local random variation at each location $\underset{\sim}{x}$ and non-stationarity in the mean is included explicitly. The methods of analysis for spatial interpolation using this approach are related to generalized least squares procedures (cf. Delfiner and Delhomme, 1975). Ord and Rees (1979) also give an introduction to this approach.

The main development of non-normal schemes is due to Besag (1974); his model for binary spatial random variables is described in Cormack (1979, section 8.1).

Relatively little has been done on the multivariate case. Kooijman (1979) has extended his work on the $I_{max}$ statistic (Section 5.2.2) to several variables while Cliff and Ord (1979) have extended the Greig-Smith method for the analysis of spatial scale to look at cross-correlation functions.

## ACKNOWLEDGEMENTS

I should like to thank both referees for their helpful comments which were of great value in revising the manuscript. Among others who read through the manuscript and gave helpful advice were Marllyn Boswell, Richard Cormack, Robert Poole and

Fred Ramsey. Several members of the Advanced Study Institutes at Texas, Berkeley, and Parma also commented upon the manuscript at various stages of its development. Of course, any errors that remain are my own particular contribution.

I am particularly grateful to David Ford for providing the data on wood cells and to Nan Duncan for introducing me to the data on phytoplankton and zooplankton.

The computing was carried out on the University of Warwick Burroughs 6700 computer.

## REFERENCES

Akaike, H. (1974). A new look at the statistical model identification. *IEEE Transactions on Automatic Control*, AC-19, 716-723.

Anderson, O. D. (1976). *Time Series Analysis and Forcasting: The Box-Jenkins Approach.* Butterworth, London.

Anderson, O. D. (1977). A Box-Jenkins analysis of the colored fox data from Nain, Labrador. *The Statistician*, 26, 51-75.

Anderson, T. W. (1971). *The Statistical Analysis of Time Series.* Wiley, New York.

Bartlett, M. S. (1960). *Stochastic Population Models.* Methuen, London.

Bartlett, M. S. (1964). The spectral analysis of two-dimensional point processes. *Biometrika*, 51, 299-311.

Bartlett, M. S. (1971). Physical nearest-neighbor models and non-linear time-series. *Journal of Applied Probability*, 8, 222-232.

Bartlett, M. S. (1975). *The Statistical Analysis of Spatial Pattern.* Chapman and Hall, London.

Bartlett, M. S. (1978). Nearest neighbor models in the analysis of field experiments (with discussion). *Journal of the Royal Statistical Society, Series B*, 40, 147-174.

Besag, J. E. (1974). Spatial interaction and the statistical analysis of lattice systems (with discussion). *Journal of the Royal Statistical Society, Series B*, 36, 192-236.

Besag, J. E. and Moran, P. A. P. (1975). On the estimation and testing of spatial interaction in Gaussian lattice processes. *Biometrika*, 62, 555-562.

Bhansali, R. J. (1974). Asymptotic properties of the Wiener-Kolmogorov predictor. *Journal of the Royal Statistical Society, Series B*, 36, 61-73.

Boswell, M. T., Ord, J. K., and Patil, G. P. (1979). Chance mechanisms underlying univariate distributions. In *Statistical Distributions in Ecological Work*, J. K. Ord, G. P. Patil, and C. Taillie, eds. Satellite Program in Statistical Ecology, International Co-operative Publishing House, Fairland, Maryland.

Boswell, M. T. and Patil, G. P. (1972). Certain chance mechanisms related to birth and death processes which have classical discrete distributions as special cases. In *Stochastic Point Processes*, P. A. W. Lewis, ed. Wiley, New York. 285-289.

Box, G. E. P. and Cox, D. R. (1964). An analysis of transformations. *Journal of the Royal Statistical Society, Series B*, 26, 211-243.

Box, G. E. P. and Jenkins, G. M. (1970). *Time Series Analysis, Forecasting and Control*. Holden-Day, San Francisco.

Bulmer, M. G. (1974). A statistical analysis of the ten-year cycle in Canada. *Journal of Animal Ecology*, 43, 701-718.

Bulmer, M. G. (1975). Phase relations in the ten year cycle. *Journal of Animal Ecology*, 44, 609-622.

Bulmer, M. G. (1978). The statistical analysis of the ten year cycle. In H. H. Shugart Jr., ed. (*op. cit.*). 141-153.

Burgner, R. L. and Rogers, D. E. (1973). IBP(PF) Data Report No. 60. Fisheries Research Institute, University of Washington, Seattle.

Campbell, M. J. and Walker, A. M. (1977). A survey of statistical work on the Mackenzie River series of annual Canadian lynx trappings for the years 1821-1934 and a new analysis. *Journal of the Royal Statistical Society, Series A*, 140, 411-431.

Chan, W-Y. T. and Wallis, K. F. (1978). Multiple time-series modelling: another look at the mink-muskrat interaction. *Applied Statistics*, 27, 168-175.

Chatfield, C. (1975). *The Analysis of Time Series: Theory and Practice.* Chapman and Hall, London.

Cliff, A. D., Haggett, P., Ord, J. K., Bassett, K., and Davies, R. B. (1975). *Elements of Spatial Structure: A Quantitative Approach.* Cambridge University Press, Cambridge.

Cliff, A. D. and Ord, J. K. (1973). *Spatial Autocorrelation.* Pion, London.

Cliff, A. D. and Ord, J. K. (1975). The comparison of means when samples consist of spatially autocorrelated observations. *Environment and Planning,* 7, 725-734.

Cliff, A. D. and Ord, J. K. (1979). The effects of spatial autocorrelation in geographical modelling (to appear).

Cormack, R. M. (1979). Spatial aspects of competition between individuals. In R. M. Cormack and J. K. Ord, eds. (*op. cit.*).

Cormack, R. M. and Ord, J. K., eds. (1979). *Spatial and Temporal Analysis in Ecology.* Satellite Program in Statistical Ecology, International Co-operative Publishing House, Fairland, Maryland.

Cox, D. R. and Lewis, P. A. W. (1966). *The Statistical Analysis of Series of Events.* Methuen, London.

Cox, D. R. and Miller, H. D. (1965). *The Theory of Stochastic Processes.* Methuen, London.

Damann, K. E. (1960). Plankton studies of Lake Michigan, II. *Transactions of the American Microscopial Society,* 79, 397-404.

David, F. N. (1971a). Measurement of diversity. *Proceedings of the Sixth Berkeley Symposium on Mathematical Statistics and Probability, Vol. 1,* University of California Press, Berkeley. 631-648.

David, F. N. (1971b). Measurement of diversity: multiple cell counts. *Proceedings of the Sixth Berkeley Symposium on Mathematical Statistics and Probability, Vol. 4,* University of California Press, Berkeley. 109-136.

Davis, J. C. and McCullagh, M. J. (1975). *Display and Analysis of Spatial Data.* Wiley, London.

Delfiner, P. and Delhomme, J. P. (1975). Optimum interpolation by Kriging. In J. C. Davis and M. J. McCullagh, eds. (*op. cit.*). 96-114.

Diggle, P. J. (1979). Statistical methods for spatial point patterns in ecology. In R. M. Cormack and J. K. Ord, eds. (*op. cit.*).

Draper, N. and Smith, H. (1966). *Applied Regression Analysis*. Wiley, New York.

Durbin, J. (1965). Discussion of Pyke (1965). *Journal of the Royal Statistical Society, Series B*, 27, 437-438.

Durbin, J. and Watson, G. S. (1950, 1951, 1971). Testing for serial correlation in least squares regression. *Biometrika*, 37, 409-428; 38, 159-178; 58, 1-19.

Elton, C. and Nicholson, M. (1942). The ten year cycle in numbers of lynx in Canada. *Journal of Animal Ecology*, 11, 215-244.

Ford, E. D. (1976). The canopy of a Scots pine forest: Description of a surface of complex roughness. *Agricultural Meteorology*, 17, 9-32.

Ford, E. D. and Robards, A. W. (1976). Short term variations in tracheid development in the early wood of *Picea sitchensis*. *Leiden Botanical Series*, 3, 212-221.

Ford, E. D., Robards, A. W., and Pinney, M. D. (1978). Influence of environmental factors on cell production and differentiation in the early wood of *Picea sitchensis*. *Annals of Botany*, 42, 683-692.

Granger, C. W. J. (1969). Spatial data and time series analysis. In *Studies in Regional Science*, A. J. Scott, ed. Pion, London. 1-24.

Greig-Smith, P. (1952). The use of random and contiguous quadrats in the study of the structure of plant communities. *Annals of Botany*, 16, 293-316.

Greig-Smith, P. (1964). *Quantitative Plant Ecology*, 2nd ed. Butterworth, London.

Griffiths, D. A. (1978). A spatially adjusted ANOVA model. *Geographical Analysis*, 10, 296-301.

Haining, R. P. (1977). Model specification in stationary random fields. *Geographical Analysis*, 9, 107-129.

Haining, R. P. (1978). The moving average model for spatial interaction. *Transactions, Institute of British Geographers*, NS3, 202-225.

Hannan, E. J. (1970). *Multiple Time-Series Analysis.* Wiley, New York.

Harrison, P. J. and Stevens, C. F. (1976). Bayesian forecasting (with discussion). *Journal of the Royal Statistical Society, Series B,* 38, 205-247.

Hengeveld, R. (1979). Analysis of spatial patterns of some ground beetles (*Col., Caribidae*). In R. M. Cormack and J. K. Ord, eds. (*op. cit.*).

Hsu, M. L. (1975). Filtering process in surface generalization and isopleth mapping. In J. C. Davis and M. J. McCullagh, eds. (*op. cit.*). 115-129.

Jenkins, G. M. and Watts, D. G. (1968). *Spectral Analysis and its Applications.* Holden-Day, San Francisco.

Jumars, P. A., Thistle, D., and Jones, M. L. (1977). Detecting two dimensional spatial structure in biological data. *Oecologia,* 28, 109-123.

Kendall, M. G. (1976). *Time Series.* Griffin, London.

Kendall, M. G. and Stuart, A. (1973, 1976). *The Advanced Theory of Statistics, Vol. 2, 3, 3rd* ed. Griffin, London.

Kooijman, S. A. L. M. (1976). Some remarks on the statistical analysis of grids especially with respect to ecology. *Annals of Systems Research,* 5, 113-132.

Kooijman, S. A. L. M. (1979). The description of point patterns. In R. M. Cormack and J. K. Ord, eds. (*op. cit.*).

Krishna Iyer, P. V. A. (1949). The first and second moments of some probability distributions arising from points on a lattice, and their applications. *Biometrika,* 36, 135-141.

Ludwig, J. A. (1979). A new quadrat variance method for the analysis of spatial pattern. In R. M. Cormack and J. K. Ord, eds. (*op. cit.*).

Matérn, B. (1960). *Spatial Variation. Meddelanden fran Statens Skogsforskningsinstitut,* band 49, No. 5, 1-144.

Matheron, G. (1971). *The Theory of Regionalized Variables and Its Applications.* Centre de Morphologie Mathematique, Paris.

Mead, R. (1971). Models for interplant competition in irregularly spaced populations. In *Statistical Ecology, Vol. 2,* G. P. Patil, E. C. Pielou, and W. E. Waters, eds. The Pennsylvania State University Press, University Park. 13-30.

Moran, P. A. P. (1951). A mathematical theory of animal trapping. *Biometrika,* 38, 307-311.

Moran, P. A. P. (1953). The statistical analysis of the Canadian lynx cycle. *Australian Journal of Ecology,* 1, 163-173, 291-298.

Ord, J. K. (1975). Estimation methods for models of spatial interaction. *Journal of the American Statistical Association,* 70, 120-126.

Ord, J. K. and Rees, M. W. (1979). Spatial processes: recent developments with applications to hydrology. In *The Mathematics of Water Resources,* Institute of Mathematics and its Applications Conference Proceedings. (to appear).

Pielou, E. C. (1975). *Ecological Diversity.* Wiley, New York.

Pielou, E. C. (1977). *Mathematical Ecology.* Wiley, New York.

Poole, R. W. (1979). The statistical prediction of the fluctuations in abundance in Nicholson's sheep bowfly experiments. In R. M. Cormack and J. K. Ord, eds. (*op. cit.*).

Priestley, M. B. (1965). Evolutionary spectra and non-stationary processes. *Journal of the Royal Statistical Society, Series B,* 27, 204-237.

Pyke, R. (1965). Spacings (with discussion). *Journal of the Royal Statistical Society, Series B,* 27, 395-449.

Quenouille, M. H. (1949). Approximate tests of correlation in time series. *Journal of the Royal Statistical Society, Series B,* 11, 68-84.

Ricker, W. E. (1958). Handbook of computations for biological statistics of fish populations. *Bulletin of the Fisheries Research Board of Canada,* 119, 1-300.

Robinson, J. E. (1975). Frequency analysis, sampling, and errors in spatial data. In J. C. Davis and M. J. McCullagh, eds. (*op. cit.*). 78-95.

Seber, G. A. F. (1973). *The Estimation of Animal Abundance.* Griffin, London.

Sen, A. (1976). Large sample size distribution of statistics used in testing for spatial autocorrelation. *Geographical Analysis*, 9, 175-184.

Shugart, H. H., Jr., ed. (1978). *Time-Series and Ecological Processes*. Society for Industrial and Applied Mathematics, Philadelphia.

Sokal, R. R. and Oden, N. L. (1978a,b). Spatial autocorrelation in biology. I: Methodology. II: Some biological applications of evolutionary and ecological interest. *Biological Journal of the Linnean Society*, 10, 199-228, 229-249.

Tanner, R. T. (1975). The stability and the intrinsic growth rates of prey and predator populations. *Ecology*, 56, 855-867.

Tobler, W. R. (1975). Linear operators applied to areal data. In J. C. Davis and M. J. McCullagh, eds. (*op. cit.*). 14-37.

Tong, H. (1977). Some comments on the Canadian lynx data. *Journal of the Royal Statistical Society, Series A*, 140, 432-436.

Watson, G. S. (1972). Trend surface analysis and spatial correlation. *Geological Society of America, Special Paper 146*, 39-46.

Whitten, E. H. T. (1975). The practical use of trend surface analysis in the geological sciences. In J. C. Davis and M. J. McCullagh, eds. (*op. cit.*). 282-297.

Whittle, P. (1954). On stationary processes in the plane. *Biometrika*, 41, 434-449.

Williamson, M. (1972). *The Analysis of Biological Populations*. Arnold, London.

Williamson, M. (1975). The biological interpretation of time series analysis. *Bulletin of the Institute of Mathematics and its Applications*, 11, 67-69.

[*Received:* *May* 1977. *Revised:* *February* 1979]

R. M. Cormack and J. K. Ord, (eds).,
*Spatial and Temporal Analysis in Ecology*, pp. 95-150. 

# STATISTICAL METHODS FOR SPATIAL POINT PATTERNS IN ECOLOGY

PETER J. DIGGLE

Department of Statistics
University of Newcastle upon Tyne, England

SUMMARY. After establishing the objectives of the statistical analysis of spatial point patterns, the available theoretical models for such data are reviewed. This is followed by a consideration of relevant statistical methods, in which a sharp distinction is drawn between exploratory analysis in the field and the more detailed approach appropriate to mapped point patterns. The paper concludes with several illustrative examples involving data on tree locations.

KEY WORDS. density estimation, goodness-of-fit, Monte Carlo test, spatial distribution, spatial point process, statistical ecology, test of randomness.

## 1. INTRODUCTION

*1.1 Objectives of Point Pattern Analysis.* Any investigation which involves the study of a population whose domain is a region of the earth's surface carries implicit spatial connotations. It follows that the environmental sciences, and in particular ecology, provide a natural area of application for the statistical methodology associated with spatial stochastic processes. Our particular concern will be with data in the form of a set of stochastically generated *point locations* of a number of *events* in some geographical region, typically trees in a forest. For such data, the statistician's first objective will be to formulate hypotheses concerning the nature of the observed pattern. If appropriate, the analysis may proceed to a formal assessment of the fit to the data of a proposed model. In either event, the objective is that the results of the statistical analysis will help the ecologist by providing additional insight into the underlying biological mechanisms.

In field-work a map of point locations may be unobtainable and we then work with distance measurements from randomly located sample points to neighboring events, or counts of numbers of events in randomly located quadrats. The objectives of any statistical analysis will now be rather more modest than above. The aim will be to provide the ecologist with an estimate of density, or mean number of events per unit area, and a qualitative description of pattern, usually expressed by the result of a test of spatial randomness. Historically, this type of analysis precedes the development of methods appropriate to a more detailed investigation of mapped data, and is now generally regarded as being of lesser importance (e.g., Bartlett, 1975, Ch. 1).

*1.2 Illustrative Examples.* To convey the flavor of the subject we now present the sets of data which will subsequently be used to illustrate the statistical methodology.

Figures 1.1 to 1.3 respectively show the locations of 929 oaks, 703 hickories, and 514 maples in a 19.6 acre square within Lansing woods, Clinton County, Michigan. The relatively uniform distribution of oaks over the square contrasts with the more patchy appearance of the hickory and maple maps.

In Figures 1.4 and 1.5 we again notice a sharp contrast in pattern, here between two rather smaller stands of 65 Japanese black pine saplings (Numata, 1961) and 62 Redwood seedlings (Ripley, 1977, extracted from Strauss, 1975). The Redwood seedlings exhibit an apparently strong tendency to cluster in groups.

Finally, Figure 1.6 shows a map of 250 Balsam fir seedlings in a 66 foot square plot, given by Ghent (1963). The plot suggests variations in local density and, in particular, a preponderance of seedlings in the lower left-hand corner.

We shall analyze each of the above maps as given, since no external information is available to us. However, we emphasize that if such information is available, for example on variations in the microtopography of the study area, it should be incorporated into the analysis if at all possible. We return to this point, with regard to the redwood data, in Section 6.3.

*1.3 Statistical Concepts.* We shall make extensive use of a number of statistical concepts which may be unfamiliar to some

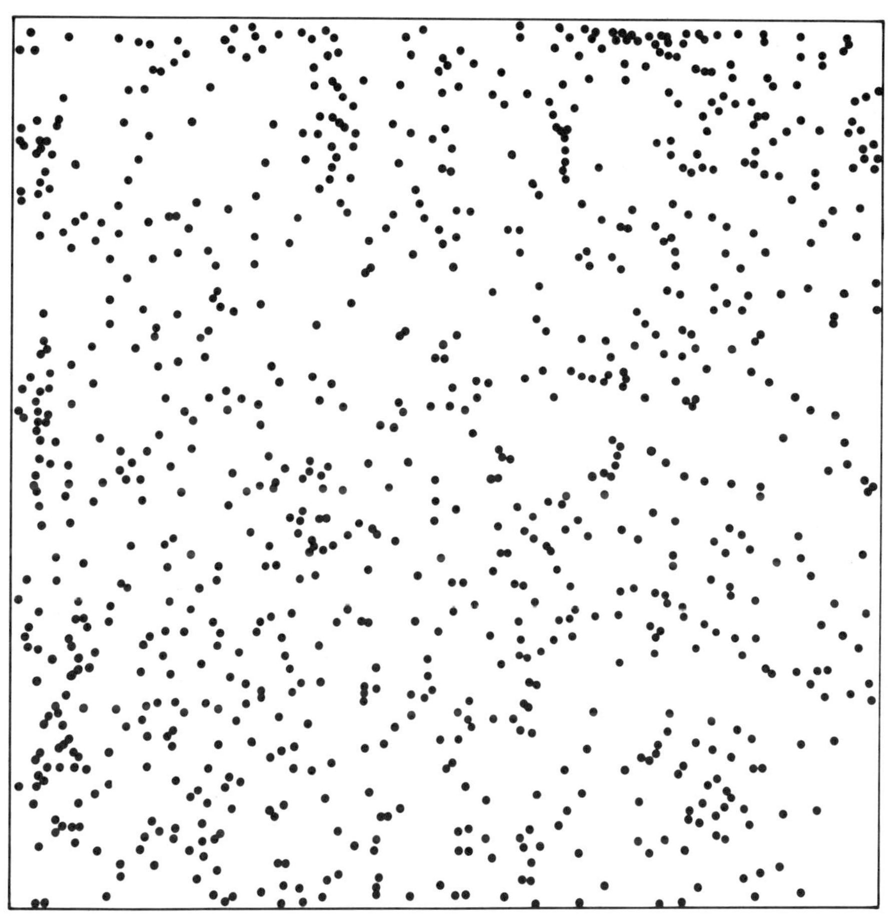

*FIG. 1.1: Oaks in Lansing Woods* (929 *trees in* 19.6 *acre plot).*

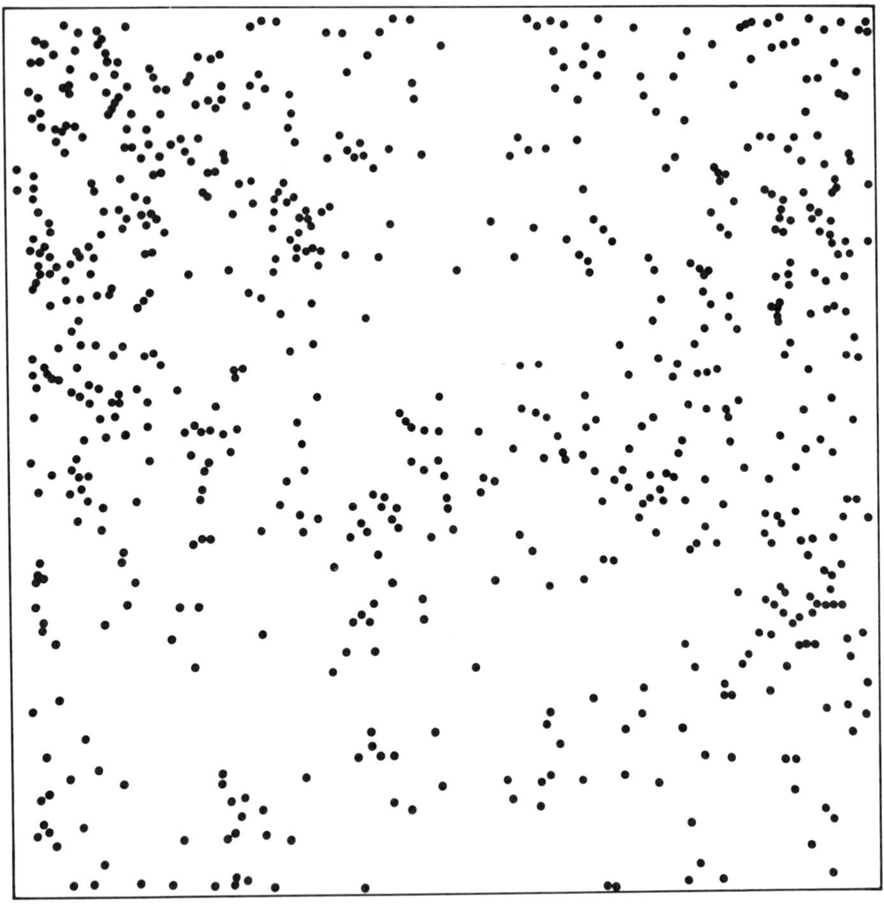

*FIG. 1.2: Hickories in Lansing Woods (*703 *trees in* 19.6 *acre plot).*

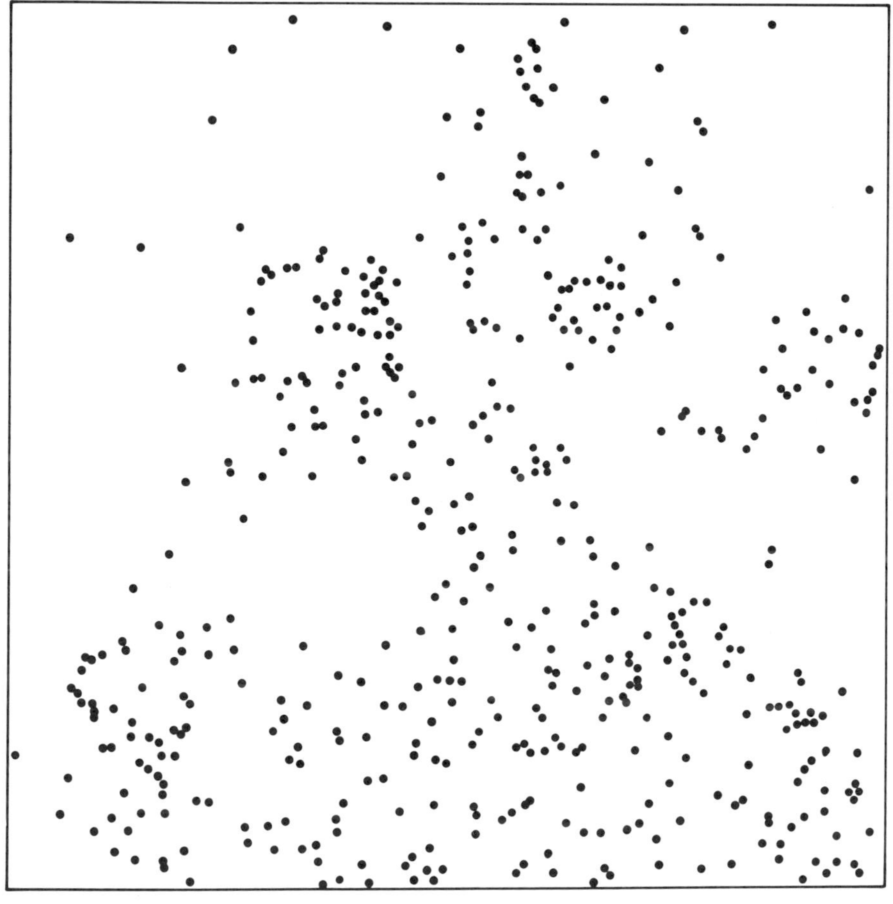

*FIG. 1.3: Maples in Lansing Woods* (514 *trees in* 19.6 *acre plot).*

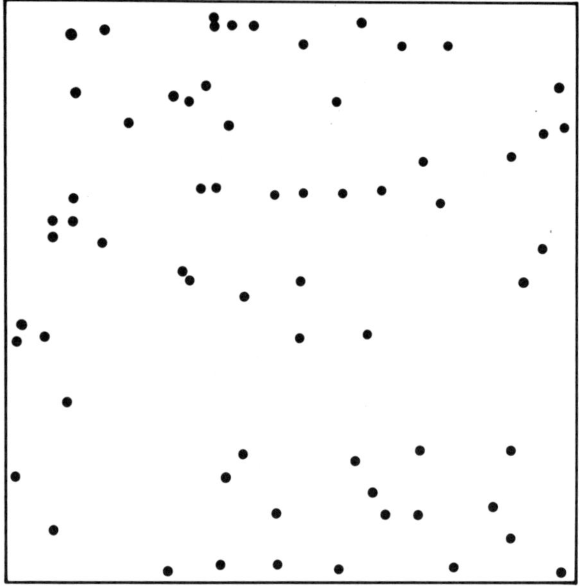

*FIG. 1.4: Japanese black pine saplings (65 saplings in 5.7 meter square plot).*

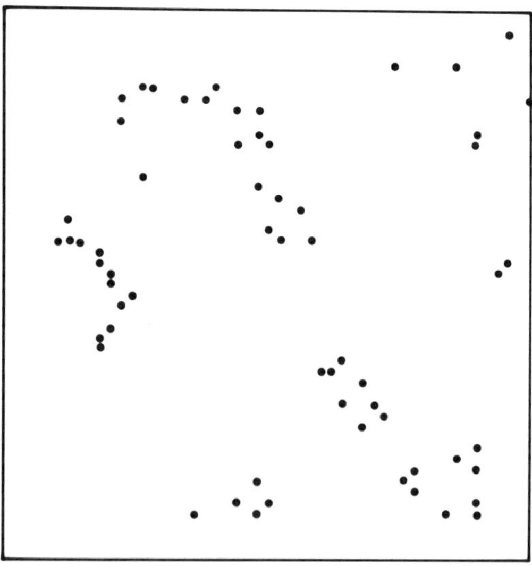

*FIG. 1.5: Redwood seedlings (62 seedlings in approximately 75 foot square plot).*

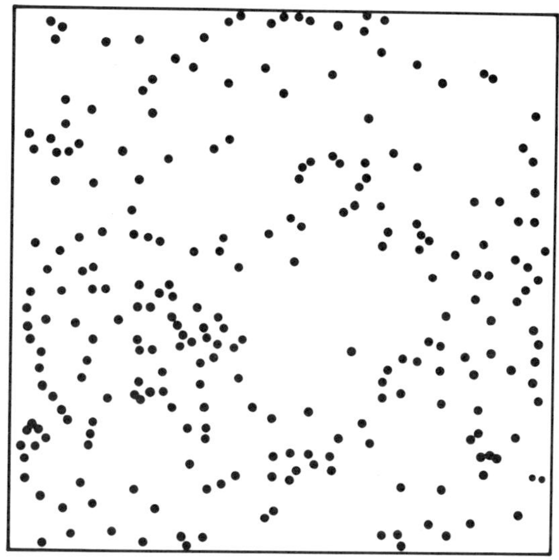

*FIG. 1.6: Balsam fir seedlings (250 seedlings in 66 foot square plot).*

readers. A description of the majority of these may be found in any one of numerous intermediate statistics textbooks, for example Mood and Graybill (1963). Two exceptions to this are the notions of *pseudo-random sampling* and *robustness*.

By pseudo-random sampling, we mean a computational device whereby a simulated sample from some distribution is generated in such a way as to have every appearance of a genuine random sample. Further details may be found in Newman and Odell (1971).

Most statistical techniques are derived on the basis of rather specific distributional assumptions about the data. For example, the well-known two-sample t-test for the equality of means assumes that the data arise as *independent* random samples from two *Normal* distributions with *common* variance. A particular technique is said to be *robust* if it continues to give reliable results when the underlying distributional assumptions are, to a greater or lesser extent, violated. Thus, robustness is a relative rather than absolute term. In the context of estimation, robustness is often summarized by the mean and variance of the estimator under various different sets of assumptions. For example, if we wish to estimate the mean number of plants per unit area in some population under study, we would like to use an estimator whose mean square error is relatively insensitive to changes in the type of spatial point pattern presented by the plants.

*1.4 Summary.* In Section 2 of this paper we review the various spatial point processes which have been suggested as possible models for point patterns in ecology. The two-dimensional homogeneous Poisson process, or Poisson forest, plays a central role in the development of such models, and is accordingly treated in some detail. We distinguish between a process defined over the entire plane and one defined only in a given finite region. This distinction may sometimes be ignored in the development of statistical techniques for 'large' populations but should always be borne in mind. Alternative classes of model are designed to reflect the qualitatively different types of spatial pattern which can arise.

In Sections 3 and 4 we draw together the various statistical techniques currently available for the analysis of point patterns. We have consciously separated methods appropriate to field-work from those designed for the analysis of mapped patterns. In the past, methods designed for the former situation have been translated somewhat uncritically to the latter. This almost invariably represents a wasteful use of scarce data and can, incidentally, be a source of unfair criticism of the particular statistical techniques. In the Appendix, we establish the simple, but extremely useful, notion of a Monte Carlo test.

Finally, in Section 5, the illustrative examples from Section 1.2 are subjected to analysis by the various methods described in the preceding two sections. These analyses indicate some of the scope for ecological applications of the methodology. In addition they should aid a critical comparison amongst rival methods of analysis, although we emphasize that authoritative conclusions in this respect must be deferred, pending fuller investigation.

Although a certain amount of overlap with Cormack's (1979) contribution to this volume is inevitable, this will be minimized by cross-referencing at appropriate points in the text.

## 2. MODELS

*2.1 The Poisson Forest.* The two-dimensional homogeneous Poisson point process, or Poisson forest, provides the simplest stochastic model for a spatial point pattern, and acts as a foundation upon which more complex models may subsequently be built. It incorporates two basic postulates, those of *uniformity* whereby the events in question exhibit no tendency to occupy particular regions of the plane, and *independence*, whereby the location of a given event is determined without reference to that of any other event. Together, these amount to the dropping of point locations completely at random on the plane. Whilst this is strictly implausible as a model for any plant population in that, for example, it allows two plants to be separated by an arbitrarily small distance, it can sometimes be a useful approximation, and is developed as an idealized standard.

More formally, let the random variable $N(A)$ denote the number of events in an arbitrary area $A$, and write $p_n(A) = \Pr\{N(A) = n\}$. Then there exists a constant $\lambda > 0$, the *density* of the process, such that for any small area $\delta A$,

$$p_n(\delta A) = \begin{cases} 1 - \lambda\delta A + o(\delta A) & (n = 0), \\ \lambda\delta A + o(\delta A) & (n = 1), \\ o(\delta A) & (n = 2,3,\cdots). \end{cases}$$

Also, for any two non-intersecting regions $A$ and $B$, the random variables $N(A)$ and $N(B)$ are statistically independent. It follows that

$$p_n(A) = e^{-\lambda A}(\lambda A)^n/n! \quad (n = 0,1,\cdots). \tag{2.1.1}$$

In other words, $N(A)$ has a Poisson distribution with mean proportional to $A$, the quadrat count distribution for the Poisson forest. Note that the mean is density times area. The proof is an exact analogue of the one-dimensional case, given for example by Cox and Miller (1965, Ch. 4).

In practice, we shall be concerned with the restriction of the process to a particular region $A$ containing a given number $n$ of events. It follows from the notions of uniformity and independence advanced above that the locations of the $n$ events will then be independently and identically uniformly distributed within $A$. A formal proof may easily be constructed along the lines indicated by Cox and Lewis (1966, Ch. 2) for the one-dimensional case. Figure 2.1 shows 200 events whose locations are so distributed on the unit square. A comparison with the data-sets illustrated in Figures 1.1 to 1.6 is instructive at this point.

The result (2.1.1) may also be used to derive various *distance distributions* associated with the Poisson forest. For example, let the random variable $X_k$ denote the distance from an arbitrary point in the plane to the $k$th nearest event in the

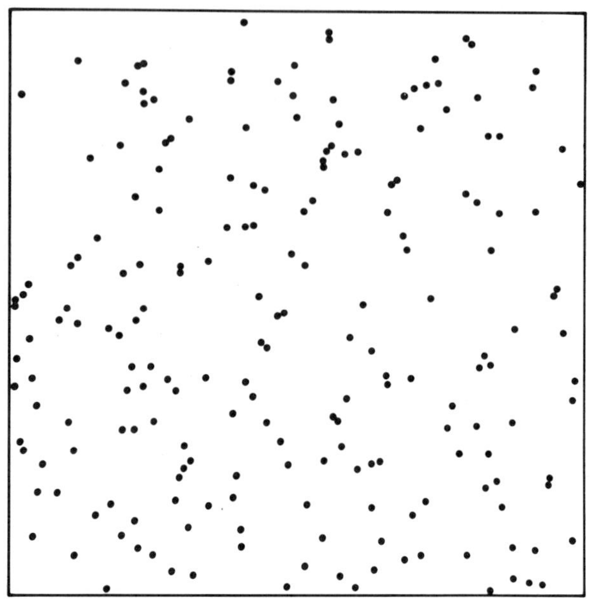

*FIG. 2.1: A partial realization of a Poisson forest:* 200 *events independently and identically uniformly distributed in the unit square.*

process. Then the observed value of $X_k$ is greater than $x$ if and only if there are at most $k-1$ events in a circle of radius $x$. If $F_k(x) = \Pr\{X_k \leq x\}$ denotes the distribution function of $X_k$, it follows from (2.1.1), with $A = \pi x^2$, that

$$F_k(x) = 1 - \sum_{n=0}^{k-1} e^{-\pi\lambda x^2}(\pi\lambda x^2)^n/n! ,$$

differentiation of which yields, after some simplification, the probability density function

$$f_k(x) = 2(\pi\lambda)^k x^{2k-1} e^{-\pi\lambda x^2}/(k-1)! \quad (x > 0). \qquad (2.1.2)$$

In other words, $X_k^2$ has a gamma distribution; $2\pi\lambda X_k^2$ is distributed as chi-squared on $2k$ degrees of freedom.

Other distance distributions based on simple geometrical constructions may similarly be derived from the basic result (2.1.1) together with the notion of the independence of numbers of events in non-intersecting regions. We defer, until Section 3, a discussion of the relevance of such distributions to inferential problems, but note briefly some of the more interesting results. First, (2.1.2) is unaffected if we choose to measure distances from an arbitrary *event* in the process, rather than from an arbitrary point in the plane. This stochastic equivalence of arbitrary point-to-event and arbitrary event-to-event distances is unique to the Poisson forest, a fact which has been turned to good effect in the construction of distance-based tests of spatial randomness. However, the selection of an arbitrary event raises problems in field-work, where a complete enumeration of the events under consideration may be impractical. A possible alternative strategy is to let $X$ denote the distance from an arbitrary point $P$ to the nearest event, at $Q$ say, and $Y$ the distance from $Q$ to its nearest neighboring event. In this case, however, $X$ and $Y$ will be neither independent nor identically distributed. It is intuitively clear that the selection procedure for $Q$ is biased in favor of the more isolated events in the population, and numerical evaluation gives $E(Y) \simeq 1.19\, E(X)$ (Kendall and Moran, 1963, Ch. 2, but note that their final expression for $E(Y)$ is incorrect by a factor of 2). The joint distribution of $X$ and $Y$ is further investigated by T. Cox and T. Lewis (1976) and by Cormack (1977). Besag and Gleaves (1973) suggest a *T-square sampling* procedure which gives rise to some rather simpler associated distribution theory. Here, $X$ is defined as above, but $Y$ is replaced by $Z$, the distance from $Q$ to the nearest event within the half-plane delimited by the perpendicular to $PQ$ passing through $Q$, and which excludes $P$ (Figure 2.2). Then, $X$ and $Z/\sqrt{2}$ are independent, and identically distributed according to (2.1.2) with $k = 1$.

The simplicity of the Poisson forest is both its strength and its weakness. On the one hand, it provides a natural starting point both for theoretical work on the development of inferential tools and for practical investigations of ecological data. On the other hand, its limitations as a plausible model for natural phenomena are only too obvious. Consequently, we proceed to a consideration of alternative classes of model.

*2.2 Poisson Cluster Processes.* An interesting generalization of the Poisson forest is obtained if we assume that randomly distributed *parents* generate random numbers of *offspring* which are spatially distributed about the corresponding parent. Formally, a *Poisson cluster process* embodies three postulates:

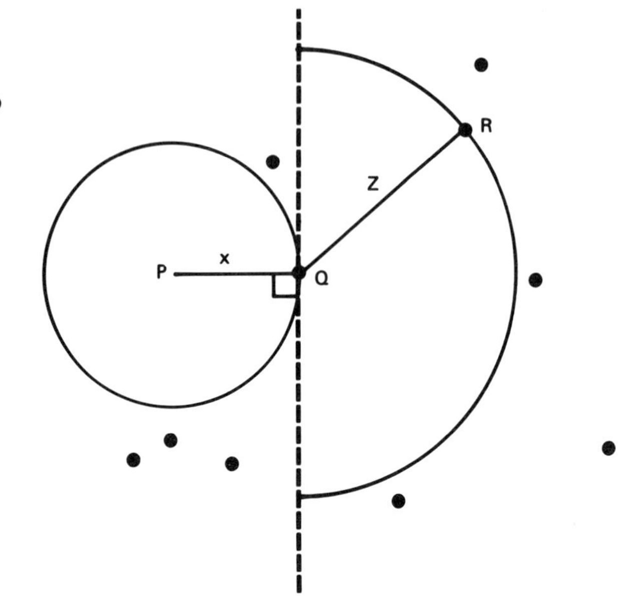


*FIG. 2.2: T-square sampling.* (P = *arbitrary point,* Q = *nearest event to* P, R = *T-square nearest neighbor to* Q).

P1: Parent events constitute a Poisson forest with density $\rho$ per unit area.

P2: Each parent produces, independently, a random number $N$ of offspring, with corresponding probability distribution $\{p_n : n = 0,1,\cdots\}$.

P3: Each offspring is, independently, spatially distributed relative to the corresponding parent. The vector displacement relative to the parent is governed by a bivariate distribution function $H(\underset{\sim}{x})$, or pdf $h(\underset{\sim}{x})$, if the latter exists.

The final pattern may consist either of offspring only, or of parents and offspring, assumed indistinguishable. It is usual to specify a radially symmetric $H(\underset{\sim}{x})$, in which case the process is both stationary and isotropic, with density $\lambda = \rho E(N)$ (add $\rho$ if parents are included in the final pattern). Figure 2.3 shows a realization of an isotropic process, parents excluded, for which $N$ is distributed as Poisson, with mean $\mu \simeq 2.3$, and the spatial dispersion mechanism is

$$h(\underset{\sim}{x}) \equiv h(r,\theta) = r/\pi\sigma^2 \quad (0 < r \leq \sigma, 0 < \theta \leq 2\pi),$$

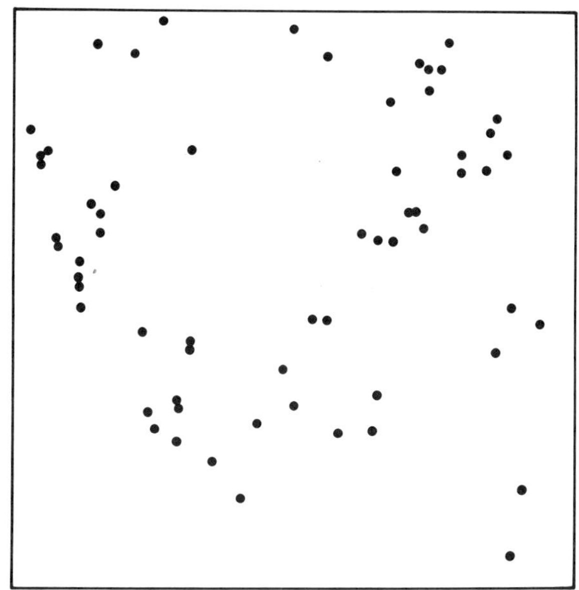

*FIG. 2.3: A partial realization of a Poisson cluster process: 62 events in the unit square.*

with $\sigma = 0.07$. The realization has been conditioned to produce 62 events from 27 parents in the unit square. Note the superficial similarity with the redwood data in Figure 1.5.

Such processes have been studied independently by a number of authors, including Neyman and Scott (1958) and Bartlett (1964). The degenerate case in which each offspring occupies the same location as its parent gives rise to various generalized Poisson, or *contagious*, quadrat count distributions. For example, if N follows a Poisson distribution, the Neyman type A results, whilst a logarithmic series distribution for N leads to the negative binomial. The assumption of randomly distributed *point* clusters implicit in the use of such contagious distributions has not always been recognized, and the fitting of such distributions to quadrat count data is best viewed simply as an empirical device which *may* provide a useful description.

Quadrat count distributions for non-degenerate cases appear largely intractable, although in principle the necessary information is available in the form of the characteristic functional of a Poisson cluster process (Bartlett, 1975, Ch. 1). Distance distributions appear to offer more scope for analytical progress.

Again, Bartlett (1975, Ch. 1) provides the general framework. Consider, for example, the isotropic case and let $P(r;x)$ denote the probability that no offspring from a parent at distance $r$ from the origin will be found within the circle, center the origin and radius $x$. Then the distance from an arbitrary point to the nearest event in the process has distribution function

$$F(x) = 1 - \exp[-2\pi\rho \int_0^\infty \{1 - P(r,x)\}rdr]$$

for the case of offspring only, and

$$F(x) = 1 - \exp[-\pi\rho \{x^2 + 2 \int_x^\infty (1 - P(r,x))rdr\}]$$

if parents are included. Similarly, the distance from an arbitrary event to the nearest other event in the process has distribution function

$$G(y) = 1 - \{1 - F(y)\}P_2(y)$$

where now $P_2(y)$ denotes the probability that no event from the same cluster as the arbitrary event will be found within a distance $y$ of that event. The probabilities $P(r;x)$ and $P_2(y)$ are in principle determined by postulates P2 and P3 of the process, although the solution presents difficulties. However, Warren (1971) and Diggle (1975) give explicit results for special cases.

2.3 *Heterogeneity*. An ostensibly different type of generative process for spatial point patterns incorporates variation in the local density, $\lambda(\underset{\sim}{x})$. If we extend the Poisson forest by replacing the previously constant density by a realization of a random function $\Lambda(\underset{\sim}{x})$, we obtain a class of heterogeneous, or *doubly stochastic* Poisson processes. Matérn (1971) considers the general formulation of such processes and offers the following explicit example: the centers of circles of given radius $\sigma$ form a Poisson forest with density $\rho$ per unit area and the realized value $\lambda(\underset{\sim}{x})$ is defined as the number of such circles which contain $\underset{\sim}{x}$; thus the marginal distribution of $\Lambda(\underset{\sim}{x})$ is Poisson with mean $\pi\sigma^2\rho$. A trivial modification is to allow an arbitrary multiplicative constant in the random function $\Lambda(\underset{\sim}{x})$, which effectively introduces the mean number of events per circle as an additional parameter. This now corresponds precisely to the isotropic Poisson cluster process used to generate Figure 2.2 and no method of analysis can therefore distinguish between these alternative representations of the same underlying process. A similar ambiguity must thwart any attempt, based solely on a spatial point pattern, to preserve a formal distinction between 'true' and 'apparent' contagion for distributions such as the negative binomial.

Despite this difficulty, the concept of heterogeneity remains an extremely useful empirical device for the description of point patterns, particularly when external evidence suggests non-uniformity of the environment. In this context it is perhaps more useful to think in terms of a *given* function $\lambda(\underset{\sim}{x})$ and to postulate a *purely heterogeneous*, or *non-uniform*, process as one for which, given a region A containing n events, the locations of the events are distributed independently and identically according to the bivariate pdf

$$f(\underset{\sim}{x}) = \lambda(\underset{\sim}{x})/\{{}_A\int\lambda(\underset{\sim}{x})d\underset{\sim}{x}\} \quad (\underset{\sim}{x} \in A). \tag{2.3.1}$$

For example, Figure 2.4 shows a realization of such a process with $n = 100$, A the unit square and

$$\lambda(\underset{\sim}{x}) = e^{-2x_1-x_2}.$$

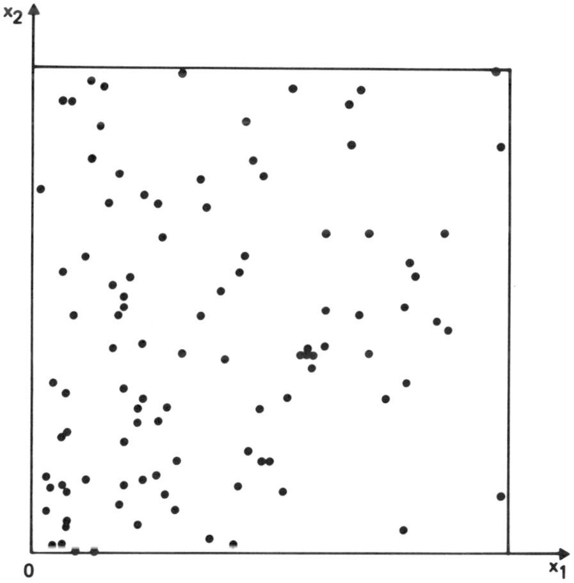

*FIG. 2.4: A realization of a purely heterogeneous process:* 100 *events in the unit square.*

We emphasize that formally, and in the absence of external information or replication, this might equally well be interpreted as a partial realization either of a doubly stochastic Poisson process, or a Poisson cluster process. However, we suggest that the meaning of pure heterogeneity is both intuitively clear and distinct from clustering, and that it might usefully be incorporated into the construction of empirical models. These might, for example, involve clustering of offspring relative to a non-uniform distribution of parents. What forms of $\lambda(\underset{\sim}{x})$ might prove useful? Kooijman (1977) uses a linear form, $\lambda(\underset{\sim}{x}) = \alpha + \underset{\sim}{\beta}'\underset{\sim}{x}$, and suggests that in practice, for data in a given region A, the possibility of $\lambda(\underset{\sim}{x})$ assuming negative values in A may be ignored. We would prefer a flexible, but reasonably parsimonious, class of non-negative functions $\lambda(\underset{\sim}{x})$. Provided that the numerical problems of fitting noted by Kooijman can be overcome, logistic polynomials of moderate degree are one possibility, whilst for data on the unit square some form of bivariate beta distribution such as Plackett's (1965) C-type might be contemplated.

*2.4 Lattice-Based Processes.* The processes described in Sections 2.2 and 2.3 above tend to generate patterns which might be described loosely as 'aggregated', in the sense that realizations of such processes will suggest a combination of relatively dense and relatively sparse sub-regions. A qualitatively different alternative to the Poisson forest would be one which constrains the events in question to be spaced out more or less regularly over the region under consideration. Such regular spatial point patterns have been given relatively little attention in the ecological literature, possibly because locally regular patterns, which might plausibly arise as a result of some form of competitive interaction between individuals, are typically obscured by aggregation at a larger scale, especially when the traditionally popular methods of analysis based on quadrat counts are used exclusively.

Extreme forms of regularity are provided by deterministic lattice structures, usually either square or equilateral triangular. A more flexible model for regular patterns would hold greater intuitive appeal, and may be obtained if we superimpose upon the lattice structure a realization of a Poisson forest with density $\rho$ per unit area. The resulting patterns range from extreme regularity when $\rho = 0$ towards realizations of a Poisson forest as $\rho \to \infty$. Diggle (1975) establishes some results for distance distributions in such a superimposed process. For example, the arbitrary point-to-event nearest neighbor distance has distribution function

$$F(x) = 1 - e^{-\pi\rho x^2}\{1 - G(x)\},$$

where $G(x)$ denotes the distribution function of the distance from an arbitrary point to the nearest lattice-node, given by Persson (1964) for the square lattice and by Holgate (1965a) for the equilateral triangular case. Brown and Holgate (1974) offer an alternative construction in which a basic square lattice structure undergoes random thinning. Lattice-based processes are essentially an artifice whereby mathematical tractability is obtained, and may seem to be of limited practical relevance. Their principal use is as a yardstick for the assessment of inferential procedures applied to regular patterns in general. Further discussion of lattice-based processes is offered by Cormack (1979).

*2.5 Inhibition and Markov Point Processes.* A more natural source of regularity in spatial point patterns is to assume an inhibitory mechanism whereby no two events may occur less than some prescribed distance, $\delta$ say, apart. If, apart from this constraint, the events are located completely at random we obtain a *simple inhibition process*, a concept which may be formalized in various ways. For example, the process termed *simple sequential inhibition* by Diggle *et al.* (1976) generates $n$ locations $\underset{\sim}{x}_i$ sequentially in a given region $A$, with $\underset{\sim}{x}_i$ uniformly distributed over the set of all points distant at least $\delta$ from all previously located $\underset{\sim}{x}_j$ $(j = 1,\cdots, i-1)$. A typical realization constitutes Figure 2.5. Note that in general the degree of regularity induced in the realized patterns will, for given $A$ and $\delta$, increase with $n$, and indeed the process may fail if $n$ is too large. Alternatively, Ripley (1977) considers the restriction of the Poisson forest to the given region $A$ and obtains a simple inhibition process by allowing only those *realizations of the process* for which no two events are closer than a distance $\delta$ apart. Further variations are described by Matérn (1960) and by Bartlett (1975, Ch. 3).

In fact, such processes fall within the general framework of *Markov point processes* introduced by Ripley and Kelly (1977). We consider a finite region $A$ and define, for any subset $B$ of $A$, the *environment* of $B$ to be the set of all points not in $B$ which are distant at most $\delta$ from some point in $B$, for a prescribed value of $\delta > 0$. A point process is Markov if the distributional properties of the configuration of events in $B$, given the configuration in the complementary region $A/B$, depend only on the configuration in the environment of $B$. Simple inhibition processes follow as a special case in which an event at $\underset{\sim}{x}$ is forbidden unless the environment of $\underset{\sim}{x}$ contains no events. In addition to their theoretical interest, such processes provide a valuable tool for the practitioner since they can be simulated as

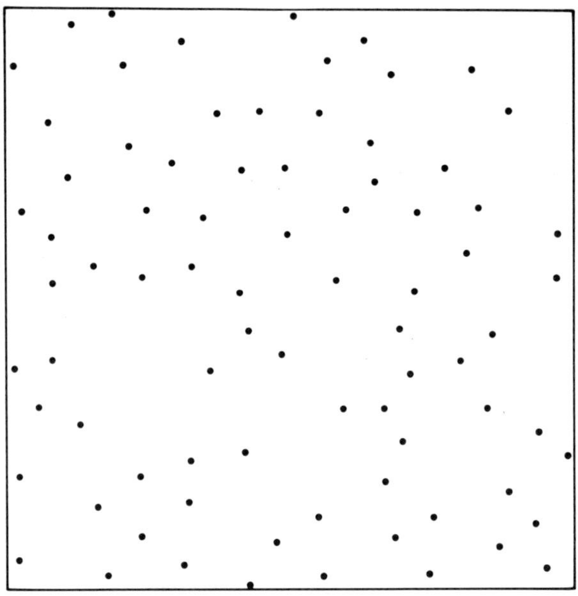

*FIG. 2.5: A realization of simple sequential inhibition:* 80 *events in the square of side* 15 *units,* $\delta = 1$.

equilibrium distributions of spatial birth-and-death processes (Preston, 1975). Briefly, from an initial distribution of n events in a region A, an event is deleted at random, then replaced by an event located according to a bivariate distribution *defined in terms of the locations of the other* n-1 *events,* and this process is repeated until equilibrium has effectively been reached. Again, simple inhibition may usefully be borne in mind as a special case. Ripley (1977) gives further details.

In principle, Markov point processes represent an extremely flexible class of models in that they can produce both regular (inhibitory) and aggregated (contagious) patterns. Figure 12 of Ripley (1977) exemplifies the latter situation. However, their principal application thus far has been to the study of regular point patterns. As such, they arise naturally in the study of competitive interactions between individuals and are again discussed further in Cormack (1979).

*2.6 Concluding Remarks.* This discussion of theoretical models may be thought unduly lengthy in the context of a review of *statistical methods* for spatial point processes, this despite the deliberate omission of almost all the underlying distribution theory. However, an understanding of the various classes of process under consideration, at least in a qualitative sense, and a similar awareness of the different types of pattern which may be generated, are pre-requisite to a critical evaluation of proposed inferential techniques, and of course to any serious attempt at model-building.

## 3. ANALYSIS IN THE FIELD

*3.1 Introductory Remarks.* When a complete map of locations is unavailable, complete spatial randomness can provide a useful working hypothesis, a test of which may be required. Rejection of the null hypothesis will usually be of relatively little interest *per se* and the aim will be rather to gain insight into the type of departure from complete spatial randomness observed. A useful dichotomy is that between *aggregation* and *regularity* in the underlying pattern, although the question of more specific alternatives will also arise. An estimate of the density of the pattern, or equivalently of the number of events in a given region, may also be useful in that it provides a measure of abundance which can form the basis for comparisons amongst different species in the same habitat, or different populations of the same species. The alternative interpretation of *inverse density* as the mean area per event may also help the decision on an appropriate area for subsequent, more detailed study, within which the population will be of reasonable size.

A basic choice of sampling method is between counts in randomly located quadrats and distances measured in various ways from randomly located points. The author's inclination is towards the use of distance methods, in part because little appears to be known about the effect on quadrat methods of the arbitrary choice of individual quadrat size. Practical considerations may also be crucial, for example in dense forest.

*3.2 Quadrat Methods.* We suppose that counts $x_i$ $(i=1,\cdots,m)$ are available of the numbers of events in $m$ quadrats each of area $A$. Under the null hypotheses, $H_o$, of complete spatial randomness, the counts form an independent random sample from the Poisson distribution (2.1.1) with unknown mean. Any goodness-of-fit test may therefore be used to test $H_o$, and a number of early workers, including Blackman (1935) and Clapham (1936) adopted Pearson's goodness-of-fit criterion, $\Sigma(O-E)^2/E$. However, this generally gives a rather weak test and requires a relatively large sample. A more satisfactory approach, now widely used but proposed originally by Fisher *et al.* (1922), is a test based on the *index of dispersion*, $I = S_x^2/\bar{X}$, i.e., the sample variance-to-mean ratio. Under $H_o$, $(m-1)I$ is distributed approximately as chi-squared on $m-1$ degrees of freedom. The approximation is satisfactory for $m > 6$ and $\lambda A > 1$ (Kathirgamatamby, 1953; see also Section 5.2 below). The mean of $I$ under $H_o$ is therefore unity, with significantly large values suggesting aggregation and significantly small values regularity in the underlying pattern. Variants of the index of dispersion are considered by Cormack (1979).

With regard to density estimation, the estimator

$$\hat{\lambda} = \bar{X}/A$$

has a natural appeal, is easily shown to be the maximum likelihood estimator in the completely random case, and is then unbiased with variance $\lambda/nA$. More generally, unbiasedness of the estimator $\hat{\lambda}$ is guaranteed for *any* underlying pattern, whilst its efficiency is clearly related to the power of the corresponding index of dispersion test.

To sum up, the index of dispersion is widely used in practice and appears generally to give a reasonably powerful test, although little objective information is available on this point. There is, however, a potential conflict between testing and estimation in that a powerful test of randomness against aggregation implies an inefficient estimator of density and *vice versa*. Also, for given total quadrat area, the effects of changes in individual quadrat size are difficult to quantify. On the other hand, the guarantee of an unbiased estimator of density is a valuable property which is not shared by rival, distance-based estimators.

*3.3 Distance-Based Tests of Spatial Randomness.* Most of the earlier references to the use of distance methods in ecological work are concerned primarily with the construction of objective sampling schemes, and the associated inferential problems are considered rather informally. See, for example, Cottam and Curtis (1949), Cottam *et al.* (1953) and Catana (1963). Holgate (1972) gives an extensive bibliography.

The test proposed by Hopkins (1954) remains the standard against which the performance of rival tests is usually assessed. The same test was suggested independently by Moore (1954). Let $X$ and $Y$ denote, respectively, the distances from a randomly located point and a randomly selected event to the nearest event in each case. Under the null hypotheses, $H_o$, of complete spatial randomness we have, from (2.1.2), that $2\pi\lambda X^2$ and $2\pi\lambda Y^2$ are each distributed as chi-squared on two degrees of freedom. Thus, for independent samples of m randomly located points and m randomly selected events, generating observations $\{(X_i, Y_i) : i=1,\cdots,m\}$, $2\pi\lambda \sum_{i=1}^{m} X_i^2$ and $2\pi\lambda \sum_{i=1}^{m} Y_i^2$ are, independently, distributed as chi-squared on 2m degrees of freedom and Hopkins' statistic

$$h = \Sigma X_i^2 / (\Sigma X_i^2 + \Sigma Y_i^2) \tag{3.3.1}$$

therefore has a beta sampling distribution, $B(m,m)$, under $H_o$. The summations in (3.3.1), and in the remainder of the present chapter unless otherwise stated, are over $i=1,\cdots,m$. As noted previously, the distributional results leading to (3.3.1) are approximate in a finite population. The intuitive idea that the approximations will be reasonable when the sample size m is small relative to population size is formalized by Holgate (1965b). As a general guide, the degree of approximation will be very slight when sampling in the field, but much more crucial when applied to mapped data, as discussed below in Section 4. The intuitive appeal of Hopkins' test, usually confirmed by comparative power studies, is that in aggregated patterns the point-to-event distances will typically be large in comparison with the event-to-event distances whilst the reverse will generally be true in regular patterns. Thus, significantly large values of h suggest aggregation whilst significantly small values indicate regularity in the underlying pattern. The objection to Hopkins' test is a practical one, in that the random selection of an event requires a complete enumeration of the population, which is precisely the operation we wish to avoid at the exploratory stage of analysis in the field. Subsequent tests have been designed to overcome this problem, at the same time admitting some loss of power.

Holgate (1965c) suggests a test based on the distances, X and W say, from each of m randomly located points to the nearest and second nearest events respectively. Here, the variate $W^2 - X^2$ assumes the role of $Y^2$ in Hopkins' test and the same distribution theory applies, giving a test statistic

$$h_B = \Sigma X_i^2 / \Sigma W_i^2 \tag{3.3.2}$$

whose null distribution is again $B(m,m)$. Because of the natural pairing of the $(X_i, W_i)$ values, a variant of this test is available, namely

$$h_N = m^{-1}\Sigma(X_i^2/W_i^2), \tag{3.3.3}$$

whose null distribution is approximately Normal, $N(0.5,(12m)^{-1})$. Holgate's tests are particularly suspect against regularity. Consider, for example, the extreme case in which an event is located at each node of a regular square lattice. The ratio $X_i^2/W_i^2$ may be arbitrarily close to unity, which would be falsely indicative of aggregation, whereas for Hopkins' test $X_i^2/(X_i^2 + Y_i^2) \leq 1/3$.

Besag and Gleaves (1973) propose two 'T-square' tests analogous to (3.3.2) and (3.3.3), but using the variates $X^2$ and $Z^2/2$ as defined in Section 2.1, rather than $X^2$ and $W^2 - X^2$. Again, the null distribution theory translates exactly to give test statistics

$$t_B = \Sigma X_i^2/(\Sigma X_i^2 + \Sigma Z_i^2/2) \tag{3.3.4}$$

and

$$t_N = m^{-1}\Sigma X_i^2/(X_i^2 + \Sigma Z_i^2/2) \tag{3.3.5}$$

with respective null distributions $B(m,m)$ and, approximately, $N(0.5,(12m)^{-1})$. The T-square tests aim to preserve the intuitive appeal of Hopkins' test, but without the tears, and in this respect they prove more successful than Holgate's test. The 'conditioned distance ratio method' of T. Cox and T. Lewis (1976) gives rise to a test whose power characteristics appear generally to be similar to those of $t_N$.

Analytical results concerning the power of the various tests are largely restricted to the extreme cases of randomly distributed point clusters and deterministic lattice structures, which are of limited practical relevance and may actually be misleading. For example, against the alternative of the Thomas (1949) process of randomly distributed point clusters, with cluster size 'one plus a Poisson variate', the Holgate and T-square Normal tests have identical power functions, and are furthermore marginally but uniformly more powerful than Hopkins' test (Holgate, 1965b, c; Besag and Gleaves, 1973). We expect neither result to extrapolate to non-degenerate Poisson cluster processes.

Diggle *et al.* (1976) describe a comparative power study, based on simulation, of the five tests (3.3.1) to (3.3.5) against two classes of alternative to complete spatial randomness. The first is the simple sequential inhibition process described in Section 2.5, which can generate a range of random-to-regular patterns. The second, chosen to produce aggregated patterns, is an isotropic Poisson cluster process as defined in Section 2.2 with parents retained, number of offspring per parent Poisson with mean $\mu$ and spatial dispersion mechanism governed by the symmetric radial Normal distribution with pdf

$$h(r,\theta) = (r/2\pi\sigma^2)e^{-r^2/2\sigma^2} \qquad (r \geq 0,\ 0 \leq \theta < 2\pi).$$

For given $\mu$, this 'modified Thomas process' embraces a continuous range of variation between the Thomas process when $\sigma = 0$ and the Poisson forest in the limit $\sigma \to \infty$.

Briefly, the results confirm the overall superiority of Hopkins' test, whilst the T-square tests tend to dominate their respective Holgate counterparts. Overall, $t_N$ appears marginally superior to $t_B$, although this latter conclusion may be affected if one has a particular alternative hypothesis in mind.

The specification of particular alternatives merits further attention, for this can remove some of the objections to a distance-based approach. For example it is not true, as has often been stated (e.g. Mead, 1974), that distance methods can detect pattern only at the smallest scale. Again using T-square sampling, consider the assumption that the variate $U_i = \pi(X_i^2 + Z_i^2/2)$ derived from the *ith* of $m$ randomly located points has pdf

$$f_i(u) = \lambda_i^2 u e^{-\lambda_i u} \qquad (u > 0).$$

Diggle (1977b) proposes a generalized likelihood ratio test of the null hypothesis $H_o:\lambda_i = \lambda$ for all $i$, against the alternative $H_1:\lambda_i$ not necessarily all equal. In the context of spatial point patterns, $H_o$ represents complete spatial randomness as before, whilst $H_1$ will be approximately true in a purely heterogeneous process, with *slowly-varying* $\lambda(\underset{\sim}{x})$. The test statistic is

$$M/C = 48m(m \log \bar{U} - \Sigma \log U_i)/(13m + 1) \qquad (3.3.6)$$

whose null distribution is approximately chi-squared on m-1 degrees of freedom (cf. Bartlett's (1937) test for unequal variances in Normal samples). As the distribution of $t_N$ is invariant with respect to differences amongst the $\lambda_i$, Diggle proposed a two-stage procedure whereby if $t_N$ yields a non-significant result the M/C-statistic is used as a test for heterogeneity. Power studies involving simulations of a doubly stochastic Poisson process of the type used to generate Figure 2.2 confirm the value of the M/C-test as a supplement to $t_N$. The combined procedure therefore effects a four-way classification of patterns as regular, random, heterogeneous, or aggregated, in which extreme heterogeneity tends, reasonably, to be classified as aggregation.

*3.4 Robust Density Estimation Using Distance Methods.* Most distance-based estimators of density, derived on the assumption of complete spatial randomness, have been found lacking in robustness (Persson, 1971). Although problems of inefficiency are seldom acute, the estimators can exhibit considerable bias when applied to non-Poisson forests. Consider, however, the problem of density estimation based on a 'Hopkins sample' $\{(X_i, Y_i) : i=1,\dots,m\}$ of random point-to-event and event-to-event nearest neighbor distances. The maximum likelihood estimator for the mean area per plant, $\gamma = \lambda^{-1}$, in a Poisson forest, is found to be

$$\hat{\gamma} = \pi(\Sigma X_i^2 + \Sigma Y_i^2)/2m,$$

which is unbiased, with variance $\gamma^2/2m$. We remark that this 'compound estimator' is the arithmetic mean of two 'simple' estimators, based respectively on the $X_i$ and $Y_i$ distances alone, which is non-Poisson forests will tend to be biased in opposite directions. The first two moments of $\hat{\gamma}$, and of a second compound estimator, $\gamma^* = \pi\sqrt{(\Sigma X_i^2 \Sigma Y_i^2)}/m$, are derived by Diggle (1975) for a particular class of Poisson cluster process and for lattice-Poisson forest superimposed processes, using explicit forms of the results quoted in Sections 2.2 and 2.4 above. As anticipated, $\hat{\gamma}$ generally exhibits smaller bias than either simple estimator and is no less efficient. The use of $\gamma^*$ leads to a further improvement in robustness, essentially because of the asymmetry inherent in the lower bound of zero for any sensible estimator of $\gamma$.

Whilst these results are of no practical significance in themselves, they do suggest the use of analogous estimators based on T-square sampling,

$$\hat{\gamma}_T = \pi(\Sigma X_i^2 + \Sigma Z_i^2/2)/2m,$$

and more particularly,

$$\gamma_T^* = \pi\sqrt{(\Sigma X_i^2 \Sigma Z_i^2/2)}/m. \qquad (3.4.1)$$

The simulation study reported by Diggle (1977a) shows that, except against regularity, $\gamma_T^*$ is rather less robust than is $\gamma^*$. Nevertheless, $\gamma_T^*$ remains tolerably robust to a wide range of aggregated departures from complete spatial randomness modelled by Poisson cluster processes. Note that $\gamma_T^*$ is a function of two, generally dependent, sample means and an empirical standard error may therefore be calculated, leading to an interval estimate for $\gamma$ in the usual way. We emphasize that associated tests of spatial randomness can and should be carried out in conjunction with the density estimation procedure. Clearly, the result of the former will influence our interpretation of the latter.

We have described one approach to the construction of robust distance-based estimators of density following Persson's (1971) demonstration of the inadequacy of the various estimators then available. An alternative approach is to use some measurable feature of the underlying pattern to correct the estimate of density. Effectively, the estimation and testing procedures are combined into a single stage. Batcheler (1973) and Cox (1976a) offer particular suggestions. Relatively little is known about the theoretical properties of the resulting estimators, although Batcheler and Hodder (1975) describe the successful application of Batcheler's technique to pine stands. They also emphasize the practical difficulties of a quadrat-based approach in this context.

## 4. ANALYSIS OF MAPPED DATA

*4.1 Introductory Remarks.* Throughout this section we assume that the data consist of a map of the locations of n events in a given area A, typically but not necessarily a square. Provided that n is sufficiently large, say of the order of 200 or so, tests of spatial randomness of the type described in Section 3 *may* help to clarify our initial impression of the data, but more sensitive tests are almost invariably available. Often, the crucial decision concerns the type of observation to be extracted from the data. The question of an appropriate test statistic may then be considered and we shall here make extensive use of the Monte Carlo approach described in the Appendix to overcome problems of analytical intractability. As noted earlier, this gives a valuable flexibility.

In the first instance we shall be concerned with formulating hypotheses about the data and would encourage the use of a variety of informative statistics, and indeed methods of analysis, at this stage. A less cavalier approach will be necessary at the later stage of fitting a model within some parametric class. Most of the methods which we shall discuss are relatively new, and comments on their comparative virtues may, of necessity, be rather speculative. On the other hand, the case studies to be reported in Section 6 will add weight to the discussion, and should assist a critical evaluation.

*4.2 Contiguous Quadrat Counts.* The material in the present section is given a more detailed treatment by Cormack (1979) under the heading 'pattern analysis'. Historically, the first attempt to systematize the analysis of a mapped point pattern is due to Greig-Smith (1952). For convenience, we assume that $A$ is a square. The method requires that the data are converted into a $2^k \times 2^k$ grid of contiguous quadrat counts and a search for *scales of pattern* is conducted, essentially by calculating and testing the index of dispersion or variance-to-mean ratio, $I$, for sets of 2×2, 4×4, 8×8, etc. *blocks* of quadrats. The combined interpretation of the tests at the various scales causes difficulties. The original practice appears to have been to interpret peaks in a plot of $I$ against block size as evidence of pattern at the corresponding scales (e.g., Kershaw, 1973, Ch. 7) although Greig-Smith emphasizes the empirical nature of this approach. More objectively, Bartlett (1975, Ch. 3) shows that for a Poisson cluster process, the plot of expected value of $I$ against block size rises to a maximum which is then maintained with further increase in block size. Bartlett also suggests that if the parent process is itself clustered, rather than Poisson, a further increase in $I$ will result and *not*, as the empirical interpretation above would imply, a double peak corresponding to the two scales of clustering.

Mead (1974) establishes a valid statistical procedure for the detection of scales of pattern from contiguous quadrat counts. His 'fours within sixteen' randomization test requires that the data are successively partitioned into 1, 4, 16, etc. blocks each consisting of 16 counts in a 4×4 grid. For each partitioning, the null hypothesis is that, within each block, the set of counts in the four associated 2×2 sub-blocks forms a random selection from the $(16!)/(4!)^5 = 2627265$ possibilities. The different partitions provide *independent* tests of pattern at the various scales. Although the full randomization procedure appears computationally prohibitive, one may compare the chosen test statistic at each scale with a pseudo-random sample from its randomization distribution. One possible such test statistic is the sum of the six absolute values of pair-wise differences between the four counts within a block, summed in turn over all blocks. For further discussion, see Besag and Diggle (1977).

Analyses of this type have a proven value for purposes of hypothesis formulation but would seem to be less appropriate for the assessment of fit between data and a proposed model. Indeed, the proponents of the method have themselves expressed the view that too definite an interpretation should not be placed on the results of the analysis (see e.g., Mead, 1974).

*4.3 Refined Nearest Neighbor Analysis.* Distance methods again provide an alternative to a quadrat-based approach for the analysis of mapped data. As noted earlier, the particular tests described in Section 3 are potentially inefficient and we propose the following refined procedure. It will again be convenient to assume $A$ a square, although the extension to other shapes will be obvious. We first overlay $A$ with a $k \times k$ square lattice of sample points and measure the distance from each such point to the nearest event in the data, generating a set of observations $X_1, \cdots, X_m$, where $m=k^2$. We further record the distance from each event to the nearest other event, generating a corresponding set of observations $Y_1, \cdots, Y_n$. This latter set will include distances recorded twice between mutual nearest neighbors. The resulting EDF's, $\hat{F}(x)$ and $\hat{G}(y)$ say, now form the basis for subsequent analysis of the data.

Note first that the choice of $k$ is arbitrary, and that $k$ large will produce a fine grid of sample points until ultimately $\hat{F}(x)$ measures the proportion of points $P$ in $A$ such that a circle of radius $x$ centered on $P$ will contain no events. Ripley (1977) refers to this as the 'test-set method.' In principle a reduction in $k$ represents a loss of information but this will be slight for a relatively fine grid since distances measured from adjacent sample points are then strongly dependent. Without as yet being able to offer any formal justification we suggest that $k$ of the order of $\sqrt{n}$ might represent a reasonable trade-off between the loss of information for small $k$ and the additional computational burden as $k$ is increased.

Note also that $\hat{F}(x)$ and $\hat{G}(y)$ convey complementary information about the data. For example, $\hat{G}(y)$ is a natural vehicle for the identification of inhibitory effects between events whereas $\hat{F}(x)$ essentially describes the 'empty spaces' within A.

With regard to choice of test statistic, the Monte Carlo EDF-test as described in Section A.3 of the Appendix may be applied to both $\hat{F}(x)$ and $\hat{G}(y)$ to assess the goodness-of-fit to a particular model. The two results will not be independent and should be suitably combined. For example, if $p_1$ and $p_2$ denote the two individual p-values, then it is always true that

$$p \leq \Pr\{\min(p_1,p_2) \leq p\} \leq 2p. \tag{4.3.1}$$

Different statistics will be appropriate for tests of spatial randomness against specific alternatives. For example, if inhibitory effects are suspected, the frequency of Y-values less than some threshold value, or even the minimum Y-value, might be used. Note also that in the completely random case, the variates X and Y have the same marginal distribution function,

$$F(x) = 1 - e^{-\pi\lambda x^2} \quad (x > 0), \tag{4.3.2}$$

bar finite population effects and where $\lambda = n/A$. This suggests a single two-sample EDF-test based, for example, on

$$s_d = \sum_{i=1}^{j} |\hat{F}(Z_i) - \hat{G}(Z_i)| \tag{4.3.3}$$

for suitably chosen $Z_1,\cdots,Z_j$. Such a test has considerable intuitive appeal since in non-random situations the two marginal distribution functions $F(\cdot)$ and $G(\cdot)$ typically deviate from the null form (5.3.1) in opposing directions, as previously noted in Section 3.3.

On a historical note, Clark and Evans (1954) have suggested a test based on $\bar{Y}$ but their derivation of the supposed null distribution of $\bar{Y}$ is strictly incorrect as it ignores the dependencies amongst the $Y_i$. Somewhat fortuitously, tests at conventional significance levels remain approximately valid (Diggle, 1975; Donnelly, 1977) but empirical investigations thus far suggest that the test is generally rather weak and cannot be particularly recommended.

Refined nearest neighbor analysis usually admits of a straightforward interpretation and therefore offers considerable promise for purposes of hypothesis formulation. In addition, the analysis leads to a natural goodness-of-fit criterion based on the combination of two Monte Carlo EDF-tests. On the other hand, the method is as yet relatively untried.

*4.4 Density Surfaces (First-Order Analysis).* The basic statistical assumption for the first-order analysis of a spatial point pattern is that the $n$ locations $\underset{\sim}{X}_i$ in $A$ are an independent random sample from a bivariate distribution over $A$, whose pdf is defined by (2.3.1) in terms of the density $\lambda(\underset{\sim}{x})$ of a purely heterogeneous process. We shall confine our attention to 'slowly-varying' $\lambda(\underset{\sim}{x})$; otherwise an interpretation in terms of a stationary cluster process would be more natural. However, we again emphasize that this distinction is one of empirical convencience and cannot formally be justified on the basis of a single set of data.

As noted earlier, complete spatial randomness corresponds to constant $\lambda(\underset{\sim}{x})$ and therefore a uniform distribution over $A$. To conduct a classical chi-squared goodness-of-fit test for uniformity, we partition $A$ into $m$ equal sub-areas containing numbers $N_i$ of events. On the null hypothesis of uniformity, the expected number of events in each sub-area is $\bar{N} = N/m$, and the chi-squared statistic is

$$\sum_{i=1}^{m} (N_i - \bar{N})^2 / \bar{N}.$$

Note that this is just $m-1$ times the quadrat count index of dispersion, $I$. This observation provides an informal derivation of the approximate null distribution of $(m-1)I$, namely chi-squared on $m-1$ degrees of freedom, quoted in Section 3.2 for a set of randomly located quadrats and applied to a grid of contiguous quadrats in Section 4.2.

When a parametric alternative to uniformity is postulated, classical generalized likelihood ratio tests may be used. For example, using Kooijman's (1977) linear form, $\lambda(\underset{\sim}{x}) = \alpha + \underset{\sim}{\beta}'\underset{\sim}{x}$, the generalized likelihood ratio test of $\underset{\sim}{\beta} = \underset{\sim}{0}$ against $\underset{\sim}{\beta} \neq \underset{\sim}{0}$ gives a test of uniformity against linear trend. As noted in Section 2.3, more flexible classes of alternative carry attendant analytical problems and we may have to fall back on sub-optimal methods of parameter estimation and testing, or content ourselves with an empirical description of $\lambda(\underset{\sim}{x})$. The latter case represents a possible analogy with the empirical estimation of a bivariate pdf as in Loftsgaarden and Quesenberry (1965). Kernel methods (Parzen, 1962) or bivariate extensions of the usual smoothing techniques for time series data present two further possibilities.

*4.5 Inter-Event Distances (Second-Order Analysis).* Bartlett (1964) suggests the conversion of the $n$ locations to a set of $m=n(n-1)/2$ inter-event distances, $t_i$ say, for subsequent analysis. The marginal distribution of $T$ in the completely random case of $n$ locations independently and uniformly distributed in $A$ is given by Bartlett for $A$ a square or circle. For a square of side $a$, the standardized variate $Z=T/a$ has distribution function

$$F(z) = \begin{cases} \pi z^2 - 8z^3/3 + z^4/2 & (0 \leq z < 1), \\ 1/3 + (4/3)(1+2z^2)\sqrt{(z^2-1)} - z^4/2 & \\ \quad - 2z^2[1 - \sin^{-1}\{(2/z^2)-1\}] & (1 \leq z < \sqrt{2}), \end{cases}$$

whilst for a circle of radius $a$,

$$F(z) = 1 + \pi^{-1}\{2(z^2-1)\cos^{-1}(z/2) - z(1+z^2/2)\sqrt{(1-z^2/4)}\} \qquad (0 \leq z < 2), \qquad (4.5.1)$$

the latter result quoted from Borel (1925). Again, goodness-of-fit tests for data in this form arise naturally and the Monte Carlo approach resolves the problem of dependencies between distances measured from a common endpoint. In principle, the recording of *all* inter-event distances incorporates more information than, say, event-to-event nearest neighbor distances alone. Against this, there is a danger that inference will effectively be based solely on small inter-event distances, since the main body of the distribution function of inter-event distance will often be relatively insensitive to departures from complete spatial randomness. Thus, EDF-tests which measure goodness-of-fit over the full range of the distribution may be relatively weak.

Ripley (1977) develops the concept of second-order analysis for isotropic, stationary stochastic processes in the plane, rather than for a given number of events in a finite area.

Although the resulting statistical procedures correspond effectively to an analysis of inter-event distances, the alternative interpretation provides useful additional insight into possible models for the data.

Rather informally, we consider a function $K(t)$ such that $\lambda K(t)$ is the expected number of further events within a distance $t$ of an arbitrary event, and $\lambda$ as usual denotes the density of the underlying process. For a Poisson forest, it follows immediately from stationarity and the independence of the individual events that

$$K(t) = \pi t^2. \tag{4.5.2}$$

Further analytical progress is possible in certain non-Poisson cases, including Poisson cluster processes as defined in Section 2.2. In this case, $\lambda K(t)$ is the sum of two contributions, the first from the same cluster as that containing the arbitrary event, the second from all other clusters. By the independence of the individual clusters, the second contribution is just $\lambda\pi t^2$ as in the completely random case. For the first, consider the case of offspring only. Let the number of offspring per parent, $N$ say, have distribution $\{p_n : n=0,1,\cdots\}$ and mean $\mu$, then $\lambda=\rho\mu$ where $\rho$ is the mean number of parents per unit area. Given that the arbitrary event lies within a cluster of size $n$, the contribution to $\lambda K(t)$ is then $(n-1)F(t)$, where $F(t)$ is the distribution function of the distance between two arbitrary events in the *same* cluster. The probability of observing $n$ is here not $p_n$, but $q_n=np_n/\mu$, since the selection of an arbitrary event corresponds to length-biased sampling from the distribution $\{p_n : n=0,1,\cdots\}$. Thus we obtain finally

$$\lambda K(t) = \lambda\pi t^2 + \sum_{n=1}^{\infty} (n-1)F(t)q_n$$

$$= \lambda\pi t^2 + \sum_{n=0}^{\infty} n(n-1)F(t)p_n/\mu$$

or, dividing throughout by $\lambda=\rho\mu$,

$$K(t) = \pi t^2 + E[N(N-1)]F(t)/\rho\mu^2. \tag{4.5.3}$$

Explicit solutions are available for $F(t)$ in useful special cases. For example, the Matérn process described in Section 2.3 corresponds to a Poisson distribution for $N$, whence

$$K(t) = \pi t^2 + F(t)/\rho$$

with F(t) as in (4.5.1).

To motivate an estimator for K(t) we revert temporarily to a consideration of n events in a given area A, and let G(t) denote the distribution function of the distance between two arbitrarily chosen events. Then, the expected number of further events within distance t of an arbitrary event is (n-1)G(t) which, with $\lambda$=n/A, suggests the estimator

$$\hat{K}(t) = n^{-1}(n-1)\ A\hat{G}(t). \tag{4.5.4}$$

Here, in the usual way, $\hat{G}(\cdot)$ denotes the empirical version of $G(\cdot)$. Now, if the n events in A are to be regarded as a *partial* realization of a process extending throughout the plane, we can correct for the edge-effects in (4.5.4) which arise from our inability to observe events outside A. We associate with each *ordered* pair of locations $(\underset{\sim}{x}_i, \underset{\sim}{x}_j)$ a weight $k(\underset{\sim}{x}_i, \underset{\sim}{x}_j)$, the reciprocal of the proportion of the circumference of the circle, centered on $\underset{\sim}{x}_i$ and passing through $\underset{\sim}{x}_j$, which lies within A as indicated in Figure 4.1. Thus, $\{k(\underset{\sim}{x}_i, \underset{\sim}{x}_j)\}^{-1}$ is the probability that an event will be recorded, given only that it is the same distance from $\underset{\sim}{x}_i$ as is $\underset{\sim}{x}_j$. The estimator $\hat{K}(t)$ now becomes

$$\hat{K}(t) = n^{-2}A\ \Sigma\Sigma k(\underset{\sim}{x}_i, \underset{\sim}{x}_j), \tag{4.5.5}$$

where the summation is over all *ordered* pairs of locations less than a distance t apart. Note that if data are relatively abundant it may be expedient to measure distances only from events within some interior sub-area, in which case (4.5.5) reduces to (4.5.4) for t less than the minimum boundary width. More generally, (4.5.4) and (4.5.5) are equivalent for sufficiently small t, but thereafter diverge progressively. Note that the weights $k(\underset{\sim}{x}, \underset{\sim}{y})$ can become arbitrarily large as t increases. For A a square of side a, the restriction $t \leq a/\sqrt{2}$ is needed to give an approximately unbiased estimator $\hat{K}(t)$.

Inference from K(t) proceeds via a comparison between the observed $\hat{K}_1(t)$ and corresponding $\hat{K}_i(t)$ obtained from simulations of a proposed model. Although goodness-of-fit criteria arise less naturally for K(t) than for the unweighted EDF G(t),

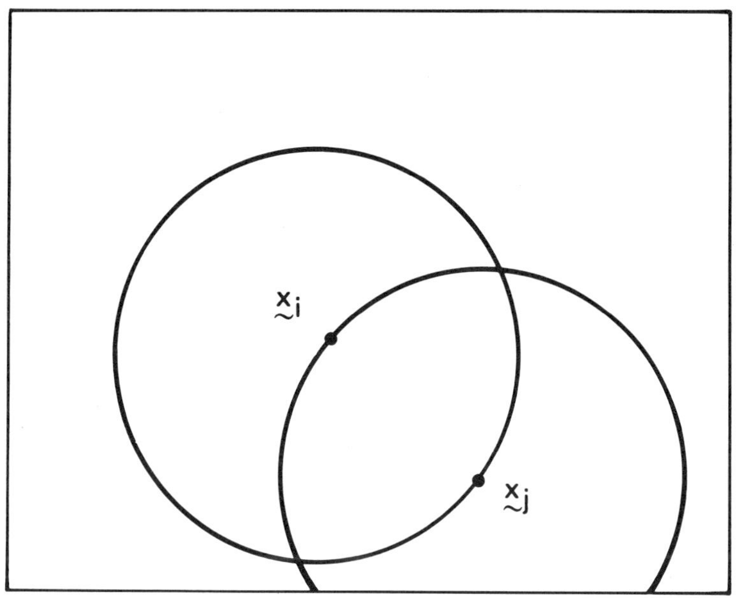


*FIG. 4.1: Estimation of the function* K(t). $(k(\underset{\sim}{x}_i,\underset{\sim}{x}_j) = 1;$ $k(\underset{\sim}{x}_j,\underset{\sim}{x}_i)>1)$.

tests may be constructed from the behavior of K(t) within prescribed ranges of t. On the other hand if, as suggested above, such inferences are effectively based on *small* inter-event distances, the unweighted G(t) might equally well be used, and is computationally simpler. Thus, the advantage of second-order analysis relative to a simpler investigation of inter-event distances would seem to lie not in the assessment of goodness-of-fit but rather in the development of plausible models. As we shall see in Section 5, this advantage can be considerable. For a more formal discussion of this approach, and alternative interpretations of the function K(t), see Ripley (1977).

Spectral analysis, also advocated by Bartlett (1964, 1975), provides an alternative method for analysis of the second-order properties of spatial point patterns. Its application to the Lansing Woods data of Figures 1.1 to 1.3 is reported by Cox (1976b).

*4.6 Plane Tessellations.* A rather different approach to point pattern analysis is to consider a partition of A into n mutually exclusive and exhaustive sub-areas, each one associated with a single event at the location $\underset{\sim}{x}_i$. In other words, each event is assigned a *territory* within A and the n territories form the basis of further analysis. Perhaps the most natural such tesselation is the *Dirichlet* or *Voronoi* tessellation whereby the territory of $\underset{\sim}{x}_i$ is the convex polygon which consists of those points in A closer to $\underset{\sim}{x}_i$ than to any other $\underset{\sim}{x}_j$.

The Voronoi tessellation has been used in ecological work by Mead (1966, 1971) to investigate relationships between plant yield and polygon area. Its application to point pattern analysis is made feasible by the publication of a remarkably efficient algorithm for its computation by Green and Sibson (1978). For the analysis of point patterns, one possible line of attack would be to consider the distribution of polygon area. For n events in a given area A the mean of this distribution is of course fixed but its variance should effectively discriminate between regular and aggregated alternatives to complete spatial randomness. Also, the spatial distribution of territory size over A has obvious potential for the empirical assessment of pure heterogeneity and as such might be regarded as a possible alternative to the use of contiguous quadrat counts. For a more detailed assessment of local geometrical structure, properties of polygon shape such as number of sides might prove useful. Further specific suggestions at this stage might be premature although contributions to the discussion of Ripley (1977) by Dr. D. Mollison and by Professor R. Sibson indicate some of the potential scope of the method.

*4.7 Concluding Remarks*. The general topic of this section is a fertile area for current research and firm conclusions cannot yet be drawn on the relative merits of the various methods. Quite possibly, no single 'optimal' method of analysis will emerge but a critical evaluation of the particular strengths and weaknesses of each technique is nevertheless clearly desirable. The analyses to be reported in the final section of this paper represent a first step in this direction.

We have emphasized the distinction between hypothesis formulation based on a variety of complementary techniques and model-fitting which requires a single objective assessment of fit. We have not so far discussed the transition from the former to the latter; indeed in particular applications we may be unable to identify a plausible parametric model. When an assessment of fit is relevant, the principal difficulty arises through the need to accommodate the effects of parameter estimation on the goodness-of-fit test. In Section A.2 of the Appendix below we note two possible solutions, cross-validation and the use of a goodness-of-fit test which is essentially unrelated to the particular estimation procedure adopted for the parameters of the proposed model.

## 5. ILLUSTRATIVE EXAMPLES

*5.1 Lansing Woods*. Dr. D. J. Gerrard (formerly University of Minnesota) has kindly made available data on the locations of 2251 trees in a square of side 924 feet within Lansing Woods, Clinton County, Michigan. This represents an unusually large body of data which provides a convenient illustration of the use of T-square smapling methods for preliminary analyses in the field. The 2251 trees divide into six species groups. We analyze three sub-sets of 929 oaks, 703 hickories, and 514 maples.

The T-square sampling procedure uses $m=50$ sample points placed randomly within the centrally located interior 'sampling square' of side 693 feet to avoid distortion due to edge-effects. To test for departures from complete spatial randomness we apply the $t_N$-test (3.3.5) followed as appropriate by the M/C-test (3.3.6). We also include the $t_B$-test (3.3.4) for comparison.

The mean area per plant, $\gamma$, is estimated using $\gamma_T^*$ as defined by (3.4.1). The value of $\gamma$ in the sampling square is known and we quote a standardized interval estimate, $\gamma^{-1}\{\gamma_T^* \pm 2S.E.(\gamma_T^*)\}$, where $S.E.(\gamma_T^*)$ denotes the empirical standard error to order $m^{-1}$, calculated by the delta technique.

The results are summarized in Table 5.1, whilst Figures 1.1 to 1.3 illustrate the three sub-populations. With regard to the interval estimates of density, we have previously suggested in Section 3.4 that these can sensibly be interpreted only in conjunction with the results of the associated tests of spatial randomness. In this respect, the positive bias in $\gamma_T^*$ associated with the detection of heterogeneity is qualitatively predictable by the following argument. Consider a population divided into $k$ sub-areas $A_i$ containing numbers $n_i$ of plants, within each of which the underlying process is completely random. Writing $A = \Sigma A_i$ and $n = \Sigma n_i$, we see that the inverse density overall is $\gamma = A/n = \Sigma(n_i/n)(A_i/n_i) = \Sigma(n_i/n)\gamma_i$, say; all summations are over $i=1,\cdots,k$. Thus, any estimator for inverse density which is unbiased in the completely random case will provide an unbiased estimator for $\gamma$ if it can be applied in such a way as to ensure

*TABLE 5.1: T-square analysis of Lansing Woods data. (Figures in brackets are p-values based on two-sided tests for* $t_N$, $t_B$ *and one-sided test for* M/C.)

| | Tests of spatial randomness | | | Interval estimate |
|---|---|---|---|---|
| | $t_N$ | M/C | $t_B$ | $\gamma^{-1}\{\gamma_T^* \pm 2\text{S.E.}(\gamma_T^*)\}$ |
| oaks | 0.447 | 48.04 | 0.512 | $1.10 \pm 0.24$ |
| | (0.194) | (0.512) | (0.811) | |
| hickories | 0.478 | 68.68 | 0.455 | $1.51 \pm 0.37$ |
| | (0.590) | (0.033) | (0.371) | |
| maples | 0.566 | 111.14 | 0.574 | $2.04 \pm 0.65$ |
| | (0.106) | (0.000) | (0.140 | |

that sampling takes place within the *ith* sub-area with probability $p_i = n_i/n$. However, T-square sampling corresponds to $p_i = A_i/A$, thus placing a disproportionate emphasis on sub-areas with relatively large $\gamma_i$. Although the argument is strictly applicable only to estimators, such as $\hat{\gamma}_T$, which combine observations from the individual sample points linearly, it remains reasonable to conclude that $\gamma_T^*$ will exhibit positive bias in this situation.

For comparison, we include in Table 5.2 the results of Mead's 'fours within sixteen' randomization test as described in Section 4.2. We use 99 Monte Carlo trials and a 32×32 grid which permits the investigation of four scales of pattern. For each of the three sub-populations there is moderate or strong evidence of pattern at the smallest scale together with a strong indication of at least one further scale of pattern.

*5.2 Japanese Black Pine.* Our second example concerns Numata's (1961) data on the locations of 65 Japanese black pine saplings in a 5.7 meter square plot (Figure 1.4). For these data, Bartlett (1964) concludes that the observed distribution of the 2080 inter-tree distances is compatible with complete spatial randomness of the trees. He compares observed and expected frequencies in successive intervals of width 0.5 meter using the classical chi-squared goodness-of-fit statistic, $\Sigma(O-E)^2/E$. Although this

*TABLE 5.2: 'Fours within sixteen' analysis of Lansing Woods data, based on 99 Monte Carlo trials in each case. (High rank suggest aggregation, low rank regularity. Where ties occur, least extreme rank is given.)*

| Rank out of 100 | Scale of pattern | | | |
|---|---|---|---|---|
| | 4/16 | 16/64 | 64/256 | 256/1024 |
| oaks | 95 | 99 | 90 | 73 |
| hickories | 99 | 100 | 100 | 90 |
| maples | 100 | 71 | 100 | 62 |

gives a value of 28.03, nominally on 13 degrees of freedom, Bartlett applies an order of magnitude correction to compensate for the dependencies in the data, and suggests that the observed value is consistent with the null hypothesis. This is confirmed by Besag and Diggle (1977) who implement a Monte Carlo version of the above test.

With a complete map of the data at our disposal other analyses are possible and in particular we compare the above use of all inter-event distances with a refined nearest-neighbor analysis. Figure 5.1 shows that the empirical distribution functions of point-event and event-event nearest neighbor distances tend, respectively, to lie above and below the null theoretical curve, which might be thought mildly suggestive of an element of regularity in the underlying pattern. However, Monte Carlo EDF-tests based on $m=99$ simulations indicate that the discrepancy between the two distributions is entirely consistent with sampling fluctuations. Formally, the two-sample EDF-test (4.3.3) gives a p-value of 0.65. The corresponding one-sample EDF-tests based on point-event and event-event distances lead to a similar conclusion with respective p-values of 0.90 and 0.60, or in combination $p \geq 0.60$, by (4.3.1).

*5.3 Redwood.* The locations of 62 redwood seedlings (Figure 1.5), extracted by Ripley (1977) from a larger plot of 199 seedlings given by Strauss (1975) provide a sharp contrast to the previous example. The spatial distribution of seedlings is markedly clustered, and clearly incompatible with complete spatial randomness. The empirical distribution functions of point-event and event-event nearest neighbor distances are shown in Figure 5.2, which may be contrasted with the corresponding Figure 5.1 for the Japanese black pine data. Formally, the two-sample EDF-test

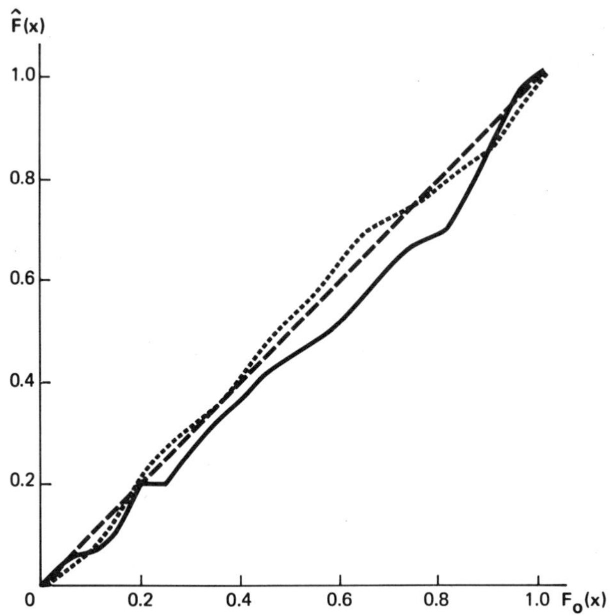


*FIG. 5.1: Empirical distribution functions of point-event and event-event nearest neighbor distances for Japanese black pine data.* ••••• *point-event;* ——— *event-event;* ----- $\hat{F}(x) = F_o(x)$ ($F_o(x)$ *denotes common marginal distribution function under complete spatial randomness).*

(4.3.3) based on 99 simulations rejects complete spatial randomness at the 1% level. The values of $s_d$ for the 99 simulations range from 0.45 to 2.20, whilst the value for the data is 7.66. In addition, both the point-event and event-event one-sample EDF-tests (A.3.1) comfortably attain significance at the 1% level. On the other hand, a formal test is redundant in this example, since the non-randomness of the seedlings is self-evident. The corresponding analysis based on all inter-event distances, illustrated in Figure 5.3, provides an interesting contrast. Superficial inspection suggests compatibility with complete spatial randomness but, as noted earlier, the standard EDF-test is quite inappropriate because the useful information about the pattern is effectively concentrated within the lower tail of the distribution.

We now search for a plausible model for the data within the class of Poisson cluster processes defined in Section 2.2. Second-order analysis now comes into its own, for we can exploit

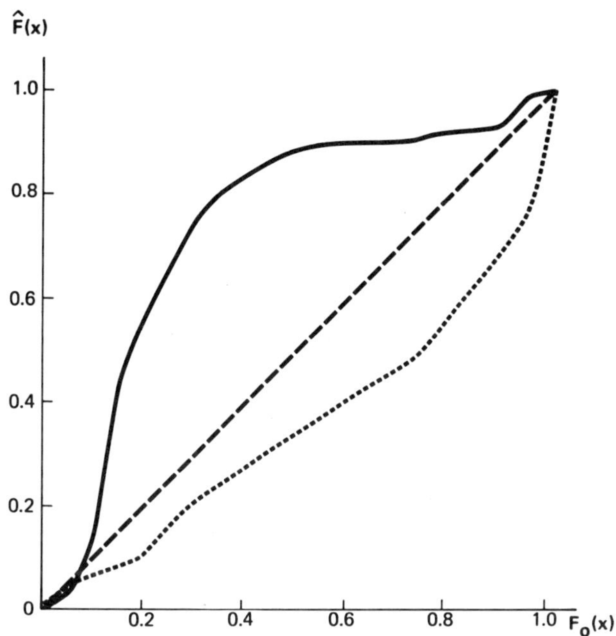


*FIG. 5.2: Empirical distribution functions of point-event and event-event nearest neighbor distances for redwood data.* ••••• *point-event;* ——— *event-event;* ----- $\hat{F}(x) = F_o(x)$ ($F_o(x)$ *denotes common marginal distribution function under complete spatial randomness).*

our knowledge of the function $K(t)$ for Poisson cluster processes, as expressed by (4.5.3). In particular, Figure 5.4 compares the estimated $\hat{K}(t)-\pi t^2$ for the data with $K(t)-\pi t^2$ for a Matérn process, for which we recall that the number of events per cluster is Poisson with mean $\mu$ and the spatial dispersion within each cluster uniform in a circle, radius $\sigma$, centered on the corresponding parent. For this process, $K(t)-\pi t^2$ first attains its maximum value of $\rho^{-1}$ at $t=2\sigma$ and thereafter assumes a constant value $\rho^{-1}$; recall that $\rho$ is the mean number of parents per unit area. Thus, the behavior of $K(t)-\pi t^2$ for large $t$ appears to be incompatible with the Matérn process, and indeed any Poisson cluster process, as concluded by Ripley (1977). Nevertheless, for small $t$ a remarkably close agreement between the data and the Matérn process is obtained by choosing $\rho=27$ and $\sigma=0.07$. This, in itself, is rather interesting for it suggests about 27 parents

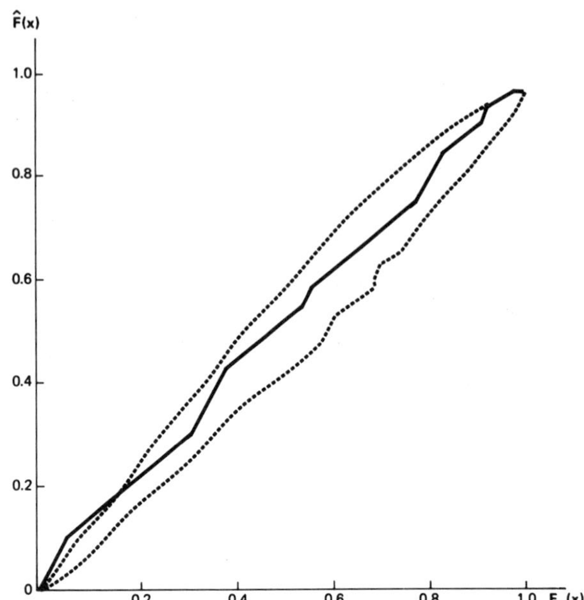


*FIG. 5.3: Inter-event distance analysis of redwood data.*
——— *data;* ••••• *envelope from* 19 *completely random simulations;*
----- $\hat{F}(x) = F_0(x)$ ($F_0(x)$ *denotes marginal distribution function under complete spatial randomness).*

in the unit square with, on average, two or three events per cluster. A visual inspection of Figure 1.5 would strongly suggest fewer, larger clusters. In addition, the envelope of $\hat{K}(t)-\pi t^2$ obtained from 99 simulations of 62 events independently and uniformly distributed over the unit square indicates the considerable sampling fluctuations which must be expected in the estimator $\hat{K}(t)$ for large $t$. In fact, the square root transformation effectively stabilizes the variance of $\hat{K}(t)$ for realizations of a Poisson forest, as demonstrated by Mr. J. Besag in the discussion of Ripley (1977).

We proceed to fit the Matérn process to the data with the parameters as indicated above, namely $\sigma=0.07$, $\rho=27$, and $\mu=62/27$, but conditioned to produce 62 events in the unit square. Figure 5.5 shows that $\hat{K}(t)$ for the data lies entirely within the envelope from 19 simulations of the Matérn process, which ostensibly suggests a good fit. However, this conclusion would be

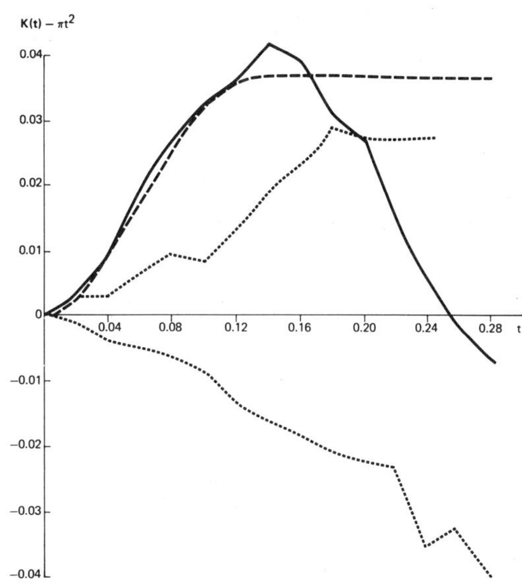


*FIG. 5.4: Second-order analysis of redwood data: the function* $K(t)-\pi t^2$. ----- *Matérn process;* —— *data;* ••••• *envelope from* 99 *completely random simulations*

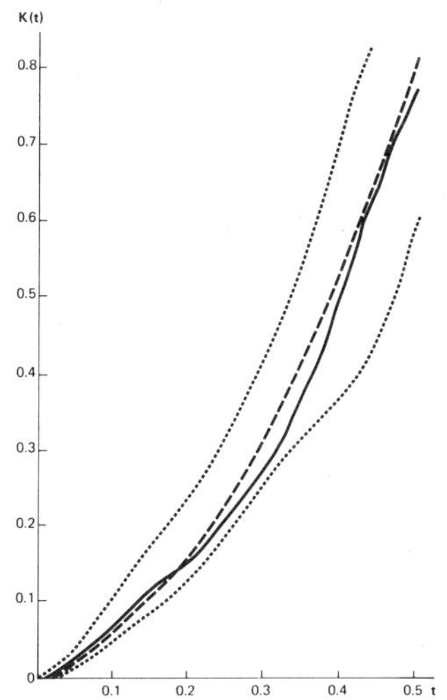


*FIG. 5.5: Second-order analysis of redwood data: the function* K(t). ----- *Matérn process;* —— *data;* ••••• *envelope from* 19 *simulations of Matérn process*

premature because the method of parameter estimation has effectively constrained the model to fit well for small $t$ whilst the wide sampling fluctuations in $\hat{K}(t)$ lead to a test which is fundamentally insensitive to discrepancies between model and data in the region of large $t$. A partial counter to this is to note that similar agreement between model and data is observed when the parameters $\sigma$ and $\rho$ are varied in the ranges 0.05 to 0.09 and 24 to 30 respectively, also for other Poisson cluster processes including an equally plausible model whereby the location of each event relative to its parent has a symmetric radial Normal distribution. It would be more satisfactory to use a goodness-of-fit statistic which is unrelated to the estimation procedure. We adopt a refined nearest neighbor analysis using the Monte Carlo EDF-test (A.3.1) for each of point-event and event-event distances, combined via (4.3.1). This is less than ideal, for we have already noted above that the parameter estimation is based essentially on the distribution of small inter-event distances and will therefore be related to the event-event nearest neighbor distribution. In this respect, the test based on the point-event nearest neighbor distribution might seem more appropriate. The respective p-values for the point-event and event-event EDF-tests, based on 99 Monte Carlo trials, are 0.86 and 0.08, or in combination $0.08 \leq p \leq 0.16$. More informatively, Figure 5.6 compares the distributions observed for the data with the envelopes from the simulations. The overall fit is satisfactory but it is noteworthy, in view of the comments above, that what suggestion there is of a lack of fit comes from the event-event nearest neighbor distribution. Recall that a realization of the fitted Matérn process is shown in Figure 2.3.

In conclusion, we have demonstrated that a Poisson cluster process appears to fit these data satisfactorily, with the important proviso that the assessment of fit has not made complete allowance for the effects of parameter estimation. Also, the proposed model represents but one of several which might fit the data equally well. The data are too sparse to give sharp discrimination between broadly similar models. Finally, the question of external information raised in Section 1.2 is relevant to our findings. Strauss (1975) states "It was felt that the seedlings would be scattered fairly randomly, except that a number of tight clusters would form around some of the redwood stumps present in the plot". Unfortunately, the locations of these stumps are not available to us. If they were, we might reasonably hope to discriminate between different clustering mechanisms by conditioning the entire analysis on the locations of the parent stumps, and the idea of a Poisson cluster process would no longer be appropriate. In addition, Strauss notes a discontinuity in the soil which is expected to be relevant. In his extraction of the

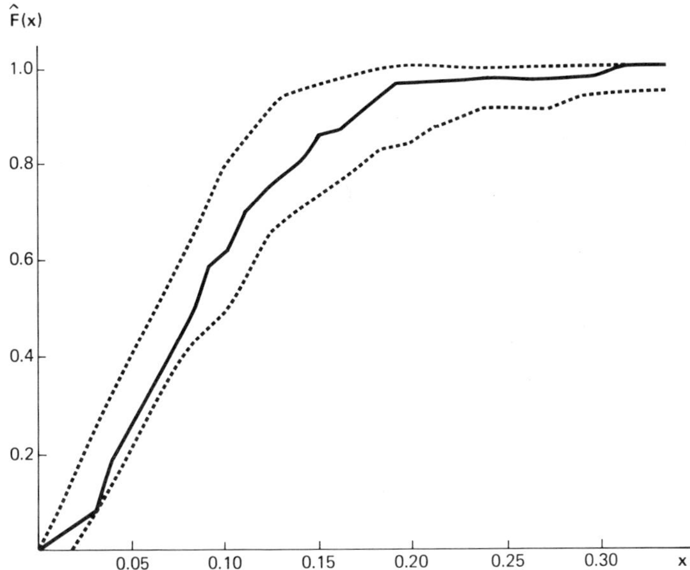


*FIG. 5.6a: Refined nearest neighbor analysis of redwood data. Point-event distances. ——— data; •••••• envelope from 99 simulations of Matérn process.*

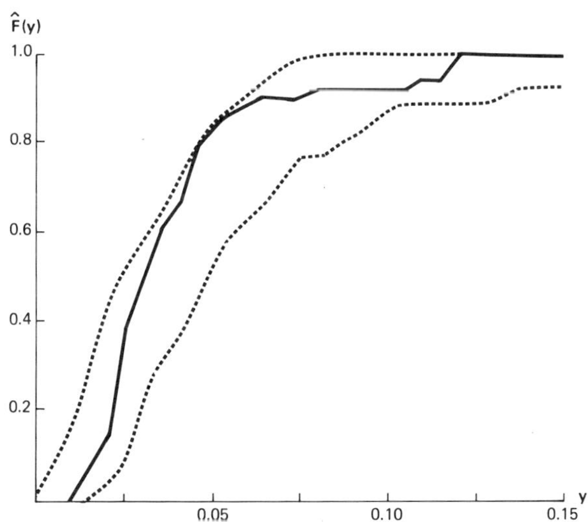


*FIG. 5.6b: Refined nearest neighbor analysis of redwood data. Event-event distances. ——— data; •••••• envelope from 99 simulations of Matérn process.*

present data from the original plot, Ripley (1977) states "for computational convenience I took a square mainly within the region of supposed clustering."

*5.4 Balsam Fir.* In our final example the concept of pure heterogeneity plays a useful role, although it does not in itself lead to an acceptable model for the data. Figure 1.6 shows the locations of 250 balsam fir seedlings in a 66 foot square plot, given by Ghent (1963). For convenience, the plot has been rescaled to the unit square. The immediate visual impression of a concentration of seedlings near the lower left-hand corner of the plot is reinforced by the counts in the four 33 foot square subplots, namely

| | |
|---|---|
| 50 | 53 |
| 92 | 55 |

The chi-squared goodness-of-fit test of uniformity gives $\chi^2=18.6$ on 3 degrees of freedom, or $p=0.0003$. Although the test can be criticized on the grounds that the $2\times2$ sub-division was chosen after inspection of the data, a highly significant result also holds for each $k\times k$ sub-division into square cells compatible with the usual requirement on null expected frequencies, say $k\leq8$, and there can be little doubt that the hypothesis of complete spatial randomness is untenable.

Given their logical equivalence to the above, it is worth recording the values of the index of dispersion applied to the counts in a $16\times16$ grid, here reproduced as Table 5.3a, and to the associated blocks of 4, 16, and 64 quadrats. The respective values of $I$ are 1.12, 1.79, 2.75, and 6.21. The first gives a nominal p-value of 0.08 using the here marginally dubious chi-squared approximation to the null distribution of $(m-1)I$. As noted above, the remaining values of $I$ are highly significant. Also, the sequence of increasing I-values is compatible with a purely heterogeneous process with slowly-varying $\lambda(\underset{\sim}{x})$. Rather surprisingly, Mead's 'fours within sixteen' randomization test applied to the same $16\times16$ grid nevertheless fails to detect pattern at any of the three available scales, although the combined p-value of the three independent *one-sided* tests, obtained by referring $-2\ \Sigma\log p_i$ to the chi-squared distribution on 6 degrees of freedom, is 0.051, mildly suggestive of aggregation. Strictly, this is an approximate device because the use of a Monte Carlo test, here based on 99 simulations, leads to a discrete null distribution of significance levels.

*TABLE 5.3: Quadrat counts in* 16×16 *grid for balsam fir data.*

a) Raw data

| 0 | 3 | 1 | 1 | 0 | 1 | 2 | 2 | 2 | 1 | 3 | 0 | 0 | 0 | 0 | 0 |
|---|---|---|---|---|---|---|---|---|---|---|---|---|---|---|---|
| 0 | 0 | 0 | 0 | 1 | 1 | 0 | 1 | 0 | 1 | 1 | 1 | 0 | 2 | 0 | 0 |
| 0 | 1 | 1 | 2 | 0 | 0 | 1 | 1 | 0 | 0 | 0 | 0 | 1 | 0 | 0 | 0 |
| 1 | 3 | 0 | 1 | 0 | 0 | 1 | 0 | 0 | 0 | 1 | 1 | 0 | 0 | 0 | 2 |
| 1 | 2 | 0 | 2 | 1 | 1 | 0 | 0 | 3 | 2 | 2 | 2 | 0 | 0 | 0 | 1 |
| 0 | 1 | 1 | 1 | 0 | 0 | 0 | 0 | 0 | 1 | 2 | 0 | 0 | 1 | 1 | 1 |
| 1 | 1 | 1 | 2 | 1 | 0 | 1 | 1 | 2 | 1 | 2 | 3 | 1 | 1 | 1 | 2 |
| 0 | 3 | 1 | 0 | 0 | 2 | 1 | 0 | 1 | 0 | 0 | 0 | 1 | 3 | 1 | 2 |
| 2 | 1 | 2 | 3 | 3 | 1 | 0 | 0 | 0 | 0 | 0 | 0 | 1 | 1 | 1 | 1 |
| 3 | 1 | 1 | 5 | 3 | 5 | 3 | 0 | 0 | 1 | 0 | 0 | 2 | 0 | 2 | 2 |
| 2 | 2 | 1 | 1 | 1 | 3 | 1 | 0 | 0 | 0 | 1 | 3 | 1 | 2 | 1 | 2 |
| 1 | 1 | 1 | 3 | 2 | 1 | 0 | 1 | 0 | 0 | 1 | 1 | 1 | 1 | 0 | 1 |
| 6 | 2 | 3 | 0 | 0 | 3 | 0 | 1 | 3 | 1 | 2 | 0 | 0 | 3 | 3 | 0 |
| 2 | 1 | 0 | 0 | 0 | 2 | 1 | 2 | 3 | 2 | 0 | 1 | 0 | 1 | 1 | 2 |
| 1 | 1 | 1 | 0 | 2 | 1 | 0 | 1 | 0 | 0 | 0 | 0 | 0 | 0 | 1 | 0 |
| 1 | 1 | 1 | 1 | 1 | 2 | 0 | 1 | 0 | 0 | 1 | 2 | 2 | 1 | 0 | 0 |

b) Spatial moving average (rounded to nearest integer)

| 1 | 1 | 1 | 1 | 1 | 1 | 1 | 1 | 1 | 1 | 1 | 1 | 1 | 0 | 0 | 0 |
|---|---|---|---|---|---|---|---|---|---|---|---|---|---|---|---|
| 1 | 1 | 1 | 1 | 1 | 1 | 1 | 1 | 1 | 1 | 1 | 1 | 0 | 0 | 0 | 0 |
| 1 | 1 | 1 | 1 | 1 | 0 | 1 | 0 | 0 | 0 | 1 | 1 | 1 | 0 | 0 | 0 |
| 1 | 1 | 1 | 1 | 1 | 0 | 0 | 1 | 1 | 1 | 1 | 1 | 0 | 0 | 0 | 1 |
| 1 | 1 | 1 | 1 | 1 | 0 | 0 | 0 | 1 | 1 | 1 | 1 | 0 | 0 | 1 | 1 |
| 1 | 1 | 1 | 1 | 1 | 0 | 0 | 1 | 0 | 2 | 2 | 1 | 1 | 1 | 1 | 1 |
| 1 | 1 | 1 | 1 | 1 | 1 | 1 | 1 | 1 | 1 | 1 | 1 | 1 | 1 | 1 | 1 |
| 1 | 1 | 2 | 1 | 1 | 1 | 1 | 1 | 1 | 1 | 1 | 1 | 1 | 1 | 1 | 1 |
| 2 | 2 | 2 | 2 | 2 | 2 | 1 | 1 | 0 | 0 | 0 | 0 | 1 | 1 | 1 | 2 |
| 2 | 2 | 2 | 2 | 3 | 2 | 1 | 0 | 0 | 0 | 1 | 1 | 1 | 1 | 1 | 2 |
| 2 | 1 | 2 | 2 | 3 | 2 | 2 | 1 | 0 | 0 | 1 | 1 | 1 | 1 | 1 | 1 |
| 2 | 2 | 2 | 1 | 2 | 1 | 1 | 1 | 1 | 1 | 1 | 1 | 1 | 1 | 1 | 1 |
| 2 | 2 | 1 | 1 | 1 | 1 | 1 | 1 | 1 | 1 | 1 | 1 | 1 | 1 | 1 | 1 |
| 2 | 2 | 1 | 1 | 1 | 1 | 1 | 1 | 1 | 1 | 1 | 0 | 1 | 1 | 1 | 1 |
| 1 | 1 | 1 | 1 | 1 | 1 | 1 | 1 | 1 | 1 | 1 | 1 | 1 | 1 | 1 | 1 |
| 1 | 1 | 1 | 1 | 1 | 1 | 1 | 0 | 0 | 0 | 1 | 1 | 1 | 1 | 0 | 0 |

To gain an informal impression of the postulated $\lambda(\underset{\sim}{x})$ we perform a spatial moving average smoothing operation whereby the smoothed value for the $(i,j)th$ quadrat is calculated as the mean of the counts in the nine (or fewer on the border) quadrats including and immediately surrounding the $(i,j)th$. As in time series analysis, the particular choice of moving average is subjective but the result, which is compared with the original data in Table 5.3b, does give a reasonable summary of the variation in local density relative to the average density of approximately one seedling per quadrat.

In an attempt to find an acceptable parametric representation of $\lambda(\underset{\sim}{x})$ we have fitted Plackett's (1965) C-type beta distribution to the data. This involves five parameters, four for the marginal beta distributions plus an interaction parameter, which we estimate by the method of maximum likelihood. Figure 5.7 shows the shape of the fitted surface, which may be compared with both the original data and the smoothed quadrat counts. The generalized likelihood ratio test of uniformity against the C-type beta alternative gives a value for $\chi^2_5$ of 9.00, corresponding to $p \simeq 0.11$. In fact, neither the null hypothesis of uniformity nor the alternative of a C-type beta distribution is tenable. Pearson's goodness-of-fit statistic, $\Sigma(O-E)^2/E$, calculated from a $6 \times 6$ grid of square cells to ensure reasonable expected cell frequencies, gives a value of 49.7 on 30 degrees of freedom $(p \simeq 0.013)$ for the C-type beta, as against 64.5 on 35 degrees of freedom $(p \simeq 0.0017)$ for the uniform. Also, we shall see below that a purely heterogeneous model is suspect on other grounds.

An analysis of event-to-event nearest neighbor distances reveals an inhibitory effect in the data. Figure 5.8 compares the distribution of nearest neighbor distances measured from each of the 139 events in the centrally-placed square of side 0.75 with the envelope obtained from 19 completely random simulations conditioned by the locations of the 111 trees in the boundary region. The plot suggests inhibition up to a distance of 0.0175, or about 1 foot. Silverman and Brown (1978) provide the necessary asymptotic distribution theory to effect tests of randomness against inhibition based on small inter-event distances. Allowing for the limited resolution in the data, a test based on the smallest of the 31125 inter-event distances gives a significance level of $p \leq 0.007$. However, a significantly large value of this test statistic will often be explainable in terms of the physical size of the individuals in question, and as such will be of limited interest to the ecologist. Here, Ghent defines the seedlings as "a useful collective term for tree regeneration between ground level and eight feet in height." A test based on

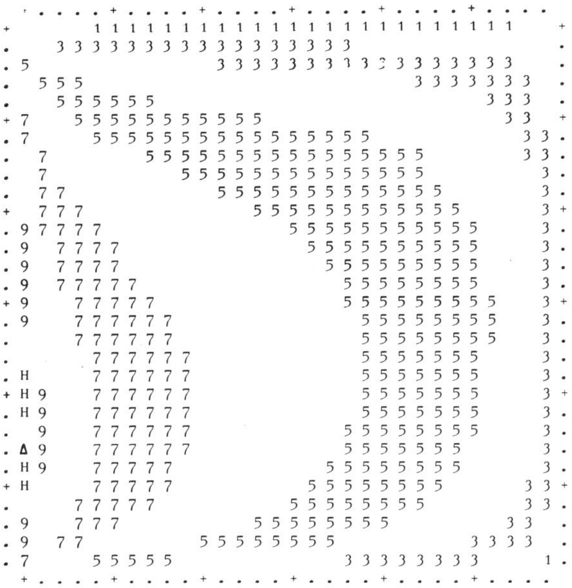

```
  ' . . . . + . . . . + . . . . + . . . . + . . . . + . . . .
+         1 1 1 1 1 1 1 1 1 1 1 1 1 1 1 1 1 1 1 1 1 1 1 1     +
.     3 3 3 3 3 3 3 3 3 3 3 3 3 3 3 3 3                       .
. 5                     3 3 3 3 3 3 3 3 3 3 3 3 3 3 3 3 3     .
.   5 5 5                                     3 3 3 3 3 3 3   .
.     5 5 5 5 5 5                                     3 3 3   .
+ 7     5 5 5 5 5 5 5 5 5 5 5                           3 3   +
. 7       5 5 5 5 5 5 5 5 5 5 5 5 5 5 5 5                 3 3 .
.   7           5 5 5 5 5 5 5 5 5 5 5 5 5 5 5 5           3 3 .
.   7               5 5 5 5 5 5 5 5 5 5 5 5 5 5             3 .
.   7 7                 5 5 5 5 5 5 5 5 5 5 5 5 5           3 .
+   7 7 7                   5 5 5 5 5 5 5 5 5 5 5 5         3 +
. 9 7 7 7 7                     5 5 5 5 5 5 5 5 5 5 5       3 .
. 9   7 7 7 7                     5 5 5 5 5 5 5 5 5 5       3 .
. 9   7 7 7 7                       5 5 5 5 5 5 5 5 5       3 .
. 9   7 7 7 7 7                       5 5 5 5 5 5 5 5       3 .
+ 9     7 7 7 7 7                     5 5 5 5 5 5 5 5 5     3 +
. 9     7 7 7 7 7 7                     5 5 5 5 5 5 5 5     3 .
.       7 7 7 7 7 7                     5 5 5 5 5 5 5 5     3 .
.         7 7 7 7 7 7                   5 5 5 5 5 5 5       3 .
. H       7 7 7 7 7 7                   5 5 5 5 5 5 5       3 .
+ H 9     7 7 7 7 7 7                   5 5 5 5 5 5 5       3 +
. H 9     7 7 7 7 7 7                   5 5 5 5 5 5 5       3 .
.   9     7 7 7 7 7 7                 5 5 5 5 5 5 5 5       3 .
. Δ 9     7 7 7 7 7 7                 5 5 5 5 5 5 5         3 .
. H 9     7 7 7 7 7                 5 5 5 5 5 5 5 5         3 .
+ H       7 7 7 7 7               5 5 5 5 5 5 5 5         3 3 +
.       7 7 7 7 7               5 5 5 5 5 5 5 5           3 3 .
. 9     7 7 7               5 5 5 5 5 5 5 5             3 3   .
. 9   7 7             5 5 5 5 5 5 5 5               3 3 3 3   .
. 7       5 5 5 5 5                   3 3 3 3 3 3 3 3       1 .
  + . . . . + . . . . + . . . . + . . . . + . . . . + . . . .
```

*FIG. 5.7: C-type beta surface fitted to balsam fir data. Note that the function,* f *say, is defined on the unit square, but illustrated only for* $0.01 \le x_1, x_2 \le 0.99$. $f(x_1,x_2) \to \infty$ *as* $x_1 \to 0$, $f(x_1,x_2) \to 0$ *as* $x_1 \to 1, x_2 \to 0$ *or* 1. *The corners* (0,0) *and* (0,1) *are points of discontinuity. The ratio* $(f-f_{min})/(f_{max}-f_{min})$ *is indicated by contour zone codes as follows:*

∇ L 1 3 5 7 9 H Δ

0% 0.1% 2.5% 5% 15% 25% 35% 45% 55% 65% 75% 85% 95% 97.5% 99.9% 100%

the numbers of distances less than some ecologically meaningful threshold is a more attractive proposition. Again, Silverman and Brown (1978) give the relevant distribution theory. In our example, it is tempting to choose a threshold of one foot, but formal significance levels based on retrospective choice of statistic are, here as elsewhere, potentially misleading.

Strictly, the detection of inhibition invalidates the previous analysis in terms of pure heterogeneity (and *vice versa*!). However, the effect of the inhibition is to induce local regularity which should *improve* the nominal fit to a purely heterogeneous model, as judged by the chi-squared goodness-of-fit test. Our rejection of both uniform and beta models therefore remains sound. Although we have been unable to find a staisfactory model

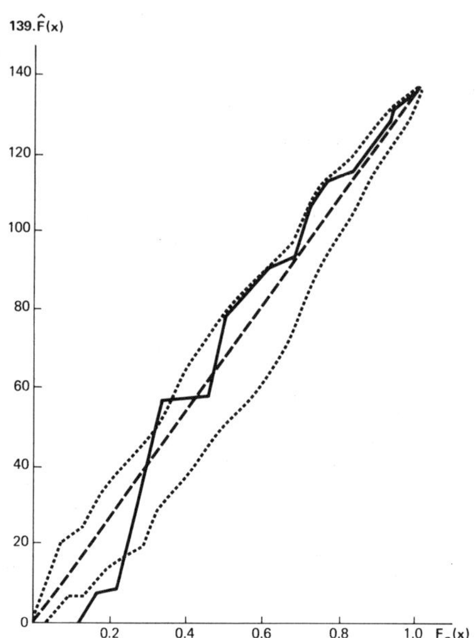


*FIG. 5.8: Event-event nearest neighbor analysis of balsam fir data.* ——— *data;* ••••• *envelope from* 19 *completely random simulations;* ----- $\hat{F}(x) = F_o(x)$ *(*$F_o(x)$ *denotes marginal distribution function under complete spatial randomness).*

for these data, the isolation of two qualitatively different non-random features is of some interest. We suggest that an adequate representation of the data would require the incorporation of both a variable inhibition effect dependent on seedling size and a slowly-varying local density.

*5.5 Concluding Remarks.* The above examples give some indication of the scope, and limitations, of the various methods currently available for the statistical analysis of spatial point patterns. Many of the methods are relatively new and untried. Further experience is needed before their usefulness to the practitioner can be evaluated accurately. In particular, comparative power studies based on simulated realizations of particular types of spatial point process would both be of considerable theoretical value and have important implications for the practitioner, provided that the chosen processes represent plausible models in the intended field of application. It would probably be unwise to look for an omnibus method of analysis. Rather, different approaches may be expected to provide appropriate tests against different types of alternative. The aim of the present paper has been to clarify the present position and to offer food for thought with regard to future developments.

## APPENDIX
## SIMPLE MONTE CARLO TESTS

*A.1 Rationale.* The idea behind Monte Carlo testing, proposed originally by Barnard (1963), is extremely simple, but nevertheless does not appear to be widely known. We therefore take this opportunity to establish the concept of a Monte Carlo test and to note the salient features of the method.

Suppose that we are given a *simple* null hypothesis $H_o$ and a set of data from which we calculate the value $u_1$ of a statistic $u$. Suppose further that values $u_2, \cdots, u_m$ are generated as a random sample from the null distribution of $u$ and that

$$u_{(1)} < u_{(2)} < \cdots < u_{(m)}$$

are the corresponding order-statistics. Under $H_o$, it is then exactly true that

$$P\{u_1 = u_{(j)}\} = m^{-1} \quad (j=1, \cdots, m), \tag{A.1.1}$$

whence the rank of $u_1$ may be used to construct an exact test of $H_o$. If the null distribution of $u$ is discrete, ties may occur. We then adopt the most conservative rank possible for $u_1$.

Hope (1968) shows that, when uniformly most powerful tests are available, the loss of power in using the Monte Carlo approach is remarkable slight for modest computational effort. For tests at conventional significance levels the use of $m=100$, or 200 if a two-sided test is appropriate, will usually be adequate and conveniently allows for retrospective calculation of a p-value in intervals of one per cent. Of course, the advantage of the Monte Carlo approach is that a test is readily available when the null distribution of $u$ is unknown, provided that it can be simulated. An important corollary is that the choice of test statistic need not be constrained by known distribution theory. As we have seen, this is particularly pertinent in a spatial context. Examples of Monte Carlo tests applied to spatial problems may be found in Cliff and Ord (1973, Ch. 2, 4), Ripley (1977), and Besag and Diggle (1977).

*A.2 Pitfalls.* We remark that, in practice, pseudo-random sampling will be used and we must assume that this device is satisfactory. A rather more serious problem is that of a non-simple null hypothesis. When the procedure for generating $u_2, \cdots, u_m$ involves the values of parameters estimated from the data, (A.1.1) does not hold. In a goodness-of-fit context we expect that failure to compensate for the estimation of unknown parameters will produce a conservative test, since the simulations are then effectively constrained to fit the data. However, the opposite effect can also arise. This phenomenon is not peculiar to Monte Carlo testing. For example, the effect on the classical chi-squared goodness-of-fit test of failing to allow for parameter estimation is to produce a conservative test, as anticipated, whereas in testing for zero mean in a Normal sample with unknown variance, the false assumption of a Normal distribution, rather than Student's t, for the standardized test statistic $\bar{x}/\sqrt{(\hat{\sigma}^2/n)}$ would lead to spuriously significant results.

One solution, seldom feasible in practice, is to condition the simulations for $u_2, \cdots, u_m$ on the observed values of sufficient statistics for the unknown parameters. When abundant data are available, a cross-validatory approach, in which one half of the data are used for parameter estimation and the other half for testing, is possible, although inefficient. A more pragmatic, albeit approximate, remedy is to choose a test statistic which is unrelated to the estimation procedure.

*A.3 A Monte Carlo EDF-Test.* Amongst the many goodness-of-fit tests available for a univariate sample $x_1, \cdots, x_n$, tests based on the empirical distribution function (EDF),

$$\hat{F}(z) = n^{-1} \times (\text{Number of } x_i \leq z),$$

have a proven value, particularly for small-to-moderate sample size and a continuous underlying distribution. Hegazy and Green (1975) give a useful list of references. Such tests assume a completely specified null distribution function $F(z)$ although critical values for particular *classes* of null distribution, say negative exponential or Normal, have been tabulated on a one-off basis, usually by simulation. See, for example, Lilliefors (1967, 1969). In our context, we shall be concerned with assessing the fit of a sample of *dependent* observations $x_i$ to a prescribed marginal distribution function and a Monte Carlo approach will certainly be required.

We now construct a Monte Carlo analogue of Finkelstein and Schafer's (1971) improvement to the classical Kolmogorov-Smirnov test. Let $\hat{F}_1(z)$ denote the EDF for the data and $\hat{F}_1(z)$ $(i=2, \cdots, m)$ the corresponding EDF's for $m-1$ simulations of the proposed model. Let $\bar{F}(z)$ denote the null marginal distribution function if known, otherwise calculate

$$\bar{F}(z) = m^{-1} \sum_{i=1}^{m} \{F_i(z)\}.$$

Further calculate

$$u_i = \sum_{j=1}^{k} | \hat{F}_i(z_j) - \bar{F}(z_j)| \quad (i=1, \cdots, m) \qquad \text{(A.3.1)}$$

from a set of values $z_i, \cdots, z_k$ such as might reasonably be used to compile a frequency table for the data. The $u_i$, although not mutually independent, are exchangeable. (A.1.1) therefore applies and an exact Monte Carlo test follows immediately. We remark that other goodness-of-fit criteria might be used, according to taste. For example, a direct analogy with the Kolmogorov-Smirnov test is given by choosing

$$u_i = \max_{j=1, \cdots, k} |F_i(z_j) - \bar{F}(z_j)|$$

rather than (A.3.1).

In Section 5 we made extensive use of this test and related procedures based on the EDF of the data. In this respect our attitude matches that of Ripley (1977) who also favors analysis in terms of a cumulative function of the data, rather than the corresponding histogram.

## REFERENCES

Barnard, G. A. (1963). Contribution to the discussion on Professor Bartlett's paper. *Journal of the Royal Statistical Society, Series B*, 25, 294.

Bartlett, M. S. (1937). Properties of sufficiency and statistical tests. *Proceedings of the Royal Society, Series A*, 168, 268-282.

Bartlett, M. S. (1964). Spectral analysis of two-dimensional point processes. *Biometrika*, 44, 299-311.

Bartlett, M. S. (1975). *The Statistical Analysis of Spatial Pattern.* Chapman & Hall, London.

Batcheler, C. L. (1973). Estimating density and dispersion from truncated or unrestricted joint point-distance nearest-neighbor distances. *Proceedings, New Zealand Ecological Society*, 20, 131-147.

Batcheler, C. L. and Hodder, R. A. C. (1975). Tests of a distance technique for inventory of pine plantations. *New Zealand Journal of Forest Science*, 5, 3-17.

Besag, J. E. and Diggle, P. J. (1977). Simple Monte Carlo tests for spatial pattern. *Applied Statistics*, 26, 327-333.

Besag, J. E. and Gleaves, J. T. (1973). On the detection of spatial pattern in plant communities. *Bulletin, International Statistics Institute*, 45(1), 153-158.

Blackman, G. E. (1935). A study by statistical methods of the distribution of species in grassland associations. *Annals of Botany*, 49, 749-777.

Borel, E. (1925). *Traité du Calcul des Probabilités et de ses Applications, I.1*, Gauthier-Villars, Paris.

Brown, S. and Holgate, P. (1974). The thinned plantation. *Biometrika*, 61, 253-262.

Catana, A. J. (1963). The wandering quarter method of estimating population density. *Ecology*, 44, 349-360.

Clapham, A. R. (1936). Overdispersion in grassland communities and the use of statistical methods in plant ecology. *Journal of Ecology*, 24, 232-251.

Clark, P. J. and Evans, F. C. (1954). Distance to nearest neighbor as a measure of spatial relationships in populations. *Ecology*, 35, 23-30.

Cliff, A. D. and Ord, J. K. (1973). *Spatial Autocorrelation.* Pion, London.

Cormack, R. M. (1977). The invariance of Cox and Lewis's statistic for the analysis of spatial patterns. *Biometrika*, 64, 143-144.

Cormack, R. M. (1979). Spatial aspects of competition between individuals. In *Spatial and Temporal Analysis in Ecology*, R. M. Cormack and J. K. Ord, eds. Satellite Program in Statistical Ecology, International Co-operative Publishing House, Fairland, Maryland.

Cottam, G. and Curtis, J. T. (1949). A method for making rapid surveys of woodlands, by means of pairs of randomly selected trees. *Ecology*, 30, 101-104.

Cottam, G., Curtis, J. T., and Hale, B. W. (1953). Some sampling characteristics of a population of randomly dispersed individuals. *Ecology*, 34, 741-757.

Cox, D. R. and Lewis, P. A. W. (1966). *The Statistical Analysis of Series of Events*. Methuen, London.

Cox, D. R. and Miller, H. D. (1965). *The Theory of Stochastic Processes*. Methuen, London.

Cox, T. F. (1976a). The robust estimation of the density of a forest stand using a new conditioned distance method. *Biometrika*, 63, 493-500.

Cox, T. F. (1976b). *Some problems in the analysis of spatial pattern*. Unpublished Ph.D. Thesis, University of Hull.

Cox, T. F. and Lewis, T. (1976). A conditioned distance ratio method for analysing spatial patterns. *Biometrika*, 63, 483-489.

Diggle, P. J. (1975). Robust density estimation using distance methods. *Biometrika*, 62, 39-48.

Diggle, P. J. (1977a). A note on robust density estimation for spatial point patterns. *Biometrika*, 64, 91-95.

Diggle, P. J. (1977b). The detection of random heterogeneity in plant populations. *Biometrics*, 33, 390-394.

Diggle, P. J., Besag, J., and Gleaves, J. T. (1976). Statistical analysis of spatial point patterns by means of distance methods. *Biometrics*, 32, 659-667.

Donnelly, K. (1977). Personal communication.

Finkelstein, J. M. and Schafer, R. E. (1971). Improved goodness-of-fit tests. *Biometrika*, 58, 641-645.

Fisher, R. A., Thornton, H. G., and MacKenzie, W. A. (1922). The accuracy of the plating method of estimating the density of bacterial populations, with particular reference to the use of Thornton's agar medium with soil samples. *Annals, Applied Botany*, 9, 325-359.

Ghent, A. W. (1963). Studies of regeneration of forest stands devastated by the spruce budworm. *Forest Science*, 9, 295-310.

Green, P. J. and Sibson, R. (1978). Computing Dirichlet tessellations in the plane. *Computer Journal*, 21, 168-173.

Greig-Smith, P. (1952). The use of random and contiguous quadrats in the study of the structure of plant communities. *Annals of Botany*, 16, 293-316.

Hegazy, Y. A. S. and Green, J. R. (1975). Some new goodness-of-fit tests using order statistics. *Applied Statistics*, 24, 299-308.

Holgate, P. (1965a). The distance from a random point to the nearest point of a closely packed lattice. *Biometrika*, 52, 261-263.

Holgate, P. (1965b). Tests of randomness based on distance methods. *Biometrika*, 52, 345-353.

Holgate, P. (1965c). Some new tests of randomness. *Journal of Ecology*, 53, 261-266.

Holgate, P. (1972). The use of distance methods for the analysis of spatial distributions of points. In *Stochastic Point Processes*, P. A. W. Lewis, ed. Wiley, New York, 122-135.

Hope, A. C. A. (1968). A simplified Monte Carlo significance test procedure. *Journal of the Royal Statistical Society, Series B*, 30, 582-598.

Hopkins, B. (1954). A new method for determining the type of distribution of plant individuals. *Annals of Botany*, 18, 213-226.

Kathirgamatamby, N. (1953). Note on the Poisson index of dispersion. *Biometrika*, 40, 225-228.

Kendall, M. G. and Moran, P. A. P. (1963). *Geometrical Probability*. Griffin, London.

Kershaw, K. A. (1973). *Quantitative and Dynamic Plant Ecology, 2nd ed.* Edward Arnold, London.

Kooijman, S. A. L. M. (1977). *Inference about dispersal patterns.* Unpublished Ph.D. Thesis, University of Leiden, Holland.

Lilliefors, H. W. (1967). On the Kolmogorov-Smirnov test for Normality with mean and variance unknown. *Journal of the American Statistical Association,* 62, 399-402.

Lilliefors, H. W. (1969). On the Kolmogorov-Smirnov test for the exponential distribution with mean unknown. *Journal of the American Statistical Association,* 64, 387-389.

Loftsgaarden, D. O. and Quesenberry, C. P. (1965). A non-parametric estimate of a multivariate density function. *Annals of Mathematical Statistics,* 36, 1049-1051.

Mardia, K. V. (1970). *Families of Bivariate Distributions.* Griffin, London.

Matérn, B. (1960). *Spatial Variation.* Meddelanden fran statens skogsforskningsinstitut, Bd. 49, No. 5, Stockholm.

Matérn, B. (1971). Doubly stochastic Poisson processes in the plane. In *Statistical Ecology, Vol. 1,* G. P. Patil, E. C. Pielou, and W. E. Waters, eds. The Pennsylvania State University Press, University Park. 195-213.

Mead, R. (1966). A relationship between individual plant-spacing and yield. *Annals of Botany,* 30, 301-309.

Mead, R. (1971). Models for inter-plant competition in irregularly distributed populations. In *Statistical Ecology, Vol. 2,* G. P. Patil, E. C. Pielou, and W. E. Waters, eds. The Pennsylvania State University Press, University Park. 13-32.

Mead, R. (1974). A test for spatial pattern at several scales using data from a grid of contiguous quadrats. *Biometrics,* 30, 295-307.

Mood, A. M. and Graybill, F. A. (1963). *Introduction to the Theory of Statistics.* McGraw-Hill, New York.

Moore, P. G. (1954). Spacing in plant populations. *Ecology,* 35, 222-227.

Newman, T. G. and Odell, P. L. (1971). *The Generation of Random Variates.* Griffin, London.

Neyman, J. and Scott, E. L. (1958). Statistical approach to problems of cosmology. *Journal of Royal Statistical Society, Series B*, 20, 1-43.

Numata, M. (1961). Forest vegetation in the vicinity of Choshi, Coastal flora and vegetation at Choshi, Chiba Prefacture IV (in Japanese). *Bulletin, Choshi Marine Laboratory*, Chiba University, 3, 28-48.

Parzen, N. E. (1962). On estimation of a probability density function and mode. *Annals of Mathematical Statistics*, 33, 1065-1076.

Persson, O. (1964). *Distance Methods*. Studia Forestalia Suecica, 15, 68 pp.

Persson, O. (1971). The robustness of estimating density by distance measurements. In *Statistical Ecology, Vol. 2*, G. P. Patil, E. C. Pielou, and W. E. Waters, eds. The Pennsylvania State University Press, University Park. 175-190.

Plackett, R. L. (1965). A class of bivariate distributions. *Journal of the American Statistical Association*, 60, 516-522.

Preston, C. (1975). Spatial birth-and-death processes. *Bulletin, International Statistical Institute*, 46, 371-391.

Ripley, B. D. (1977). Modelling spatial patterns (with discussion). *Journal of the Royal Statistical Society, Series B*, 39, 172-212.

Ripley, B. D. and Kelly, F. P. (1977). Markov point processes. *Journal of the London Mathematical Society*, 15, 188-192.

Silverman, B. and Brown, N. T. (1978). Short distances, flat triangles and Poisson limits. *Journal of Applied Probability*, 15, 815-825.

Strauss, D. J. (1975). A model for clustering. *Biometrika*, 62, 467-475.

Thomas, M. (1949). A generalization of Poisson's binomial limit for use in ecology. *Biometrika*, 36, 18-25.

Warren, W. G. (1971). The center-satellite concept as a basis for ecological sampling. In *Statistical Ecology, Vol. 2*, G. P. Patil, E. C. Pielou, and W. E. Waters, eds. The Pennsylvania State University Press, University Park. 87-118.

[*Received:* *June* 1977. *Revised:* *December* 1978]

R. M. Cormack and J. K. Ord, (eds).,
*Spatial and Temporal Analysis in Ecology*, pp. 151-212. 

# SPATIAL ASPECTS OF COMPETITION BETWEEN INDIVIDUALS

R. M. CORMACK

Department of Statistics
University of St. Andrews
North Haugh, St. Andrews
Fife, Scotland KY16 9SS

SUMMARY. The main aim of this paper is to summarize, without going into mathematical detail, recent developments in statistical models for spatial interactions between individuals, to discuss how such interactions can be observed and how experiments to make them more apparent can be designed, and to suggest further developments needed to unify these statistical models which are spatial and individual-based with current ecological models which are aspatial and population-based.

KEY WORDS. automodels, beehive designs, competition circles, lattice processes, pattern analysis, plant competition experiments, quadrat sampling, spatial point processes, statistical ecology, tests of randomness, Voronoi polygons.

## 1. INTRODUCTION

Several of the papers given at the International Symposium on Statistical Ecology at Yale in 1969 (Patil, Pielou, and Waters, 1971) and at the conference on Stochastic Point Processes at Yorktown Heights in 1971 (Lewis, 1972) comment on the lack of applicable models for spatial pattern. In the intervening years has occurred a surge of development so that a variety of models, founded on rigorous mathematics but with considerable intuitive appeal, are now available. Together with this development has come a conviction that many of the classical forms of statistical analysis are strictly invalid, and can provide misleading results, when applied to such data. New investigatory tools, impracticable before the computer revolution, have been developed for the analysis of spatially observed data and the exploration of the new models.

Competition is the determinant of much, some might say all, of the natural living world, and there has also been in the seventies a surge of development of mathematical descriptions and of quantitative field investigations of these processes. The index of the Macarthur Memorial Volume (Cody and Diamond, 1975) lists multiple references to the effects of competition on community structure, niche breadth, phenotypic traits, population density, reproductive strategy, and species distributions. Most of this development concerns competition between populations, not between individuals, and most avoids any quantitative description - if any cognizance at all is taken - of the relative positions in ordinary physical or geographical space of the competing units. Connell (1975) considers the problem of how to detect and measure competition, and lists one of the methods which has been most generally used as "to describe the pattern that exists at a point in time to see whether it does or does not conform to the predictions of the model."

This paper provides a review of recent research on statistical models for spatial competition and points the way for future developments.

## 2. COMPETITION AND SPACE

*2.1 Competition.* Ecologists commonly say that there are as many definitions of competition as there are ecologists. One bone of contention is whether to define any interaction by the action or its effect; another is whether the definition should refer to the individual or the population; a third, whether it should be verbal or mathematical; a fourth, whether or not it should be restricted to situations in which the effect is injurious to both competitors.

To discuss spatial aspects of competition we require a definition which can apply to whatever units have a definite spatial location. Since these are usually individuals, but may be clones, colonies, or groups, reference to "a biological population, or any part of it" (Pielou, 1974) is needed - organisms alone (as in Milne, 1961; Donald, 1963) or populations alone (as in Williamson, 1972) will not serve our purpose. If the units involved are groups or clones, we shall regard them as indivisible and continue to refer to them as individuals. Since most of the models to be discussed are primarily models for interaction in general, the more general the definition we make the wider will be the applicability of the model. The idea of interaction implies that more than one individual is involved, a requirement which does not seem to be imposed by Pielou's definition. We shall also adhere to the idea inherent in Donald's and Milne's definitions that individuals compete when some factor on which both depend is, at least potentially, in short supply. Both direct interference and competition through exploitation of the environment are envisaged. We shall not, however, impose Williamson's requirement that both individuals are affected by the competition, although we shall discover later that some models are not valid when competition is one-sided. We shall formulate specific mathematical models, consistent with these verbal concepts, which represent the effect of one individual on another, be it of the same or a different type (species, variety or any other biological category), in terms of restriction in its location or reduction in some aspect of its performance.

As stated above, any attempt to investigate specifically spatial aspects of competition forces one to consider competition between individuals as the basic element of any theory. The widening of our definition of individual to include any spatially indivisible unit of population overcomes Williamson's argument that such a definition cannot apply, for example, to the classical examples of competition between species of Tribolium. In such examples the spatial relationships between individual organisms change much faster than interactive effects occur. But I would argue that there is more reason than this semantic sidestepping of the basic difference between motile and sessile organisms for emphasizing competition between individuals even if the struggle for existence between two populations is the subject of study. Harper (1967) quotes Darwin: "the structure of every organic being is related, in the most essential yet often hidden manner, to that of all the other beings with which it comes into competition for food and residence, or from which it has to escape, or on which it preys." Taking this as a text we may hope to understand fully the effect of competition on populations only by examining competition between individuals. The remark made above about the time span in which animals change their location and interact suggests an analogy with thermodynamics and indeed there is a very close correspondence between several of the models to be discussed and the theory of statistical mechanics. This may justify the concentration on population studies with animals although, if we continue the analogy, much of Newtonian 'population physics' is illuminated by atomic or molecular 'individual models'. It does not justify the study of population dynamics of plants other than as one observable outcome of individual interactions.

2.2 *Space*. Competition is one concept with which we shall be concerned; space is the other. The space that we shall be discussing is two-dimensional physical space, the distance between individuals being the physical distance in cms. or whatever other unit is appropriate. Many of the types of model which will be considered apply equally well in 3-dimensional space, and we shall ignore the changes of detail which may be necessary. It is of course rare for physical space to be the factor directly limiting the potential development of an individual. If however, we make the assumption of a reasonably homogeneous environment - and this imposes a serious restriction on the applicability of our models - then physical area will be highly correlated with availability of resources. We use physical space as a mapping of available resources. This will be incorrect if there is, for example, a habitat gradient between parts of the study area. Such a gradient will undoubtedly affect the performance of individuals, but I do not think it is profitable to consider it as directly affecting the interaction between individuals. It is therefore assumed that if such environmental heterogeneity exists, it has been removed from the data by appropriate trend surface methods before spatial analysis is attempted. Caution is then necessary in making inferences, since such trend removal induces statistical dependence among the residual observations. In the representations in Section 5 of an individual plant by a circle or polygon, the shape is drawn in physical space but its area represents the resources available to the plant. In this there is a further assumption that individuals are equally good at converting available resources for their use. The lack of realism in this assumption has been at the root of some dissatisfaction with the use of physical space in these models. However, although there are technical difficulties in detailing the correspondence, it is simple conceptually to envisage a strong competitor who makes good use of available resources as occupying a larger physical space than does a weak competitor who does not. Thus we can still hope to map competitive strength on to physical space.

Instead of this double distortion to retain physical space as the basic representation of the location of an individual, it might seem simpler to represent the individual directly in its niche in a space whose axes represent the resources being competed for. This has become very popular among ecologists. If there is a single limiting factor for which individuals are competing this can be a satisfactory representation. If, however, the resource space cannot be reduced satisfactorily to a single dimension, it suffers from the severe disadvantage that the axes are not independent in the way that the compass directions of physical space necessarily are. Euclidean distance is the natural distance in physical space: it is not in resource space. Moreover, the appropriate distance is dependent on the correlation between the resources, and there does not seem to me any prior reason why this should remain the same in different sections of resource space. Although this dependence between variables has been acknowledged in the literature (May, 1975), it is often forgotten.

The word 'competition' is used in what follows to designate any process - physical inhibition, direct interference, or overlapping niches - which has a negative influence on an individual's location, viability, or strength. Such competition will be evidenced by the spatial pattern of individuals or measurements made on them.

## 3. DATA, ANALYSIS, AND MODELS

Different types of data require different forms of statistical description and analysis and will require different approaches to statistical modelling. Since competition between biological units can occur and can be studied in highly disparate ways, a wide variety of situations needs to be considered. These range from a forest, with the exact position of each tree mapped and recurrent measurements taken on all or a sample of individuals, to a flower pot in which seed of a mixture of species has been sown, allowed to grow, harvested in bulk at one instant of time, and a single measurement made on each species. Since the population at any time has developed as a result of competition which has acted continuously over a period of time, possibly in different ways and usually at different strengths at different times, data on the same individuals seem likely to be most informative biologically. There are three reasons why the whole of this review is not devoted to such spatio-temporal data. The first is simply the effort and time necessary to collect such observations and the volumes of data which then require analysis. The second is the difficulty of formulating an adequate, yet precise, model for such a variable and complex process. The third is a version of Heisenberg's uncertainty principle: in many situations the act of measuring the competing individuals will influence the process of competition.

Reliable sets of data on the spatial relationships between the same individuals at different times are rather rare, and largely restricted to forestry studies. Although I have no doubt that in most situations, and all plant situations, any mechanistic model ought to be a spatio-temporal one, we shall be concerned principally with static descriptions of spatial pattern at one time. If these patterns are investigated at different times, they will refer to different sets of individuals at each time.

*3.1 Point Data.* Consider first the study of individuals. Of prime importance is their spatial location, and this may be the only information available. If the position of individuals is naturally determined and if all points of some region in the plane are potentially capable of containing an individual, the resulting pattern is a realization of a two-dimensional point process: the data are the location vectors $\underset{\sim}{r}_i$ referred to some arbitrary origin. Statistical models for competitive point processes will be studied in Section 7. Other variables, specifying the particular type of individual or some qualitative, usually binary (e.g., large or small, diseased or healthy, nesting successfully or not), or quantitative (height, weight, grain yield, number of eggs), aspect of its performance may be observed at each individual location: $\underset{\sim}{Y}(\underset{\sim}{r})$. Most such models refer either to binary or normally distributed data, usually to the latter, and measurements may require initial transformation to help them to fit these statistical distributions. Models for mixed distributional forms are not yet available.

*3.2 Lattice Data.* In other situations the locations $\underset{\sim}{r}_i$ may be extrinsically determined, as in a field crop or plantation. There may also be value in examining the process of competition between naturally located individuals by studying only the spatial pattern of measurements on these individuals conditional on the observed locations. In the redevelopment of a natural forest, after fire for example, the major factor in the location of seedlings will have been the location of parent trees, so that competition between the seedlings is made manifest, and should be studied, by their subsequent individual development given the presence of the other individuals at pre-determined points. We may then speak of a lattice process, whether or not the locations are regularly spaced. Variables $\underset{\sim}{Y}_i$ defined at the lattice points may again be binary - including presence/absence of an individual if unoccupied lattice points are observable - qualitative, discrete, or continuous.

One of the first decisions to be taken in modelling or analyzing a lattice process is the choice of a set of neighbors $\eta_i$ of each point $P_i$ - those individuals $P_j$ upon whose values $\underset{\sim}{Y}_j$ the variable $\underset{\sim}{Y}_i$ will depend. If a sample of individuals is observed then typically measurements will be made on these individuals and on their neighbors. On a regular square lattice the decision is usually restricted to the inclusion or exclusion of second-order neighbors (x on the diagonals)

```
X  0  X
0 (X) 0
X  0  X
```

as well as the first-order neighbors (on the rectangular grid). The regular triangular lattice does not present a comparable problem, the six neighboring lattice points determining the first-order neighbors.

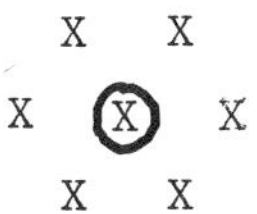

Such models are referred to as nearest neighbor models on the lattice. More distant points can be included within $\eta_i$ but seldom are. Reasons for this are largely simplicity of mathematical framework and directness of biological interpretation. All situations must display first-order competition before they display competition at greater distances, and first-order competition will be the most intense and is likely to be crucial. At larger distances, moreover, environmental heterogeneity may dull the picture. But we should be warned by the observation that at high density the roots of 32 different corn plants were found under any point on the row (Haynes and Sayre, 1956). High-order competition exists. Statistical models for processes on regular lattices will be described in Section 8. On an irregular lattice such a categorization of neighbors is not so easy. The usual definition is to include in $\eta_i$ any individual $P_j$ whose distance $r_{ij}$ from $P_i$ is less than some threshold, which can be varied to be a function of the size of individuals. Alternatively $\eta_i$ may be taken to contain all individuals whose cells in a Dirichlet tessellation (see Section 5.2) border on $P_i$'s cell.

The determination of a set of neighbors $\eta_i$ is merely to delimit which individuals are assumed to be affecting $P_i$. It does not force us to assume that every $P_j \in \eta_i$ will affect $P_i$ in the same way: usually the effect will be modelled as being dependent on the distance $r_{ij}$.

*3.3 Quadrat Data.* In many practical situations the point locations of individuals are not recorded, either directly or in the form of a set of inter-individual distances. This may be because no point can be assigned to an individual, as is the case with plants spreading vegetatively, or because the ecologist has made summary measurements on a selected area, which we shall refer to as a quadrat, whatever its actual shape. If the location of each quadrat is known, in particular if the quadrats form a contiguous grid, then such data can be treated as a lattice process on the points defined by the centroids of the quadrats. Strictly they form a derived process and care is needed in applying methods of analysis to such data (Bartlett, 1975). The important problem of large-scale continuous spatial structures such as mosaics require different methods of study (Pielou, 1969, 1974, and many papers referenced therein). It has recently been suggested (Murray, 1976) that such patterns can be described as a stable travelling wave obtained as the solution of a nonlinear diffusion model incorporating a time lag. Data of this type will not be discussed further in this review.

*3.4 Restrictions on Inference.* One question, mentioned briefly earlier, concerns the nature of the data as a sample or as a population. The sad truth is that in any situation involving spatial interaction, no individual is strictly independent of any other. Even if it is assumed that interaction occurs only between nearest-neighbors, the effect of such dependence can be such as to create large-scale patterns. This is recognized in the study of mosaics but is often forgotten in the study of individuals. In tests of randomness, discussed in Section 4, points, distances, and measurements are independent provided the areas involved do not overlap when the null hypothesis of randomness is true: not so under any alternative. To ensure that a sample is effectively independent, we must ensure that the individuals are sufficiently far apart for the direct correlation between them to have died away and that the points are not related by any systematic large-scale spatial pattern. The assumption of a random sample of observations from a spatial process is thus extremely insecure. What is under observation is a single realization of an enormous multivariate distribution. Without strong assumptions of structure, classical formal statistical inference is impossible from a single observation, a fact that forces both statistician and biologist away from attempts at statistical description towards biological modelling.

The above strictures do not of course refer to experimental situations in which the outcome of the spatial process is summarized by a single datum such as the yield of Species A from a pot or field in which it was grown in mixture with Species B. Genuinely independent observations result from replicates of this experimental situation repeated with different mixtures or spacings and appropriate classical statistical analyses can be performed on such data. The strictures do however apply to data in the form of frequency distributions of weights, heights, or yield of the surviving individuals from a single competitive process. Such distributions, described in Section 6, again constitute a single multivariate observation with specific but limited value in the simulation or assessment of stochastic models.

## 4. TESTS OF RANDOMNESS VERSUS REGULARITY

Early approaches to spatial pattern consisted largely of testing the hypothesis that individuals are randomly distributed. If this hypothesis were rejected, some phenomenon existed which required explanation. The explanation was usually non-statistical except that the value of the statistic used in the test of randomness was used as a guide to the ecological argument. The hypothesis of random distribution has two components, independence, and homogeneity. If either fails to hold, spatial pattern will be observed; but the causal factors are different in the two cases. Heterogeneity implies an extrinsic cause - lack of uniformity in the environment - while dependence is intrinsic to the process itself. Although the environment is obviously of the greatest biological importance, its effect is best studied by standard statistical procedures. We are concerned here with dependence, its intrinsic causes, and how it can be measured and assessed. Most spatial methods implicitly or explicitly assume homogeneity in the environment or, at least, that trends have been eliminated before the spatial analysis begins.

Although the aim of testing the null hypothesis of randomness is a very limited one, the methods of achieving it remain of interest because the statistics devised for that purpose can also be used to measure pattern and to guide attempts to model the non-randomness. Let us consider first point process data - data purely on the location of individuals. The null hypothesis of no pattern is then that the individuals form a Poisson forest. Four different types of approach have been devised - quadrat counts, nearest neighbor distances, spectral analysis, and the application of test sets. Some of these are discussed in the companion articles by Ord (1979) and by Diggle (1979) - let us give an overview here.

*4.1 Quadrat Counts.* Exploration of spatial pattern by counting the numbers of individuals $Y_i (i=1,\cdots,Q)$ in $Q$ randomly sited quadrats is limited by the fact that information is available only on a single scale of pattern chosen by the experimenter. Of course this can be of benefit if only one scale is of interest. However, the complexity of ecological systems makes this unlikely. Moreover the suitable systematic choice of quadrats can provide information on a range of scales of pattern. Most interest in quadrat counts has lain in aggregation as an alternative to randomness, and work has developed along the lines of describing this by a Poisson distribution with additional variability superimposed. The only test statistics whose properties have been explored when the pattern has a degree of regularity to it, as would be expected of a competition process, are the sample variance to mean ratio

$\Sigma(y_i - \bar{y})^2/[(Q-1)\bar{y}]$, Lloyd's index of crowding $\overset{*}{m} = \Sigma[y_i(y_i-1)]/(Q\bar{y})$, and Morisita's index of aggregation $I_\delta = \Sigma[y_i(y_i-1)]/[\bar{y}(Q\bar{y}-1)]$. In fact for any formal test of randomness, only one of these is relevant since $\overset{*}{m}$ and $I_\delta$ have to be converted to the form $\Sigma(y_i-\bar{y})^2/\bar{y}$ which has approximately a $\chi^2_{Q-1}$ distribution for Poisson data provided $\bar{y}$ is not too small. A recent variant of Morisita's index (Smith-Gill, 1975) seems to offer no advantages in this context. If the pattern is regular, a small value of the $\chi^2_{Q-1}$ statistic will be observed. Use of these statistics to investigate scales of pattern will be described later. There has been one study of the properties of Morisita's index for regular patterns (Stiteler and Patil, 1971). They showed that its population value was less than 1 for a number of lattices when sampled by rectangular quadrats with the same orientation as the lattice.

*4.2 Nearest-Neighbor Methods.* Observations from randomly sited quadrats seem of limited value in assessing competition. Nearest-neighbor distances are potentially more valuable, and a list of such statistics is provided in Table 4.1. The quoted distributions refer to an independent sample of observations, and will not be appropriate otherwise, although there is some evidence that the effect of dependence is relatively slight. The first three require an independent estimate of density to be available: the effect of using as estimate the density over the whole area from within which the distance measurements are made is assumed to be bad but has not been studied in detail. Full discussion of these tests is given by Diggle (1979).

*4.3 Pattern Analysis.* Spatial pattern of individuals may occur at a number of different scales, of which nearest-neighbor methods can be expected only to reveal the smallest. Of course this is the scale at which competition between individuals will be manifest, so that our concern will be largely with such methods. However, competition between clumps will give rise to regular pattern at a larger scale, and moreover, methods designed to investigate pattern at all scales will provide information *inter alia* on competition between neighbors. Most of these explore the second-order structure of the process - the correlation between observations at points or areas a given distance r apart. Spectral analysis is the most statistically sophisticated of these methods; in that it is usually used for the exploration of models other than randomness, we shall defer consideration of it until Section 7.2. Much more widely used is Greig-Smith's pattern

*TABLE 4.1: Distance tests of randomness of a set of locations.*

| Statistic | Distribution under $H_o$ | Reference |
|---|---|---|
| $2\sqrt{\hat{\lambda}}\,\Sigma r_i/n$ | approx $N[1, (4-\pi)/(\pi n)]$ | Clark and Evans (1954) |
| $\pi\hat{\lambda}\,\Sigma u_i/n$ | approx $N[1, (\lambda A + n + 1)/(n\lambda A)]$ | Pielou (1959) Mountford (1961) |
| $2\pi\hat{\lambda}\Sigma u_i$ | approx $\chi^2_{2n}$ | Skellam (1951) |
| $\Sigma u_i/\Sigma(u_i + u_i^*)$ | $B(n,n)$ | Hopkins (1954) |
| $\Sigma u_i/\Sigma u_{i2}$ | $B(n,n)$ | Holgate (1965a) |
| $\Sigma(u_i/u_{i2})/n$ | approx $N(1/2, 1/12n)$ | Holgate (1965a) |
| $\Sigma u_i/\Sigma(u_i + .5v_{iT})$ | $B(n,n)$ | Besag and Gleaves (1973) |
| $\Sigma\{u_i/(u_i + .5v_{iT})\}/n$ | approx $N(1/2, 1/12n)$ | Besag and Gleaves (1973) |
| $-.5\Sigma\ell n\{u_i/(u_i + .5v_{iT})\}$ | $\chi^2_{2n}$ | Besag and Gleaves (1973) |
| $[2\pi + \sin B_i - (\pi + B_i)\cos B_i]^{-1}$ | $U(1/4\pi, 1/\pi)$ | Cox and Lewis (1976) Cormack (1977) |

where $B_i = 2\sin^{-1}(\sqrt{v_i/2u_i})$

a) $r_i$: distance from a random point to the nearest individual; b) $u_i$: $r_i^2$; c) $u_{i2}$: squared distance from a random point to the second nearest individual; d) $u_i^*$: squared distance from a random individual to its nearest neighbor; e) $v_i$: squared distance between the nearest individual to a random point and its nearest neighbor; f) $v_{iT}$: as $v_i$ with the nearest neighbor restricted to T square sampling.

analysis and its variants. This was introduced initially (1952) to provide some indication of scales of heterogeneity. Data, counts or measurements of frequency or of cover, are collected on the basis of a regular grid of contiguous sample areas, square, rectangular, or line segments. Most commonly the grid is a linear transect of unit areas (Kershaw, 1957). If $2^m$ units of area are observed, they are regarded successively as 2 areas, each comprising $2^{m-1}$ contiguous units, each of which is split into 2 areas, each comprising $2^{m-2}$ contiguous units, the split continuing down to the smallest scale of $2^m$ unit areas considered as $2^{m-1}$ pairs of contiguous units. At each level of this hierarchy, a measure of the variability among the area totals is calculated and used as a description of the pattern at the scales determined by groups of $2^k$ units $(k=0,1,2,\cdots,m-1)$. If the data are counts and the individuals are randomly distributed then the index of dispersion $\Sigma(y_i-\bar{y})^2/[(n-1)\bar{y}]$ among $n$ values should always be about unity, and the mean square of the observations $(n-1)^{-1}\,\Sigma(y_i-\bar{y})^2$ should be proportional to the block size. If at each level in the hierarchy the mean square is taken as the sample variance among the $2^{m-k}$ observations on the areas formed by $2^k$ units (taking $n = 2^{m-k}$), successive mean squares will clearly be statistically correlated. To avoid this the principle of a hierarchical analysis of variance was adopted, the mean square at each level being obtained *within* the pairs which are combined to form the units at the next higher level. Three different statistical test procedures have been applied to such a hierarchy:

i) for counts, to use the mean squares from the hierarchy as estimates of the variance in the $\chi^2$ test of the index of dispersion;

ii) to assess each mean square by an F-test against the 'residual' mean square among single units (Greig-Smith's own proposal in 1952);

iii) to assess each mean square by an F-test against the mean square at the next smallest level in the hierarchy.

These tests are only approximate and assumptions for the F-tests in particular are unlikely to be satisfied (Thompson, 1958).

The major difficulty is that these distributions are no longer valid once any non-randomness has been established. This has been resolved by Mead's (1974) randomization and rank tests. If a formal test is required of the existence of pattern at any scale or scales in a transect, this is the only recommendable procedure. It is based on the same hierarchy as before, examining at each level sets of four consecutive observations and assessing the way in which they are spatially combined in pairs against the randomization distribution of all possible combination in pairs. With data $y_1,y_2,y_3,y_4$ there are three possible differences between pair totals $|y_1+y_2-y_3-y_4|$, $|y_1+y_3-y_2-y_4|$, and $|y_1+y_4-y_2-y_3|$ of which the first pairing has been observed. Regularity will be revealed by a pattern Large, Small, Large, Small in which the observed pair-difference will be less than at least one of the other differences; clumping by the opposite pattern - Large, Large, Small, Small. Making allowance for ties, the three possible differences must be of one of the forms (a,a,b), (a,b,c), or (a,b,b) where the possible differences a,b,c are such that $a<b<c$. If the data yield $n_1,n_2,n_3$ respectively of these three types in which we observe respectively $n_{1a},n_{2a},n_{3a}$ a's, $n_{1b},n_{2b},n_{3b}$ b's and $n_{2c}$ c's, then the rank test takes the form of testing

$$\frac{2(n_{1b} - n_{1a}) + 3(n_{2c} - n_{2a}) + 2(n_{3b} - 2n_{3a})}{\sqrt{(8n_1 + 6n_2 + 8n_3)}}$$

as a standard normal variable. A misprint in Mead's original formula (repeated in Bartlett, 1975) should be noted. The lower tail of the distribution provides a test of regularity. The normal distribution is as usual an asymptotic approximation and, when few observations are available as at the larger scales, the exact randomization distribution should, and can easily, be evaluated. Extensions to square lattices and to triangular lattices are in progress.

Pattern analysis was introduced as an exploratory technique. As such, much straining after formal statistical tests may be thought irrelevant. Scales of pattern can be visually assessed from a graph of variance among block totals against block size, a peak indicating contagion, a deep trough regularity. Numerous variants (e.g., Hill, 1973; Usher, 1975) are discussed and compared by Ludwig (1979).

An alternative statistic for investigating pattern at different scales has been much used by Japanese ecologists. This consists of plotting Lloyd's index of crowding $\overset{*}{m}$, which is algebraically related to Morisita's $I_\delta$ as $\overset{*}{m} = \bar{y} I_\delta$, against the sample mean $\bar{y}$ for different quadrat sizes. Since the statistic $\overset{*}{m}$ is a function of the sample mean and variance, exactly the same information as in Greig-Smith's analysis is being presented in a different form. Iwao (1972) shows the range of forms taken by this relationship as the size of sampling unit changes, for a variety of theoretical patterns of individuals and colonies, and illustrates these with a wide range of real examples. In other papers Iwao (1970a, b) discusses a linear relationship $\overset{*}{m} = \alpha+\beta\bar{y}$ which holds, with values of $\alpha$ and $\beta$ which characterize the pattern, for various theoretical distributions and a range of empirical data. The manner in which $\alpha$ and $\beta$ change with change in area should reveal the nature and scale of any pattern. Regularity is thought to occur in one of two ways. Direct competition for resources, which limits the maximum number of individuals possible in a discrete unit, will be shown by $\alpha=0$, $\beta=(k-1)/k$, where $k$ is the carrying capacity of the unit. Mutual repulsion without any effective limitation on capacity will be shown by $\alpha$ lying between 0 and -1, $\beta=1$.

The method is claimed to be superior to other types of pattern analysis, not least in doing away with the necessity for a contiguous grid of quadrats. Although the methods are tried on several data sets taken from the literature in addition to various new examples, and biologically meaningful conclusions are reached, assessment of the variability in the estimates is not made, nor is guidance given as to situations which are suited to these analyses. Since pattern analyses are considered very useful by ecologists, a detailed comparison of the statistical properties of the methods is urgently needed.

*4.4 Lattice Processes.* Tests of randomness of lattice processes were reviewed and generalized by Cliff and Ord (1973). The statistics are summarized in Table 4.2. Weights $w_{ij}$ are to be pre-assigned to represent the spatial interaction between $P_i$ and $P_j$, observation $y_i$ being made at $P_i$. With binary A, B, $y_i = 1$ if $P_i$ contains a B and zero otherwise. In the original references, which developed the theory for regular lattices, $w_{ij} = 1$ or 0 according as $P_i, P_j$ are or are not neighbors. Cliff and Ord evaluate the means and variances of the randomization distribution of these statistics, and of the generalized Moran and Geary statistics under the hypothesis that the

*TABLE 4.2: Tests of randomness for a lattice process.*

| Data | | Test statistic | Original reference |
|---|---|---|---|
| Binary ($y_i = 0$ or 1): | BB | $\sum_{i \neq j} w_{ij} y_i y_j^2 / 2$ | Moran (1948) Krishna Iyer (1949) |
| | BA | $\sum_{i \neq j} w_{ij} (y_i - y_j)^2 / 2$ | Moran (1948) Krishna Iyer (1949) |
| Quantitative | | $\dfrac{n \sum_{i \neq j} w_{ij} (y_i - \bar{y})(y_j - \bar{y})}{\sum_{i \neq j} w_{ij} \sum_i (y_i - \bar{y})^2}$ | Moran (1950) |
| | | $\dfrac{(n-1) \sum_{i \neq j} w_{ij} (y_i - y_j)^2}{2 \sum_{i \neq j} w_{ij} \sum_i (y_i - \bar{y})^2}$ | Geary (1954) |

$y_i$ are independent $N(\mu, \sigma^2)$, and show that for reasonable choice of $w_{ij}$ they are asymptotically normally distributed while warning that in small samples they can depart considerably from normality. For normal data, they consider the power of the test against the alternative of a first-order Markov process

$$y_i = \rho \sum_j d_{ij} y_j + u_i \quad \text{with} \quad u_i \text{ i.i.d. } N(0, \sigma^2),$$

and conclude that for $\rho > 0$ Moran's test is asymptotically more powerful than Geary's and that the optimal choice of weight $w_{ij}$ is the $d_{ij}$ of the alternative of interest. By contrast, with binary data the joins count between opposite kinds is found to be more efficacious than that between similar individuals. It should be emphasized that these studies by Cliff and Ord were against alternatives of positive spatial autocorrelation whereas competition will reveal itself through negative autocorrelation.

Further work by Strauss (1975a, 1977) on join-count statistics on a square lattice will be discussed in Section 8.

A different statistic was considered by Verhagen (1971), the number of individuals of one type, say B, all of whose neighbors are of the same type. The approach is similar to that of Roach (1968). Expectations and variances are given for the triangular lattice, for the square lattice with both 4 and 8 neighbors, and for these lattices with a few well-separated nodes missing. Verhagen is particularly interested in clustering and claims that his statistic is more sensitive than a join count. This certainly holds for the example he gives, although in that situation with A a very uncommon occurrence the distribution of his statistic under the hypothesis of randomness is very close to Poisson with a very small mean rather than the normal distribution he implies. Normality will only be approached if clusters of the common type of individual are sought. The behavior of this statistic could repay further study. It will clearly be very sensitive against total regularity.

An alternative range of test statistics was given by David (1971) for the random mixing of a number of different types on a square lattice. These are obtained by assigning a score to each (overlapping) square block of four lattice points, summing the scores and standardizing the total S with respect to its theoretical mean and variance under the null hypothesis of random mixing. Ideally the probability that S was no greater than the observed value would be the recommended index, but this would require knowledge of the whole distribution of S, not currently available. One scoring system used by David for each block is 6 if all 4 lattice points have the same type, 3 if 3 are identical, 2 if there are two pairs of identical types, 1 if 2 points are the same and the others different, 0 otherwise. David mentions also an alternative scoring scheme in which diagonally opposite lattice points are excluded. Except for edge effects on the boundary of the lattice the first scheme is the same as the join count including diagonal neighbors between points of the same type (Strauss, 1975a), and can be made equivalent to the BB joins only by slight rescoring. Strauss showed that this statistic has a very high efficiency compared to the more usual rectangular join count for large lattices and recommends its use even when the alternative is restricted to rectangular neighbors.

The derivation of the distribution of S depends on the symmetry of a regular lattice. The generality of the weights $w_{ij}$ allows sensible application of the statistics listed in Table 2 to irregular lattices. One possible choice of weights is to have $w_{ij} = 1$, if $P_j$ is a neighbor of $P_i$, and 0 otherwise. The most restrictive case is to include only nearest neighbors.

*4.5 Two Species.* Consideration only of nearest neighbors is the basis of Pielou's test for segregation between two species, a $\chi^2$ test of the 2×2 table of counts of number of individuals of species A,B whose nearest neighbor is of species A,B. Although Pielou (1974) recommends this test for data in which each individual in a study area has been examined and the identity of its nearest neighbor established, the bugbear of dependence in the complete census makes the standard contingency table test inappropriate. A correct analysis must include the lattice structure, the number of pairs for which $w_{ij} = w_{ji}$. At the opposite extreme, in the sense of giving equal weight to all lattice joins, is a test of segregation proposed recently by Delfiner (1976) which is appropriate whether or not the lattice points are regarded as given. Label the lattice points $A_i$ or $B_j$ according to the type of individual, A or B located at them. For each $A_i$, rank the distances $r(A_i,A_k)$ and $r(A_i,B_\ell)$ in increasing order of distance and obtain $R_{i(AA)}$ as $\sum_{k \neq i}^{a} r(A_i,A_k)$, $R_{AA}$ as $\sum_i R_{i(AA)}$. This is a spatial analogue of the usual two-sample Wilcoxon test of the null hypothesis that the given number of A's and B's have been randomly allocated to the observations, in this case the lattice points. The ranks can be regarded as a particular set of weights $w_{ij}$ in Cliff and Ord's formulation. A weighted combination $[(b-1)R_{AA} + (a-1)R_{BB}]/(a+b-2)$ has minimum variance but, although both $R_{AA}$ and $R_{BB}$ are asymptotically normal, the combined statistic is not. Interspecific competition will result in large values of $R_{AA}$. Delfiner's statistic suffers from the disadvantage of any test which incorporates the complete lattice in that it confounds different scales of pattern. It cannot readily be restricted to one scale because of the randomization over the given number of A's and B's; any restriction turns the numbers a and b into random variables.

## 5. GROWING SPACE MODELS

The most natural representation of the physical area required by an organism for development is a circle centered at the organism. In a homogeneous environment the space occupied represents also the resources available. This is an ideal and may, even where there is no competition, be distorted by local variation in the environment: the upper parts of a plant may respond to the direction of the illumination, roots to water gradients, animal behavior to the availability of cover. With even more limitations, a circle can be taken to represent the competitive strength of a freely growing organism. This need not be the same as, but is usually related to, the physical circle occupied by the organism.

There have been two distinct approaches to the problem of representing geometrically what happens when neighboring plants interact:

i) As growth takes place each individual in the population continues to be represented as a circle: the circles overlap and the amount of overlap is taken to describe the competition pressure imposed by one individual on another.

ii) The circles are not allowed to overlap but are distorted into polygons which then represent the space available to the organism for development.

*5.1 Competition Circles.* The first technique has been much used in forestry and a number of elaborate specific formulations have been given, differing in the details of the correspondence between the model and its application. This correspondence has three elements; for each individual we must define

1) the radius $y_i$ of its competition circle as a function of its measurable aspects;

2) an index $I_i$ of the competitive effect of its neighbors as a function of the radii of the circles;

3) the effect of a competition index on the measurable aspects of an individual.

In forestry $y_i$ is principally expressed as a function of diameter at breast height $D_i$, and of other variables such as basal area, height, and average crown radius, closely related to it. The index $I_i$ is taken as $\sum_{j \in N(i)} w_{ij} z_{ij} / \pi y_i^2$, where $z_{ij}$ is the area of overlap of circles $i$ and $j$, and $w_{ij}$ is a weighting factor, either constant (Opie, 1968; Gerrard, 1969; Keister and Tidwell, 1975), or a function of $D_i$ and $D_j$ such as $(D_j/D_i)^p$ (Bella, 1971). Subsequent growth of an individual, for example the increment in $D_i$ or its logarithm, is then expressed as a regression model in terms of $I_i$ or some function of it (e.g., a polynomial (Bella, 1971) or a logistic (Hegyi, 1973)), or of it and $D_i$ (Opie, 1968; Beck, 1973), unknown parameters in the index $I_i$ being chosen to give the best least-squares fit to data. Empirical data-dredging by multiple regression analysis abounds in the forestry literature, different forms being tried by different workers for different ages of different species in different stands. Few such studies give any thought to the statistical assumptions required for, and the statistical dangers inherent in, these analyses.

The extreme result of intense competition is death. By replacing the assumption implicit in all the above models that competition occurs simultaneously from all neighbors by the alternative assumption that competition from neighbors occurs independently, Diggle (1976) has developed an applicable spatially stochastic model for the pattern of mortality at one instant of time. As in the models described above in which the combined effect of the set of neighbors is the sum of their individual effects, so here competitors are assumed to have independent effects, the probability that an individual survives the competition from a set of neighbors being the product of the probabilities that it survives the individual competitions. The model assumes that individuals $P_i$ are situated at given points in the plane whose distances apart, $r_{ij}$, are fixed and that the sizes $y_i$ of the individuals are independent realizations of a random variable with probability density functions $f(y)$.

Denote by $q_{ij}(y_i,y_j)$ the probability that the individual $P_i$ of size $y_i$ will be eliminated in competition with an individual of size $y_j$ located at $p_j$. To survive, an individual must survive independent competitions with all other individuals, so that: $1 - q_i(y_i,\underset{\sim}{z}) = \prod_{j\neq i} \{1 - q_{ij}(y_i,y_j)\}$, where $\underset{\sim}{z}$ is used to denote the sizes of all individuals other than $P_i$. Thus $P_i$ survives with probability

$$1 - q_i(y_i) = \prod_{j\neq i} \{1 - \int q_{ij}(y_i,y_j)\ f(y_j)\ dy_j\}.$$

Specification of any competition function $q_{ij}(y_i,y_j)$, which in a homogeneous environment would be assumed to depend on the location of $P_i$ and $P_j$ only through the interpoint distance $r_{ij}$ so that $q_{ij}(y_i,y_j) = q(y_i,y_j,r_{ij})$, would enable the probability of survival of any individual to be evaluated, at the worst by numerical integration.

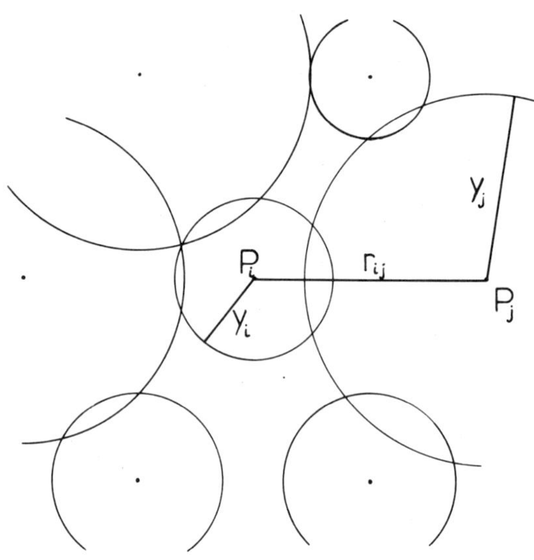


*FIG. 5.1: Overlap model for individuals on a regular lattice.*

When the $P_i$ lie on a regular lattice, so that no individual occupies a favored position and we may assume $q_i(y) = q(y)$ for all $i$, the p.d.f. of sizes of survivors can be found:

$$g(y) = \{1 - q(y)\}\ f(y)/\{1 - \int q(y)\ f(y)\ dy\}.$$

The function $q$ chosen by Diggle for the most extensive further development is the proportional linear, not areal, overlap between the competition circles, radii $y_i, y_j$, of the two individuals at $P_i, P_j$. This is piecewise linear in $y_j$:

$$q(y_i, y_j) = \begin{cases} 0 & (y_j < r_{ij} - y_i), \\ (y_i + y_j - r_{ij})/2y_i & (r_{ij} - y_i \le y_j \le r_{ij} + y_i), \\ 1 & (y_j > r_{ij} + y_i); \end{cases}$$

the first case occurring when the circles at $P_i$ and $P_j$ do not intersect, the third when the circle at $P_j$ completely includes that at $P_i$. The degree of competition induced can be adjusted by the scale factor in defining the radii relative to the physical distance between the plants, and this can be chosen to make the model approximate to competition among first-, second-, or *nth*-order neighbors.

When competition is for light, as in the specific situations which Diggle is attempting to model, it may be thought appropriate to assume that a smaller individual has no competitive influence on a larger one. Such a one-sided scheme can readily be included in Diggle's model: it can give rise to bimodality in the survivor size distribution as observed experimentally by Ford (1975) - and does so whether linear or areal overlap is used. Gates (1978) discusses a different model in which again one-sided competition can yield a bimodal size distribution.

In this form the model requires an initial size distribution to be specified for some stage of the time development of the population (Diggle chose a range of gamma distributions) and it makes the implicit assumption that the eventual outcome of life or death for any individual is determined by the state of the system at that stage. As Diggle points out the model can be adapted to furnish a dynamic representation of the development of the population. The assumption of the initial form of the size distribution can be made more acceptable by initiating the process before competition begins. A sequential discrete time process can then be formulated and simulated, the size of an individual at any stage being zero if the individual has previously been eliminated, otherwise modified by an additive or multiplicative increment from its value at the previous stage, the increment being subjected to competition from neighbors according to some appropriate mechanism. The approach is then the same as that of much of the forestry work referred to above.

More recently Gates and Westcott (1978) have formalized six basic minimal requirements which appear natural and realistic for an overlap model of competition. The resulting mathematical structure is found to be remarkably rich, but sufficiently tight to permit the rigorous demonstration of certain well-known empirical relations between typical measurements and survival frequencies.

*5.2 Dirichlet Tessellations.* Given a set of points $P_i$ in the plane, the surface of the plane can be partitioned into contiguous areas $A_i, A_i$ containing all those points nearer to $P_i$ than to any other of the given $P_j$. This procedure, shown in Figure 5.2, has many names - Voronoi polygons, Dirichlet cells or tessellations,

S - mosaic. The statistical properties of these polygons have been the subject of much work, detailed references to which may be found in Kendall and Moran (1963), Moran (1966, 1969), and Little (1974). The properties of the polygons provide a description of the stochastic process according to which the points $P_i$ were distributed. However, analytical results are in short supply even in the null case of a random Poisson forest. Simulation results have been given for that case by Crain (1972): an efficient computer program for the construction of Voronoi polygons from a given set of points has been developed recently by Green and Sibson (1978).

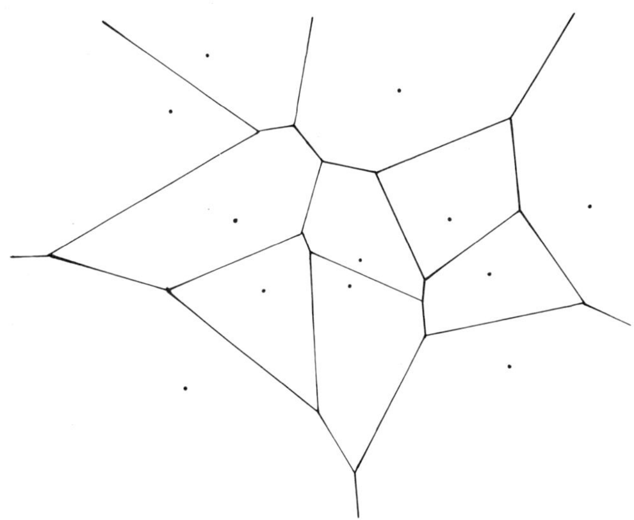

*FIG. 5.2: Dirichlet tessellation of* 12 *individuals.*

For a given set of points, for example the locations of tree seedlings or the nest sites of a strongly territorial bird species, in an area of uniform environment the polygons represent the resource available to the individuals if this is restricted by contact inhibition with immediate neighbors and if the individuals have equal competitive strength. Even if this last unrealistic proviso is accepted, it is unlikely that the area of a polygon will be the sole determinant of the strength of its occupant. Some parts of the polygon will be less accessible to the occupant than others despite the plasticity of plant material. Considerable notice is taken of this in the study of field crops to determine the inter- and intra-row spacing. The only serious attempt to model this effect is due to Mead (1969). Define a polygon by its vertices $V_j$ with interior angles $\alpha_j$ and centroid C, distance $r_j$ from $V_j$, the individual being situated at P distance r from C (see Figure 5.3). Then in addition to the area A, Mead considered the eccircularity of the polygon $D\sqrt{\pi/A}$, where D is a weighted average of the $r_j$, $\Sigma(\pi-\alpha_j)r_j/\Sigma(\pi-\alpha_j)$, and the abcentricity of the plant $\nu = r/D$ as variates on which the strength of the individual might depend. A linear dependence of carrot diameter on the two new variates and on log A was suggested. Although it is intuitive that an individual situated at the centroid of its polygon is well situated to utilize all the available resources, it has proved impossible to make any general statement about the effect on the area of a polygon of a single individual moving. In considering the accessibility of an individual to capture, either by a predator or by a mate, Cannings and Cruz Orive (1975) were able only to conclude that a move by a single individual away from a regular lattice arrangement reduced the area of its polygon.

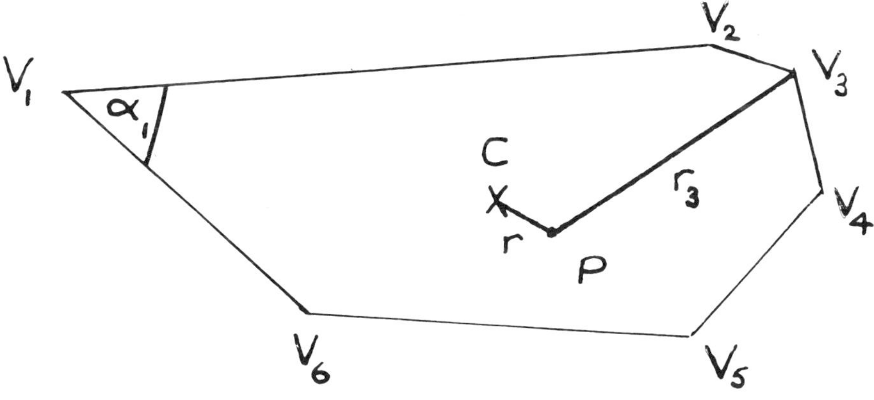


*FIG. 5.3: Components of eccircularity and abcentricity for one of the polygons in Figure 5.2.*

*5.3 Other Tessellation Models.* The main disadvantage of this representation is its inability to accommodate variation in the strength of individuals. If instead of bisecting the line joining two neighboring individuals we divide the line in some other proportion according to the relative strengths of the individuals, we obtain polygons which are not necessarily convex, and regions claimed by more than one individual. For an example of the type of ad hoc decision which requires to be made, see Adlard (1974). The idea of $\varepsilon$-tolerant or $\varepsilon$-inhibitory mosaics (Lewis and Rogers, 1974), in which polygons either overlap or are separated by a designated amount, seems likely to meet similar difficulties. A profitable line of development might come from a modified interpretation of a crystallographic model introduced by Johnson and Mehl (1939) to incorporate a time dimension for the arrival of an individual. Time is included as an extra dimension: if $P_i$ arrives at time $t_i$ then $A_i(t)$ is a circle of radius $v(t-t_i)$ if no other competing individual exists. The area $A_i$ ultimately includes all points $Q$ first covered by it: if $QP_i = r_i$, $Q$ will be in $P_i$'s polygon if $t_i+r_i/v<t_j+r_j/v$ for all other $P_j$. An individual arriving in an occupied area fails to take root. Individuals of different strengths could be modelled by an appropriate distribution of arrival times, or by allowing different individuals to have different rates of growth, or both. The first corresponds geometrically to the intersection of cones all with the same angle of revolution but with vertices which are not coplanar, the latter to cones of unequal angle. Although discussion of the Johnson-Mehl process has been restricted to random arrival times (Gilbert, 1962), there is no difficulty in simulating the behavior under biologically more reasonable assumptions.

In the growth of annual crops, for example, different times of germination of individual seeds may make a major contribution to the later success or otherwise of the plants. Currah (1977) has demonstrated such a relation with final weight of individual vegetable crops although there is a considerable scatter about the relationship because of genetic variability in the seed. Fischer and Miles (1973) used the Johnson and Mehl tessellation as a basis for their model of competition between a crop, planted on a regular lattice, and weeds, regarded as a spatial Poisson process starting later but growing faster than the crop plants.

As a standard of randomness, an L-mosaic formed by the intersection of random lines has been preferred to the S-mosaic described above "because of its greater mathematical tractability" (Pielou, 1969). For modelling non-random patterns, particularly competition, this advantage no longer pertains. Although one can visualize an L-mosaic in which segments of lines are subsequently displaced, the concept seems self-contradictory and there is no convincing mechanistic correspondence between model and practice, as I find to exist with the Dirichlet tessellations. To model the continuing development of a stand in which neighbors are interacting would require further assumptions about displacement of boundaries, but acceptable formulations can readily be imagined, and development of simulation models along these lines is quite possible as an alternative to competition circles.

## 6. SIZE FREQUENCY DISTRIBUTION

Competition alters the degree of variability which exists between individuals in a population. One aspatial phenomenon resulting from any competition process is the frequency distribution over the individuals of any appropriate measure of productivity. This is described by Harper (1967) as the emergence of a hierarchy with a few large individuals and a large number of small ones. Koyama and Kira (1956) noted that the distribution changed during the period of growth from a bell-shaped to a highly skewed and even L-shaped form. They argued that populations which become skew at an earlier stage of growth display evidence of a greater degree of competition. According to Donald (1963) evidence was then inconclusive as to whether a skew distribution does develop in an even-aged monoculture in the absence of any competition, although there are good theoretical reasons why any multiplicative process would have such an outcome. The lognormal distribution was introduced by Koyama and Kira to describe their data, although they did not attempt to test its fit. It has been used often, sometimes successfully (Bliss and Reinker, 1964; Harper, 1967), other times unsuccessfully. There are many other standard skew distributions which can be fitted to such data - gamma (Hoel, 1943), Weibull (Clutter and Allison, 1974), beta scaled to the appropriate range (Burkhart and Strub, 1973). However, none of these distributional forms can represent the apparently bimodal distributions obtained by Ford (1975) under conditions of severe competition.

At this point a note of statistical caution should be interposed, and questions asked about the aim of fitting a particular form of distribution to such data. The dangers of trying to infer the mechanism which causes a particular distribution from the statistical form of the distribution have been widely publicized in the case of the negative binomial distribution fitting numbers of individuals in disjoint areas (see Cane, 1977), but are equally applicable here. If, however, a particular biological theory of competition or a probabilistic model for a process predict that a resulting frequency distribution should have one form rather than another, particularly if there are qualitative differences between the distributions, then a lack of fit of data to a distribution casts doubt on the validity of the model or theory in that situation. The resulting frequency distribution is known for too few models.

A bimodal distribution is qualitatively different from a skew one, and hence Ford's data must cast doubt on any theory which generates a lognormal distribution. Unfortunately many data sets which are samples from unimodal populations give the appearance of bimodality, and a sensitive statistical test of bimodality is sadly lacking. However, the bimodality recorded by Ford at later stages of growth appears as a natural continuation of the pattern of growth at earlier times. Moreover, he argues that the construction of the leaf canopy in the species considered will lead to a discontinuity in the quantity of light received by different individuals and hence to a discontinuity in the distribution of relative growth rate over individuals. As was noted earlier, Diggle's competition circle models can result in a similar bimodality in the size distribution of surviving individuals provided competition is restricted to be one-sided in the sense that an individual has no competitive effect on one larger than itself. Diggle was unable to generate bimodality from an initial gamma distribution without this restriction. One-sidedness is of course a possible mathematical representation of the biological arguments adduced by Ford. It would be of equal interest to discover whether any of the other statistical models of the competition process can generate bimodality in the size distribution of the population. One further statistical warning needs to be given. In that the frequency distribution of measurements has been generated by competition between neighbors, the measurements on neighbors are not independent. It is therefore strictly invalid to test the fit of a particular form of distribution by standard methods such as $\chi^2$ or Kolmogorov-Smirnov. If a model for the underlying process has been formulated, its outcome, in this respect as in others, can be assessed against Monte Carlo repetitions of the process (see the Appendix to Diggle, 1979). There is no statistical procedure for, as well as no biological value in, testing the fit of a statistical description of the outcome, as distinct from a statistical model for the process.

## 7. POINT PROCESS MODELS

In order to study the performance of statistical analyses of competition processes it would be useful to have available a library of models which possess some degree of mathematical tractability. The first set of models, the fiction shelves if you like, attempt to mimic the outcome of competition without seriously attempting a mechanistic representation of the process. The first such point processes by which distance methods were assessed were the regular lattices: square, triangular, and hexagonal. These are too regular to be of much practical relevance and attempts have been made to modify them to make them more realistic.

One form of modification is to effect an independent displacement of each point on the lattice. In an investigation of the robustness of distance estimators of density, Holgate (1972) considers a fixed small displacement of $\varepsilon$ along a randomly chosen one of the six lattice edges through each point of a triangular lattice. If $\varepsilon$ is small enough relative to the scale of the lattice for $\varepsilon^2$ to be neglected, the expectation of the squared distance from a random point to its nearest individual can be found by exhaustive enumeration of possible cases. A more general displacement of a triangular lattice was used by Dacey (1966) as a model of the effect of the heterogeneous environment on the siting of towns according to central place theory. The size of displacement has to be such that each individual remains nearer its own original lattice point than any other: one specific model examined assigned a circular normal distribution to the displacement. Under the possibly dubious assumption that the distances from a random individual to its six nearest neighbors could be regarded as approximately independent observations from some distribution, and the distances from a random point to the three nearest individuals similarly as independent observations from another distribution, Dacey evaluated the order statistics of these distributions by numerical integration and found a close correspondence with results from simulated data. The final, and most realistic, model of this type is the thinned plantation studied by Brown and Holgate (1974), a regular square lattice from which each point independently has the same probability of being deleted.

Another form of distribution is the displacement of alternate lines in a rectangular lattice, equivalent to the superposition of two lattices, used by Stiteler and Patil (1971) in their investigation of the behavior of the variance to mean ratio of quadrat sampling in the presence of regularity. Another device they adopted was the random distribution of a fixed number of individuals within the equal areas of a regular lattice, thereby obtaining a considerable flexibility of data while retaining a pattern which is exactly regular at one particular scale of quadrat sampling with correct choice of origin and orientation. A more general version of this process in which each quadrat can independently contain one or no individual was used by Dacey (1971). The strictness of a regular lattice may also be diluted by superposition of a Poisson forest with different density (cf. Fischer and Miles, 1973). Some distance properties of such a mixture were investigated by Diggle (1975).

*7.1 Point Processes.* The procedures just described mimic the outcome instead of modelling the process, the important question. Since modelling point processes is covered by Diggle (1979) and is also the subject of an important paper by Ripley (1977), we shall avoid detailed methodology and merely describe briefly some types of model for competition.

The most severe form of competition is direct exclusion from a physical space, whether by complete occupancy, allelopathy, or aggressive territorialism. Models for this situation are termed hard-core models: individuals form a Poisson forest of independently located points except that there is a hard-core of radius R round each point which cannot overlap with the hard-core round any other point. Thus any individual has a circular neighborhood of radius 2R within which no other individual can exist. Mathematically there are several distinct ways in which the exclusion can be generated and these give rise to different properties in the realized process. As usual, biological realism varies inversely with analytical tractability. We can

1) start with a Poisson forest and simultaneously delete all points within a distance 2R of another point (Matern, 1960);

2) label the points in a Poisson forest with a time of 'birth' and delete points within a distance 2R of a point born earlier,

   a) whether or not that point has already been deleted (Matern, 1960; Paloheimo, 1971; Bartlett, 1974);

   b) if the earlier point has not already been deleted - simple sequential inhibition (Diggle, Besag, and Gleaves, 1976);

3) generate realizations of Poisson forests and accept any realizations which contain no pair of points less than 2R apart (Ripley, 1977).

Of these models 2a) and 2b) seem most realistic as mechanistic models of some ecological processes. Model 1), regarded literally, assumes that all competitions are fights to the death - of both parties; model 3) that only one of many systems tried has survived. The only properties of these processes which have been found analytically are the density and the distribution of nearest-neighbor distances, although the latter is not known for model 2b). A variant of model 2), relaxing the strict requirement of a fixed radius of inhibition R by giving it a hypothesized probability distribution, has been used by Bartlett (1974) to model the spatial distribution of nests in gulleries. Properties of the models have to be explored by simulation. Realizations of models 1) and 2a) are generated by deletions from Poisson forests, whereas model 2b) must be generated sequentially. Simulation of model 3) is less practicable since only a minute fraction of simulations give an acceptable realization if any effective, observable, degree of inhibition is present.

Hard-core models have been used in the theory of liquids (see, for example, Temperley, Rowlinson, and Rushbrooke, 1967), and summaries of many simulations of such models are to be found in the literature of chemical physics. For example, tables of probabilities of inter-point distances are given for various models by Chae, Ree, and Ree (1969). They differ in two respects from the stochastic process models being considered here. Firstly, they are based on chemical theories which postulate attractive forces operating outside the hard-core. Secondly, they depart from the probabilist's sacred canon of stationarity, that the interaction between two individuals depends only on their relative and not on their absolute positions, by explicitly introducing boundary conditions (Ripley, 1976). However these differences might be an advantage. In the first place clustering of individuals with a physical, inhibitory, size is a common biological phenomenon which could well be represented by an attractive potential. In the second place the extent of an ecologically homogeneous area in which stationarity is a reasonable assumption is usually very limited. Moreover, the fact that a finite sampling area is observed itself introduces edge effects which require awkward maneuvers in statistical analyses of data. The device by which these difficulties are avoided in stochastic theory, that of wrapping the sampling area round a torus so that opposite boundaries of the area are superimposed, always seems to me one of those supremely elegant, effective, mathematical tricks whose practical consequences cannot be visualized. The assumption of stationarity however contains one necessity for successful analysis of spatial data, which would have to be replaced if stationarity were abandoned - that of providing inbuilt replication within one realization of the process, since the process is assumed to present the same appearance wherever it is observed. It is not clear how this could be obtained from an alternative model.

A hard-core inside which there is complete inhibition and outside which there is no interaction seems too extreme for any biological organism which displays plasticity of size and shape. Models with fixed-range interactions are less extreme, in that within the neighborhood $\eta$ of fixed radius $R$ around each individual inhibition is not complete, but a competitive effect is experienced. The degree of competition is not a function of the distance $r$ between the individuals - if $r<R$ competition takes place; if $r \geq R$ it does not. If the distance between $P_i$ and $P_j$ is less than $R$, $P_i$ and $P_j$ are neighbors of one another. The total competitive effect experienced at a point in the plane is a function only of the number of neighbors it has.

This is a natural extension to a continuous space of the nearest-neighbor Markov lattice processes to be discussed in the next section. Subject to certain conditions which should always hold in ecological contexts, a valid mathematical formulation of this situation exists in the competitive situation and leads to a joint density function $f(\underset{\sim}{r}_1, \cdots, \underset{\sim}{r}_n) \propto \alpha^n e^{-\beta s}$ dependent solely on $s$ the number of neighboring pairs, conditional on $n$ individuals having been observed (Strauss, 1975b; Kelly and Ripley, 1976). Strauss relates the probability density of $s$ given a particular value of the competition parameter $\beta$ to its distribution for a Poisson forest $(\beta = 0)$, and discusses approximations to the distribution of $s$ which are useful in estimating $\beta$. The model requires an *a priori* choice of $R$, the distance to which this constant competition extends around the individual. Direct information could be obtained on this by, for instance, measuring the size of tree crown or the extent of a root system, or tracking animal movements. It is likely to be more informative to operate in the other direction by estimating $R$ from data, although the only methods currently operable is trial of various values, and drawing conclusions from that estimate about the utilization of the environment by the individual. I know of no investigations of this kind.

The nearest approach to ecological reality would come from allowing competitive effects to depend on the distance between individuals rather than purely on the number of neighbors. There are mathematical difficulties in such a general formulation. These are avoided by Ripley (1977) by imposing a hard-core center on this process. He considers the joint density $f(\underset{\sim}{r}_1, \cdots, \underset{\sim}{r}_n) \propto \alpha^n \prod_{i \neq j} h(r_{ij})$ which is zero if $r_{ij} < R$. No general method of estimating the function $h$ from data is yet known, and again we are forced to make assumptions about its specific form and discover whether such a model can generate observations which resemble the data.

We noted earlier that many observed spatial patterns have been generated over time. We noted also the hazards of trying to reach biological conclusions from spatio-temporal data, and the richness of the purely spatial models described above should reinforce these warnings. Nevertheless it would be comforting to know that these spatial processes can be the stable outcome of a reasonable process developing over time. Ripley (1977) provides this reassurance.

All these models describe the competitive influence of an individual in terms of a circular region around the individual.

The superficial correspondence with the competition circles described in Section 5 is obvious but has not been explored.

*7.2 Fitting and Testing Models.* Classical methods of statistical inference are based on a random sample of independent, identically distributed, observations. From these would be calculated a few summary, if possible sufficient, statistics representing most or all of the information provided by the observations about the assumed model. As we saw in Section 3 this approach is of doubtful validity in the context of a spatial process, although properties of quadrat counts and nearest-neighbor distances may remain a valuable guide to the class of appropriate models. Techniques are now required which assess the complete realization of the process.

One of these, largely developed and advocated by Bartlett (1964, 1974, 1975), is two-dimensional spectral analysis. Unfortunately, although the spectra of some clustering processes are known, no spectra are known for any of the above competition models. Some investigation of the appearance of the spectrum of a competition process has been carried out by Cox (1976), calculating the smoothed isotropic periodogram for a simulated simple sequential inhibition process, but he did not compare it with that from any other type of competition process.

The isotropic spectrum depends on all inter-point distances. An alternative, also proposed by Bartlett (1964), is to examine directly the frequency distribution of all inter-point distances. A refined version of this procedure has been developed by Ripley (1976, 1977). He summarizes any stationary point process by the density of points $\lambda$ and the function $K(t)$ which is such that $\lambda K(t)$ is the expected number of further points within distance $t$ of any point. For a regular spacing of points $K(t)$ will be smaller than the Poisson forest value $\pi t^2$ for small values of $t$. The derivative of $K(t)$ is proportional to the joint density for the occurrence of pairs of points distance $t$ apart. This density has been evaluated analytically for the hard-core models defined on the infinite plane.

The development of these ideas as tools for the analysis of a map of locations of all individuals in an area is discussed fully in Diggle (1979, Section 4).

## 8. LATTICE MODELS

*8.1 Binary Data.* To fix ideas let us imagine a plantation of conifers so closely planted on a square lattice that some have died. Imagine further that you have been transmuted into one of these trees in the interior of the plantation. You can look around and observe which of the other trees near you are alive or dead, and you ask yourself, given that information, whether you are alive or dead. Clearly, if a number of your immediate neighbors have been killed by competition, there must be a successful competitor and the chance that you are it is fairly high: by contrast if all your neighbors are alive then there is a greater probability that your obituary has already been written. If we denote the random variable at the $(r,s)th$ point of the lattice by $Y_{rs}$ and the two states of this binary process by 0 and 1, we want to model $P(Y_{rs}=0)$ given that $Y_{ij} = y_{ij}$, the states at all the other lattice points. In the simplest situation the dependence will be only on the states of immediate neighbors - first order interaction.

In the corresponding situation in a non-spatial time process we can formulate any model we choose for $P(Y(t) = 0 | Y(t-1) = y_{t-1})$, since in time the state at time $t$ does not causally affect the state at time $(t-1)$. In space, even in one-dimensional space, there is no such causal ordering, and in two or more dimensions there is no spatial ordering either. (Barnett (1976) surveys proposals for avoiding this latter difficulty in statistical analyses.) If in space $Y_{rs}$ is dependent on the state of $Y_{r(s+1)}$, so is $Y_{r(s+1)}$ dependent on the state of $Y_{rs}$. This interlocking dependence imposes stringent restrictions on the possible form of

$$P(Y_{rs} = 0 \mid y_{(r-1)s}, y_{r(s-1)}, y_{r(s+1)}, y_{(r+1)s}).$$

Indeed Besag (1972a) has shown that, when interaction affects only the four nearest neighbors on a square lattice and does so isotropically equally in all four directions, the only possible distribution is the autologistic defined, in a less cumbersome notation, by:

$$P(Y=y \mid t,u,v,w) = \frac{e^{-y[\alpha+\beta(t+u+v+w)]}}{1+e^{-[\alpha+\beta(t+u+v+w)]}},$$

where $t$, $u$, $v$, and $w$ denote the given values for four nearest neighbors. Expressed in terms of the example above, this means that the probability that a tree with $n$ living neighbors (out of 4) is alive is $\exp(\alpha+\beta n)/[1+\exp(\alpha+\beta n)]$. In this formulation competition requires $\beta$ negative; in the absence of interaction $\beta$ is zero.

Equal dependence on all four first-order neighbors of a square lattice is the simplest form of interactive model, and extensions are readily made to include directional effects, to allow more distant neighbors to interact, or to other regular and to irregular lattices. Directional effects are incorporated by allowing a different $\beta$ parameter for the E-W neighbors from that for the N-S neighbors. The effect of the four diagonal neighbors is incorporated in an isotropic autologistic second-order model by extending the exponent from $\alpha+\beta n$ to $\alpha+\beta n_1 + \gamma n_2$, where $n_1$ and $n_2$ are respectively the number of living first-order and second-order neighbors. As can be seen from its form, this extension assumes that pairwise interactions are themselves independent and that the effect of the four diagonal neighbors is additive, in the logistic scale, to that of the rectangular neighbors - the same type of assumption made in the competition circle models of Section 5. Interaction among triples

```
  X
X X
```

and quadruples

```
X X
X X
```

can be modelled by adding to the exponent $\xi n_3$ and $\eta n_4$ respectively where $n_3$ is the number of the twelve pairs of neighboring neighbors of the subject individual that are both alive, and $n_4$ the number of the four triples that are all alive.

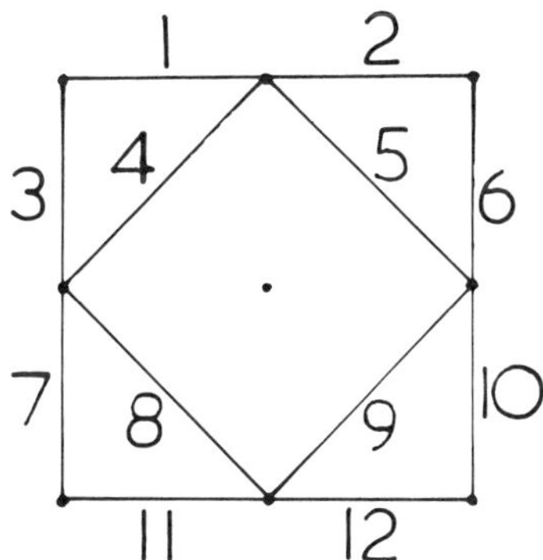


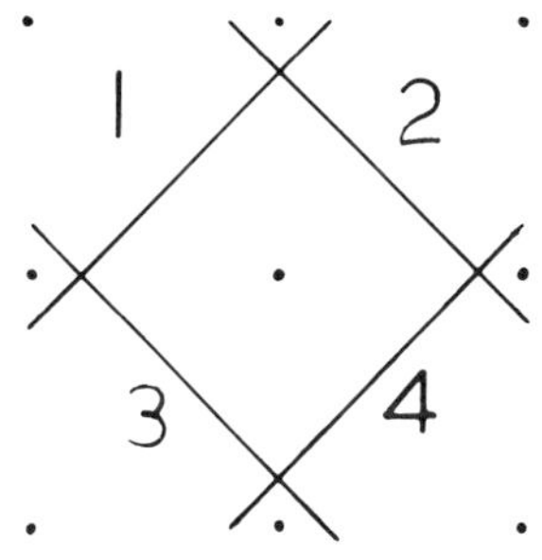


*FIG. 8.1: Extended sets of interactions on a square lattice.*

These extra terms correspond to the second and third virial coefficients in the expansion of the radial distribution function in liquid theory. Again directional effects could be included by assigning different parameters to different orientations of triplets and quadruplets, but ecological data will seldom be sufficiently precise to estimate efficiently the increased number of parameters and to test whether they are necessary. The regular triangular lattice has six first-order neighbors. The isotropic model with only pairwise interaction is autologistic, the probability that a tree with $n$ (out of 6) living neighbors is alive being $\exp(\alpha+\beta n)/[1+\exp(\alpha+\beta n)]$ as before. The extended model allowing interaction among triples has exponent $\alpha+\beta n_1+\gamma n_2$ where $n_2$ is the number of the six pairs of neighboring neighbors with both alive. This model could readily be fitted, and the significance of $\gamma$ tested, from a well-designed experiment. Such an investigation of the additivity of competitive effects would be valuable in view of the number of empirical models for forest stands which make that assumption.

Given a set of binary data on a lattice, we wish to test which order of competition is required and to estimate the parameters in the appropriate model. We can categorize each interior point of the lattice by the number of live neighbors (0 to 4 for a square lattice, 0 to 6 for a triangular one) and by whether the point itself contains a live or dead individual. It would, however, be wrong to analyze the resulting $2\times5$ or $2\times7$ contingency table by a conventional $\chi^2$ test because of the lack of independence of the observations, exactly as was noted above in the discussion of Pielou's test of segregation of two species. The test can be carried out on any selection of *independent* individuals. The limited range of dependence assumed by these models allows regular selection of such sets of individuals according to certain coding schemes introduced by Besag (1972b). The points labelled x in Figure 8.2 are independent under the stated models. If we are going to utilize respectively only 1/2, 1/4, or 1/3 of the interior observations and divide them into 10, 10, and 14 categories respectively, it is clear that large lattices will be required for any analysis of these models. Besag found a $24\times24$ lattice too small for a second-order model. Of course more than one coding scheme is available for any set of data. Thus, for a first-order model on the square for example, we could carry out a second analysis on the o points in the lattice conditional on thei neighbors, the x points. Four different analyses can be computed on the second-order square lattice, three on the first-order triangular lattice. Although these analyses are likely to be dependent so that it is invalid to combine them formally, it is still valuable to carry out all the analyses and examine them informally for consistency.

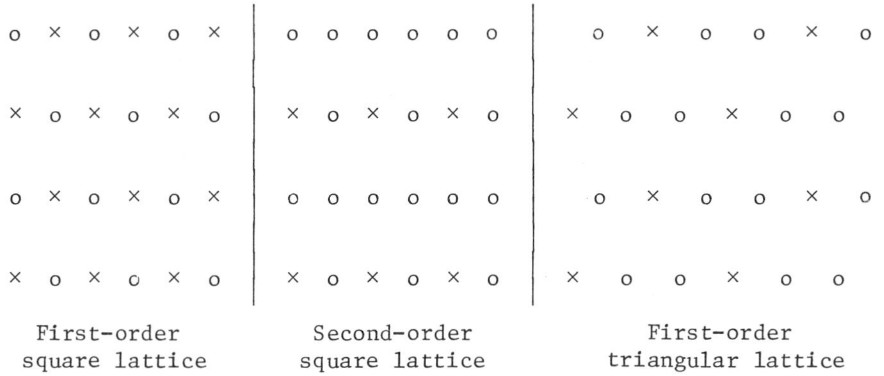

| | | |
|---|---|---|
| o × o × o × | o o o o o o | o × o o × o |
| × o × o × o | × o × o × o | × o o × o o |
| o × o × o × | o o o o o o | o × o o × o |
| × o × o × o | × o × o × o | × o o × o o |
| First-order square lattice | Second-order square lattice | First-order triangular lattice |

*FIG. 8.2: Coding schemes for automodels.*

For each anslysis a $\chi^2_4$ test of randomness can be performed, if thought necessary, on the 2×5 contingency table. If randomness is rejected, estimates of $\alpha$ and $\beta$ are found by maximizing the likelihood over the (independent) coded sites:

$$\prod_{i \text{ coded}} \exp[x_i(\alpha+\beta n_i)]/[1+\exp(\alpha+\beta n_i)],$$

where $x_i = 1$ or $0$ according as the individual at site $i$ is alive or dead, and $n_i$ is the number of its live neighbors. Expected values for the cells in the contingency table are calculated from these estimates $\hat{\alpha}$, $\hat{\beta}$, and a $\chi^2_2$ can be carried out of the fit of the first-order model to the data. However, this 2×5 contingency table relies on the independence of the coded sites, which will not hold if the first-order model is inappropriate. A non-significant value of this $\chi^2_2$ should not therefore be taken as conclusive evidence of the fit of the first-order model. Instead, second-order coding should be used if the lattice is large enough to permit this, both first- and second-order models fitted, and the standard likelihood-ratio test between them assessed as a $\chi^2_1$ to determine the acceptability of the first-order model. A second-order model partitions the 2×5 contingency table into a 2×5×5 table to which a $\chi^2$ test can, in turn, be applied, after suitable pooling of classes to make expected numbers sufficiently large, in order to test the fit of the second-order model. The details can be followed in an example of a contagious process on a $10 \times 940$ lattice given by Besag (1974).

On a triangular lattice the distinction between first-order and second-order competition is greater than on a square lattice. If second-order competitors are considered to be the six individuals at distance $d\sqrt{3}$ from the subject plant, then the appropriate coding scheme is the same as for the square lattice observing one-quarter of the plants, but if they include also the six individuals at distance $2d$ the most efficient coding scheme utilizes only every seventh plant. The dangers of accepting a first-order model without testing it against a second-order model remain.

A possible alternative method of fitting automodels is to forget the dependence between neighbors and to maximize what could be called the pseudo-likelihood obtained by multiplying together the conditional probabilities for all sites. This seems to break all canons of statistical behavior by treating the data set as if it contained much more information than it in fact does. However it seems to work, although considerably more experience is needed of its behavior with known systems before it can confidently be recommended.

Another method of estimating $\beta$ for a square lattice is given by Strauss (1975a) using the join count statistics described in Section 4. He gives three approximations to the distribution of the number of first-order BB joins, and provides estimates from functions of the first three cumulants of the distribution. Since the approximations diverge from the exact distribution in different ways, confidence in their validity is generated if the three estimates of $\beta$ are reasonably close.

The restriction of conditional binary lattice processes to the autologistic is a consequence of the Hammersley-Clifford theorem discussed in detail by Besag (1974). This theorem has similar consequences for conditional models for other types of lattice data. Binomial and Poisson automodels have been defined by Besag. The latter, in particular, seems potentially a useful model for a lattice of small areas. It is defined only for competition processes, the observation at any point having, in the homogeneous and isotropic case, a Poisson distribution with mean $\alpha\beta^{\bar{x}}$ where $\beta<1$ and $\bar{x}$ is the average of the values at the neighboring sites. An interesting link between lattice and point process competition models is that the auto-Poisson tends to Strauss's fixed range model (see Section 7.1) as the area of the lattice grid tends to zero (Besag, 1977).

*8.2 Normal Data.* The other important form of conditional distribution is for normal data, for which the Hammersley-Clifford theorem imposes no restriction on the general linear model for pairwise interaction: that the observation $Y_{rs}$ at a particular lattice point, conditional on the values of $y_{uv}$ at all other lattice sites $(u,v)$ is normally distributed with mean $\mu_{rs} + \sum_{u,v} \beta_{(rs)(uv)}(y_{uv}-\mu_{uv})$ and variance $\sigma^2$. The notation with double suffices is cumbersome, and also has an air of restriction to a regular lattice, whereas, as with the lattice test statistics of Section 4.4, the structure of any lattice can be incorporated in the choice of the dependency parameters $\beta$. For these reasons we shall now refer to the *ith* or *jth* point on the lattice, the spatial relationship between points being subsumed in the parameters:

$$E[Y_i \mid \text{all other } y_j] = \mu_i + \sum_{j \neq i} \beta_{ij}(x_j-\mu_j), \quad \text{where } \beta_{ji}=\beta_{ij}.$$

Unlike the other conditional models, for the normal the joint distribution of all the observations on the lattice, can be written down:

$$P(y) = (2\pi\sigma^2)^{-n/2} \mid \underset{\sim}{I} - \underset{\sim}{B} \mid^{1/2} \exp\{-(\underset{\sim}{y}-\underset{\sim}{\mu})'(\underset{\sim}{I}-\underset{\sim}{B})(\underset{\sim}{y}-\underset{\sim}{\mu})/2\sigma^2\},$$

where $\underset{\sim}{B}$ is the matrix of dependencies $\beta_{ij}$ with $\beta_{ii} = 0$. The means $\mu_i$ can be made to reflect environmental trends: for a homogeneous process, $\mu_i=\mu$. Clearly $\underset{\sim}{B}$ contains more elements than there are observations. Within this framework, modelling the competition process is reduced to the choice of form of the $\beta_{ij}$, whose values will be negative or zero if all direct interactions are competitive. In addition to the symmetry restriction $\beta_{ij} = \beta_{ji}$, we should note that $(\underset{\sim}{I} - \underset{\sim}{B})$ has to be positive definite.

Even for the isotropic regular lattice with first-order competition, so that there is only a single competition parameter $\beta$ (restricted for a square lattice to $|\beta| < 1/4$ by the positive definite requirement), maximum likelihood estimation from the joint density is complicated by the form of the matrix $(\underset{\sim}{I} - \underset{\sim}{B})$. For an $r\times s$ square lattice this matrix is $rs\times rs$ with 1 on the main diagonal, $\beta$ on four other diagonals except where distorted by edge effects, and 0 otherwise. Thus for a $3\times 3$ square lattice:

$$\underset{\sim}{I} - \underset{\sim}{B} = \begin{bmatrix} \underset{\sim}{A} & -\beta\underset{\sim}{I} & 0 \\ -\beta\underset{\sim}{I} & \underset{\sim}{A} & -\beta\underset{\sim}{I} \\ 0 & -\beta\underset{\sim}{I} & \underset{\sim}{A} \end{bmatrix} \quad \text{where } \underset{\sim}{A} = \begin{bmatrix} 1 & -\beta & 0 \\ -\beta & 1 & -\beta \\ 0 & -\beta & 1 \end{bmatrix}.$$

For a very small lattice, such as a $3\times3$, the likelihood function can be evaluated directly. For a large lattice, asymptotic estimates can be obtained by neglecting the distortions in the matrix caused by edge effects, or alternatively the edge effects can formally be eliminated by considering opposite edges of the lattice to be joined up, thus making $(\underset{\sim}{I} - \underset{\sim}{B})$ a circulant matrix. The properties of such processes are discussed by Moran (1973): the method of obtaining the maximum likelihood estimate of $\beta$ and its variance is given by Besag and Moran (1975).

Alternatively the coding procedure defined for the autologistic model can be used. The procedure is exactly the same, but estimates are much simpler than for the autologistic model since the conditional models are classical regression models, with one independent variable for the isotropic first-order model, more than one if competition is permitted with more distant neighbors. With the same notation as before, if $x_i$ is the mean of the observations neighboring the *ith* of $r$ independent lattice points at which observation $y_i$ is made,

$$\tilde{\beta} = \sum_{i=1}^{r} y_i \, (x_i - \bar{x}) / \sum_{i=1}^{r} (x_i - \bar{x})^2 .$$

The variance of $\tilde{\beta}$ over all possible realizations of the process is not of the classical form because of the variability in $x_i$. It is shown by Besag and Moran to be approximately $\beta\sigma^2/r\nu\gamma$ when $\beta \neq 0$, and $1/r\nu$ when $\beta = 0$, when $r$, the number of coded points, is large, $\nu$ being the number of neighbors of each site, $\sigma^2$ the conditional variance of an observation and $\nu$ the covariance between $y_i$ and any of its neighbors. The coding technique is shown to be fully efficient as a test of randomness only for first-order competition on a square lattice of those we have considered: the efficiency is 2/3 on a triangular lattice, and 1/2 for second-order competition on a square. Its efficiency in the non-random case is shown to be reasonable only when $|\beta|$ is fairly small, i.e. when competition is not very intense.

The formulation of lattice models in terms of conditional models derives its impulse from consideration of one particular individual and the question of how that one depends on the others. In many ways it is more natural for statisticians, with a body of experience in stochastic processes and time series analysis, to model the process as a whole. In many applications the lattice is not a lattice of points but a lattice of observational areas on which the aggregate of a continuous process has been observed. In such cases the idea of dependence on nearest neighbor individuals is less natural. For all these reasons the first models of spatial lattice processes were joint models describing the whole system by its variance and autocorrelation structure, which of course completely determines a multivariate normal distribution after the mean has been eliminated by some process for removal of trend. The seminal paper on such processes is that by Whittle (1954).

The model is a simultaneous autoregressive one linking the random variables: $Y_i = \mu_i + \sum_{j \neq i} \beta_{ij}(Y_j - \mu_j) + \varepsilon_i$ not restricted to $\beta_{ji} = \beta_{ij}$, the $\varepsilon_i$ being assumed independent $N(0,\sigma^2)$. With the mean suppressed, this can be written as $\underset{\sim}{Y} = \underset{\sim}{B}\underset{\sim}{Y} + \underset{\sim}{\varepsilon}$. The corresponding joint probability distribution is

$$P(\underset{\sim}{y}) = (2\pi\sigma^2)^{-n/2} \mid \underset{\sim}{I} - \underset{\sim}{B} \mid \exp\{-(\underset{\sim}{y}-\underset{\sim}{\mu})' (\underset{\sim}{I}-\underset{\sim}{B})' (\underset{\sim}{I}-\underset{\sim}{B}) (\underset{\sim}{y}-\underset{\sim}{\mu})/2\sigma^2\}.$$

The matrix $(\underset{\sim}{I} - \underset{\sim}{B})$ of the conditional model is replaced here by $(\underset{\sim}{I} - \underset{\sim}{B})' (\underset{\sim}{I} - \underset{\sim}{B})$. In this formulation the dependence has the nature of a correlation coefficient, whereas in the conditional model it is more akin to a regression coefficient. The only model which has been much studied is that in which competition is determined by a single unknown parameter $\beta$, the matrix $\underset{\sim}{B}$ being $\beta\underset{\sim}{W}$ where $\underset{\sim}{W}$ is a known matrix of weights (cf. the lattice statistics in Section 4.4). If we denote $\underset{\sim}{I} - \beta\underset{\sim}{W}$ by $\underset{\sim}{A}$, the maximum likelihood estimators of $\sigma^2$ and $\beta$ are related (Mead, 1967; Ord, 1975a) by:

$$\hat{\sigma}^2 = \underset{\sim}{y}' \, \hat{\underset{\sim}{A}}' \, \hat{\underset{\sim}{A}} \, \underset{\sim}{y} \, / \, n$$

$\hat{\beta}$ the value that minimizes $\hat{\sigma}^2 \, |\underset{\sim}{A}|^{-2/n}$.

Computational methods for obtaining these estimates have been given for medium-sized lattices by Ord (1975a) and Cliff and Ord (1975) based on finding the eigenvalues $\lambda_i$ of $\underset{\sim}{W}$ and noting that $|\underset{\sim}{A}| = \prod_{i=1}^{n} (1 - \beta\lambda_i)$. For large lattices approximate estimates can be obtained by approximating the awkward term $-\ln|\underset{\sim}{A}|/n$ by a term in the power-series expansion of the spectral function: a method due to Whittle (1954) and discussed further by Besag (1974) and Bartlett (1975). The same techniques can be applied to the conditional autonormal, save that $|\hat{A}|^{2/n}$ is replaced by $|\underset{\sim}{I} - \underset{\sim}{\hat{B}}|^{1/n}$.

The estimate $\hat{\beta}$ is a measure of the degree of competition. A proper study of biological competition will explore how $\hat{\beta}$ changes with aspects of the competition such as spacing. Analysis by $\hat{\beta}$ of experiments on vegetable crops grown in triangular lattices, described by Mead (1968), give rather ambivalent results. Earlier Monte Carlo studies on small lattices (Mead, 1967) had shown the distribution of $\hat{\beta}$ to be highly skew. A family of transformations, analogous to Fisher's transformation of the sample correlation coefficient, was introduced to convert $\hat{\beta}$ to an approximation of a normal variable: two parameters in the family were fitted for a given lattice pattern by goodness-of-fit of the transformed Monte Carlo results to a normal distribution. For the transformation to be successful the distribution of plant weights has to be reasonably symmetric, so that from the discussion of Section 6 some preliminary transformation of the variate may be necessary. With experimental data the estimates of competition, while showing the anticipated pattern of change with spacing, were highly variable, although this may be partly due to the inefficient method of estimation used by Mead. Further studies by Ross-Parker (1975) have shown that for small lattices $\hat{\beta}$ can have a bimodal distribution, even when $\beta$ is close to zero, under either the conditional or joint model: the choice of mode can usually be readily determined on empirical grounds, and a transformation made from $\hat{\beta}$ to $\beta^*$ which has approximately a normal distribution and hence allows statistical comparison of competition effects.

The question of which type of model - joint or conditional - to use for continuous normal data is a vexed one. Efficient parameter estimation requires knowledge of the joint distribution, as does knowledge of the marginal distribution of a single individual and of the overall covariance structure, and generation of simulated realizations. On these grounds a model which purports to describe the global properties of the system is attractive. On the other hand the conditional distributions are based on quasi-mechanistic models of the local biological interactions which generate the system. There exists in a natural way hierarchical families of conditional models allowing interaction to take place at greater and greater distances. These facts give conditional distributions great intuitive appeal. From any spatio-temporal normal model,

$$\underset{\sim}{Y}(t) = \sum_r \underset{\sim}{B}(r)\, \underset{\sim}{Y}(t-r) + \underset{\sim}{\varepsilon}(t)$$

both a conditional model (Bartlett, 1971) and some kind of joint model (Ord, 1975b) can be derived for the purely spatial relation at a given time. However, the joint model corresponding to a first-order conditional model need not be of nearest-neighbor type. Ord shows that the conditional formulation is the one which arises naturally when the value taken by an individual depends on the past history, and not directly on the present state, of its neighbors. It seems likely that this will apply to all plant studies. The joint formulation is the natural model when the interaction is strictly contemporaneous. This seems appropriate to interaction between animals.

## 9. EXPERIMENTAL DESIGN

Left to herself Nature provides us with situations in which a plant is affected by competition from different sizes and species of neighbor. Field observations always have the advantage of referring to the situation of direct practical interest, but Nature does not usually conduct her experimentation with a view to revealing her own secrets most efficiently. During the natural development of a regularly planted pine forest for example the death of some trees reduces the degree of competition experienced by their neighbors, but not in any balanced way.

Many experiments, some involving considerable ingenuity in the construction of inter- or partially-connected boxes, have been performed to investigate the contribution of root competition for water and nutrients and of canopy competition for light. For an early survey see Donald (1963): a recent example that appeals to me for its construction is described by Fitter (1976). Many of these are designed and analyzed in standard statistical ways. Two types of design problem are however specific to this area and should be discussed here.

The first of these comprises spacing experiments in monocultures in which the effect of different spacings between plants is investigated within a single block of land. In such experiments plants are situated on a rectangular grid, and the factors of interest are the area $A$ round each plant and the rectangularity $B$ of that area. A class of systematic designs was introduced by Nelder (1962) in which the grid of points at which plants are situated form the intersections between radii and circumference of concentric circles. In the first two the angle between the radii is kept constant and the distance between circles either increases with increasing radius $r$ in such a way as to keep the shape of the region round each plant approximately square $(B=1)$ while $A$ increases with $r^2$, or decreases with increasing $r$ so that $A$ remains constant while $B$ increases with $r^2$. In two other designs the angle between radii changes so that both $A$ and $B$ change systematically over the area. Another design suggested by Nelder consists of a log-log grid set diagonally at 45° to the sides of the block so that area increases from one side to the other of the block and retangularity from bottom to top. These blocks can be replicated in different orientations, parameters of a yield-density-rectangularity relationship estimated from each block, and the estimates from different blocks treated as a random sample, which will be justified provided no periodic variation occurs in the environment to bias the estimates from systematic sampling.

The restrictive assumptions required by the systematic designs led Lin and Morse (1975) to propose a randomizable design on a rectangular lattice. A $3\times3$ design has replicate blocks of 9 plots laid out in 3 rows of 3 columns, rows having constant vertical spacing between plants, columns constant horizontal spacing. These plots are of different sizes so that the crucial assumption to be tested in this case is that of equal variances. If it holds, the interaction of each factor with blocks provides the error appropriate to the test of that factor in a standard analysis of variance.

The other type of design involves the measurement of individual plants surrounded by different numbers of neighbors or of different proportions of neighbors of different types. Some experiments do precisely that, harvesting only the plant on the central 'hill'. Sakai (1955) surrounded single rice plants with six neighbors on a triangular lattice, $m$ being of a weakly competitive variety, $(6-m)$ strongly competitive. Initially he used only the values $m=0,3,6$, but later (1957) he experimented with sets of 14 plants in a design in which each species in turn was measured surrounded by all values of $m$ from 0 to 7: yield of both species increased linearly with $m$. Schutz and Brim (1967) suggested using only four types of unit on a square lattice:

```
A A A    B B B    A A A    B B B
A A A    B A B    A B A    B B B
A A A    B B B    A A A    B B B
```

provided competitive effects are assumed linear with respect to the number of competitive hills. With such experiments only one-seventh or one-ninth of the experimental material is actually measured. The design problem is to overlap such units so that balance is retained while as large a proportion as possible of hills become testable. Since competition is assumed to be contributed solely and equally by a specified set of neighbors experimentation on a triangular lattice has advantages, and the honeycombing of the lattice must be utilized.

Beehive designs were developed by Martin (1974) and explored further by Veevers and Boffey (1975). These designs comprise two disjoint regular hexagonal arrays of hills, each hill occupied by a plant either of species A or B, the arrays being complementary under exchange of the symbols A and B, the margins acting as guards and not being measured. The hexagons investigated had 6-fold symmetry, being constructed from six replicates of the same basic triangle rotated about the central axial, non-testable, point of the hexagon. The number of hills $r$ on the side of each triangle defines the 'radius' of the design; the number of testable hills $k_i$ in each triangle which are surrounded by $i$ hills of the opposite species defines the (vector) profile $\underset{\sim}{n} = 6\ (k_0,k_1,\cdots,k_6)$ of the design, $n_i$ being the total number of testable hills with $i$ different neighbors. Ideally one would like a balanced design $(n_i = n)$ with equal information on each degree of competition. This is only possible, and need not even then be attainable, if $r = 7j$ or $7j + 1$: a design which minimizes $\sum_{i=0}^{6} n_{\underset{\sim}{i}}^2$ with $n_i = 6k_i$ and $\sum_{i=0}^{6} n_i = 3r(r-1)$ is said to be levelled. Martin gave levelled designs of radius 4 and 5 and asserted that none such existed of radius 6. Veevers and Boffey confirmed Martin's assertion and showed that no levelled design existed for radii 7 and 8. They tabulated the nearest to levelled designs for radii 4 to 8.

Six-fold symmetry is a very stringent condition and Veevers and Boffey (in preparation) have searched for balanced designs in beehives with 2- or 3-fold symmetries. Although this freedom allows designs nearer to balance to be achieved no balanced beehive has yet been found.

Since balance is the main property being sought, a better approach might be to seek a small balanced array, with the property of interlocking with repetitions of itself in such a way as to retain balance. A simple pattern of rows of two alternating patterns achieves this (Boffey and Veevers, 1977; Veevers and Boffey, in preparation) and no other way of building elementary arrays has been found. Together with its complement such a pattern provides a design which is simple to lay out, and is highly efficient in its proportion of testable hills despite the fact that single plants in each row between the basic pattern and its complement are not harvested - unharvested plants are bracketed (A). Twelve such designs exist: one example is given below, generated by the two strings 0 0 0 1 1 1 1 0 and 0 1 1 1 1 1 0 1 1 and their complements. This yields the design whose first rows are:

```
  (B) (B) (B) (A) (A) (A) (A) (B) (A) (A) (A) (B) (B) (B) (B) (A) (B)
(B)  A   A   A   A   A   B   A  (A)  B   B   B   B   B   A   B  (B)
  (B)  B   B   A   A   A   A   B  (A)  A   A   B   B   B   B   A  (B)
(B)  A   A   A   A   A   B   A  (A)  B   B   B   B   B   A   B  (B)
  (B)  B   B   A   A   A   A   B  (A)  A   A   B   B   B   B   A  (B)
```

each pair of rows forming one replicate of all patterns. Choice of number of replicates is open to the experimenter: the smaller the experiment the higher the proportion of plants used as guards and not harvested. One very minor disadvantage of this layout is the imbalance in arrangement of the two types of plant between the basic pattern and its complement: in the example above 10 A's are harvested for every 4 B's in the left-hand section, vice versa in the right.

These elegant designs can be used both for inter-species competition experiments and for intra-species competition experiments regarding B as an unplanted site. In the latter case we do not need to ensure that an equal number of holes exist with each number of plants as neighbors, and a slightly more compact design can be constructed.

Robertson and Ford (in preparation) carried out such an experiment using replications of the basic pattern given below read alternatively from left to right and right to left separated by empty 'guard rows' required for balance.

```
G   G   G   G   G   G   G
  G   G   G   G   G   G   G
G   A   A   A   B   A   A
  A   B   A   A   B   B   G
G   B   A   A   A   A   B
  A   A   A   A   A   A   G
G   B   A   A   A   B   B
  B   B   A   A   B   B   G
G   A   A   A   B   A   A
  G   G   G   G   G   G   G
G   G   G   G   G   G   G
```

Eight replications arranged as 4 'rows' by 2 'columns' of the above design gives 224 testable out of 513 hills. This can be improved on by a design (Veevers, personal communication) formed by repetition of the arrangement

(B) A A A B B A B B B A A A (A)

(A) A B B B A B B A A A A A (B)

guard rows at top and bottom being repetitions of these rows and at the sides being alternatively A and B as indicated, the alternation of the rows being reversed in any replicate to the side of that shown. This arrangement gives 224 testable hills out of 484. Either design can be readily laid out by use of a template; Veevers' design is the more suitable for field experimentation since a single key for vertical rows specifies every point in the design. Fascinating combinatorial problems remain to be solved, but are unlikely to provide much increase of efficiency in designs for testing competition between nearest neighbors. To achieve balance of further neighbors would be much more difficult, as would balance over a set of coding schemes of independent lattice points. Developments on the square lattice are reported by Street and Macdonald (1979).

## 10. YIELD MODELS

Competition between individuals will influence the total output from an area, an effect which is of direct practical importance to agriculturalists and foresters. Many annual crops are normally grown as monocultures: the farmer can influence the yield obtained by increasing the availability of growth-limiting factors and also by modifying the spatial distance of one plant from its neighbors. The natural type of experiment by which to investigate the latter effect is to sow different plots in different patterns and at different densities, and to model the yield from the plot as a function of the density and spacing of individuals. The 'yield' of a plant is in reality multivariate with competition occurring within the plant for available resources, and detailed specific models have been proposed for different univariate components of the yield. When the density is very low there is no competition between plants and the yield per plant, $w$, is independent of the density $\lambda$: as density increases yield per plant decreases until the total yield from the area remains constant or even decreases when density is further increased.

The model:

$$w^{-\theta} = \alpha + \beta\,\lambda^{\phi} \qquad (1)$$

proposed initially by Bleasdale and Nelder (1960) has been found empirically to fit data from a wide variety of crops. With four parameters $\alpha,\beta,\theta,\phi$, this is a very flexible model, and includes as special cases earlier models:

$$w = K\lambda^{A} \quad \text{(Kira, } et\ al.\text{, 1953)} \tag{2}$$

$$w^{-1} = \alpha+\beta\lambda \quad \text{(Shinozaki and Kira, 1956).} \tag{3}$$

These simpler models have a restricted but useful range of applicability particularly for total crop yield. The power law cannot describe a situation in which plants, whether spatially dependent or not, reach a ceiling. The linear reciprocal relation is based on the concept of a ceiling yield but is restricted in shape and cannot describe a situation in which yield per unit area may decline at high densities.

The general model (1) gives this when $\theta<\phi$. Bleasdale (1967), arguing that sufficient data seldom exist to allow accurate determination of both $\theta$ and $\phi$, and that only the ratio $\theta/\phi$ is important, proposed a simplified equation $w^{-\theta} = \alpha+\beta\lambda$. In either case the parameters $\alpha$ and $\beta$ dominate the behavior of response curve respectively at low and high plant densities.

General comparison between data sets is meaningful only when the shape parameters $\theta,\phi$ are constant in the different sets. If this does not hold, attention must be given to specific features of interest, such as the density at which a certain proportion of maximum yield is obtained. A good discussion of these and other models, and of their biological validity or lack of it is given by Willey and Heath (1969).

Methods of fitting general inverse polynomials were discussed by Nelder (1966). If $w$ has expectation $W=1/(\alpha+\beta\lambda)$ and variance proportional to $W^2$ then he recommends the use of weighted least squares. With the additional assumption that log W is normally distributed with constant variance maximum likelihood estimation can be used. For this model Nelder suggests that weighted least squares will provide a good approximation to, and involve much less computing than, maximum likelihood. Mead (1970) gives similar consideration to fitting Bleasdale and Nelder's general model and makes the same recommendation for fitting $\alpha$ and $\beta$. The shape parameters $\theta$ and $\phi$ must first be fitted by a general minimization routine.

Simpler methods of fitting these curves have sometimes been adopted. Considerable caution is however needed. The response curve $W = 1/(\alpha+\beta\lambda)$ is of essentially the same form as the Michaelis-Menten equation of enzyme kinetics $y = Vx/(K+x)$ for which a great amount is known about the efficiency and robustness of different fitting techniques. The obvious inverse linear plot of $1/y$ against $1/x$, recommended by de Wit (1960) and used for example by Fery and Janick (1971) for fitting the model: $w^{-1} = \alpha+\beta\lambda$, is known to be virtually uniformly the worst possible method (Colquhoun, 1969). Weighted least squares and maximum likelihood are known also to be lacking in robustness against departure from the assumptions under which they are derived. An intriguing non-parametric method has been proposed by Eisenthal and Cornish-Bowden (1974) and discussed by Porter and Trager (1977). The corresponding method for model (3) would consist of using each pair of observations $(\lambda_i, w_i)$, $(\lambda_j, w_j)$ to obtain a pair of estimates $(\alpha_{ij}, \beta_{ij})$ by solution of the pair of equations: $1/w_i = \alpha+\beta\lambda_i$: $1/w_j = \alpha+\beta\lambda_j$. The estimates $\alpha^*, \beta^*$ are the medians of the sets of $\alpha_{ij}, \beta_{ij}$ as every observational point is paired with every other. This estimate has been shown by Monte Carlo simulation to be very robust over a wide range of values of $\sigma^2$ and of error distributions (Lamb, Cormack and Henderson, in preparation). The principle is that of the Theil estimator for linear regression (Theil, 1950).

These models are static descriptions of the final outcome and permit investigation of the time development of competition only by the empirical comparison of the estimated parameters from experiments grown at different spacings and harvested at different times. A dynamic model can be built up from the static logistic model by postulating for example that

$$\frac{dw}{dt} = r\,w\,(F - w - \phi_c)$$

where $F$ is the constant effect of growth-limiting factors on isolated plants, and $\phi_c$ is a competition function dependent on the spacings. This is best as a model for the development of an individual but could be adapted as a yield model.

Much of the development of this global approach to the study of competition has been influenced by the work of de Wit (1960) which relates the effects of inter-species competition to that of intra-species competition. The basic assumption is that the yield from a species is proportional to the 'area' available to it. Although this is described as physical area the principle is essentially the same if 'area' is taken to be a measure of volume of resources. Without inter-specific competition the area $A_i$ available to the *ith* species $S_i$, and hence its yield $Y_i$, will be in proportion to the number of seeds sown or plants planted $X_i$. If instead a plant of $S_i$ occupies an effective area $b_i$, then the yield $Y_1$ from the first species in a mixture of two will depend on the relative crowding coefficient $k_{12} = b_1/b_2$ of $S_1$ with respect to $S_2$ and on the proportion $x_1 = X_1/(X_1+X_2)$ of seeds that are of $S_1$ according to:

$$Y_1 = \frac{M_1 k_{12} \, x_1}{[k_{12}-1]x_1+1}$$

where $M_1$ is the yield of a monoculture of $S_1$ planted at the same density as the mixture. From an experiment with different plots sown at the same density with differently constituted mixtures, estimates can be obtained for $M_1, M_2, k_{12}, k_{21}$. A case of special interest occurs when $k_{12}k_{21} = 1$, when the species are directly excluding each other. Maximum likelihood estimates, under the assumption that the error structure of $(Y_1, Y_2)$ is dependent bivariate normal with variances some specifiable functions of the mixing level $x_1$ (but not directly of the yield), are obtainable by numerical solution of an explicit set of equations (Machin and Sanderson, 1977). de Wit's model applies directly to a mixture of healthy and diseased plants of the same species and hence a limiting form should apply to a mixture of present and absent plants - i.e., a spacing experiment of the type previously discussed. The concept of relative crowding of a small square with a plant, relative to one which is empty, has lost the natural interpretation of relative effective areas which is so appropriate for the multi-species situation. If however the concept is formally applied, monoculture yield from plants at density $\lambda_1$ is related by the model to the yield from plants at density $\lambda_2$ by

$$\frac{M_1}{M_2} = \frac{\beta + 1/\lambda_2}{\beta + 1/\lambda_1}$$

so that, if $M_\infty$ is the hypothesized yield at infinite density,

$$M_\lambda = \frac{\beta \lambda M_\infty}{\beta \lambda + 1}$$

or, expressed as the yield per plant $w \propto M/\lambda$,

$$w^{-1} = A\lambda + B$$

the relationship proposed by Shinozaki and Kira described above.

Extended experiments to investigate competition between several varieties or genetic strains have been performed by growing replicate pots of each monoculture and of each 50-50 mixture. Parameters $\alpha_{ij}$ represents the yield from variety i when grown in mixture with variety j, competition being summarized by $(\alpha_{ij} - \alpha_{ii})$. This is related to a type of experimental design used in breeding programs known as the diallel cross, except that yields $y_{ij}$ and $y_{ji}$ from the same pot are not independent. The appropriate analysis of variance is described by McGilchrist and Trenbath (1971). Analyses of experiments with mixtures in unequal proportions are given by Gleeson (1978).

de Wit established a correspondence between this logistic relation and the Pearl-Verhulst equation for the growth of a resource-limited population:

$$\frac{dM_t}{dt} = \frac{rM_t\ (K - M_t)}{K}$$

by regarding $M_{t+1}$ as the number of seeds obtained in the (t+1)*th* year by resowing the harvest $M_t$ of the t*th* year.

## 11. FUTURE OUTLOOK

Most of the models described in Sections 7 and 8, and the methods available for their analysis, are so new that they have not been subjected to the critical test of performance in a variety of real ecological situations. The lack of methods of analysis and the apparent complexity of inferring population properties from interacting individuals has meant that few ecologists have collected data in a form suitable for such an approach. I hope that such data sets are now forthcoming so that we can see whether these new tools are mathematical playthings or as useful as some of us think likely. Even if some of the mathematical models are rejected by the acid practical test, the revolution in approach offers a freedom for development and investigation of new models which was not available before.

Historically the development of theoretical models in ecology has been hampered by limitations of appropriate statistical models. It could continue to be hampered in the future by limitations in ecologists' knowledge of the statistical models now available. I am not advocating that ecologists avidly pursue every latest development in statistical theory as some photographers or hi-fi addicts purchase every new technical gimmick, but I am suggesting that they be aware of what is in the current catalogues before committing themselves to cheap, shop-soiled, remaindered items for which technically superior, relevant replacements have been found.

As examples of what I regard as out-of-date products I would include:

a) the use of the normal distribution as a distribution in space to model the appearance, increase, climax, decline and disappearance of a species on a habitat gradient;

b) the use of the normal distribution, or descriptive measures of departure from it, to investigate change with time in the frequency distribution of measurements over individuals in a competing population;

c) attempts to describe a yield-density curve by the 3/2 power law when the observations at high density fall consistently below the linear log-log plot;

d) continuing attempts to improve the fit of a multiple regression model by adding yet another variable, particularly when that variable is so highly correlated with those already in the model that considerable numerical instability is likely and it is likely that the 'model' will not be validated by any other set of data.

All of these have appeared in recent publications.

Another aspect of awareness required of ecologists is awareness of theoretical invalidity and of practical bias in methods of fitting models to data. To say, as some have done, that such methods do not matter because models can be generated whose variability dominates sampling and experimental variability seems to me to display an arrogance bordering on the self-destructive.

What of the new products, and what further developments would be valuable? There are many exciting possibilities, some of which have been mentioned earlier. General developments might include:

1) A closer relationship between models for competition processes and observations on the outcome of competition. What are the crucial, observable differences between realizations of different models? What frequency distribution of yields can be

generated by them? Is the correspondence between competition circles and point-process models more than superficial?

2) The development of corresponding multi-dimensional models in resource space, together with methods of overcoming problems of dependence between the axes.

3) Attempts to incorporate such models in the essentially aspatial Gause-type formulation to provide it with applications in situations where spatial relations are important.

4) Incorporation of the time domain in these models, either by allowing the parameters of automodels or competition circles to change with time, or by the formulation and analysis of spatio-temporal models which have these static models as stable equilibria.

5) Extension to mosaics.

6) Modelling invasion processes and multi-species communities, possibly by Johnson-Mehl cells.

## REFERENCES

Adlard, P. G. (1974). Development of an empirical competition model for individual trees within a stand. In J. Fries, ed., (*op. cit.*). 22-23.

Barnett, V. D. (1976). The ordering of multivariate data. *Journal of the Royal Statistical Soceity, Series A*, 139, 318-355.

Bartlett, M. S. (1964). The spectral analysis of two-dimensional point processes. *Biometrika*, 51, 299-311.

Bartlett, M. S. (1971). Physical nearest neighbor models and non-linear time-series. *Journal of Applied Probability*, 8, 222-232.

Bartlett, M. S. (1974). The statistical analysis of spatial pattern. *Advances in Applied Probability*, 6, 336-358.

Bartlett, M. S. (1975). *The Statistical Analysis of Spatial Pattern*. Chapman & Hall, London.

Beck, D. E. (1974). Predicting growth of individual trees in thinned stands of yellow poplar. In J. Fries, ed., (*op. cit.*). 47-55.

Bella, I. E. (1971). A new competition model for individual trees. *Forest Science*, 7, 364-372.

Besag, J. E. (1972a). Nearest-neighbor systems and the auto-logistic model for binary data. *Journal of the Royal Statistical Society, Series B*, 34, 75-85.

Besag, J. E. (1972b). On the statistical analysis of nearest-neighbor systems. E.M.S. Budapest Transactions, European Meeting of Statisticians, Budapest.

Besag, J. E. (1974). Spatial interaction and the statistical analysis of lattice systems. *Journal of the Royal Statistical Society, Series B*, 36, 192-236.

Besag, J. E. (1975). Statistical analysis of non-lattice data. *The Statistician*, 24, 179-195.

Besag, J. E. (1977). *Bulletin of the International Statistical Institute*, (to appear).

Besag, J. E. and Gleaves, J. T. (1973). On the detection of spatial pattern in plant communities. *Bulletin, International Statistical Institute*, 45, 153-158.

Besag, J. E. and Moran, P. A. P. (1975). On the estimation and testing of spatial interaction in Gaussian lattice processes. *Biometrika*, 62, 555-562.

Bleasdale, J. K. A. (1967). The relationship between the weight of a plant part and total weight as affected by plant density. *Journal of Horticultural Science*, 42, 51-58.

Bleasdale, J. K. A. and Nelder, J. A. (1960). Plant population and crop yield. *Nature*, 188, 342.

Bliss, C. I. and Reinker, K. A. (1964). A lognormal approach to diameter distributions in even-aged stands. *Forest Science*, 10, 350-360.

Boffey, T. B. and Veevers, A. (1977). Balanced designs for two-compartment competition experiments. *Euphytica*, 26, 481-484.

Brown, S. and Holgate, P. (1974). The thinned plantation. *Biometrika*, 61, 253-262.

Burkhart, H. E. and Strub, M. R. (1974). A model for simulation of planted loblolly pine stands. In J. Fries, ed., (*op. cit.*). 128-135.

Cane, V. R. (1977). A class of non-identifiable stochastic models. *Journal of Applied Probability*, 14, 475-482.

Cannings, C. and Cruz Orive, L. M. (1975). On the adjustment of the sex ratio and the gregarious behaviour of animal populations. *Journal of Theoretical Biology*, 55, 115-136.

Chae, D. G., Ree, F. H., and Ree, T. (1969). Radial distribution functions and equation of state of the hard-disk fluid. *Journal of Chemical Physics*, 50, 1581-1589.

Clark, P. J. and Evans, F. C. (1954). Distance to nearest neighbor as a measure of spatial relationships in populations. *Ecology*, 35, 445-453.

Cliff, A. D. and Ord, J. K. (1973). *Spatial Autocorrelation.* Pion, London.

Clutter, J. L. and Allison, J. L. (1974). A growth and yield model for *Pinus radiata* in New Zealand. In J. Fries, ed., (*op. cit.*). 136-160.

Cody, M. L. and Diamond, J. R., eds. (1975). *Ecology and Evolution of Communities.* Harvard University Press, Cambridge, Massachusetts.

Colquhoun, D. (1969). A comparison of estimators for a two-parameter hyperbola. *Applied Statistics*, 18, 130-140.

Connell, J. H. (1975). Some mechanisms producing structure in natural communities: a model and evidence from field experiments. In M. L. Cody and J. R. Diamond, eds., (*op. cit.*). 460-490.

Cormack, R. M. (1977). The invariance of Cox and Lewis's statistic for the analysis of spatial patterns. *Biometrika*, 64, 143-144.

Cox, T. F. (1976). *Some problems in the analysis of spatial pattern.* Ph.D. thesis, University of Hull.

Cox, T. F. and Lewis, T. (1976). A conditioned distance ratio method for analysing spatial patterns. *Biometrika*, 63, 483-491.

Crain, I. K. (1972). Monte Carlo Simulation of random Voronoi polygons: preliminary results. *Search*, 3, 220-221.

Currah, I. E. (1974). Crop plant population density and vegetative yield. *Land*, 1.

Dacey, M. F. (1966). A probability model for central place locations. *Annals of the Association of American Geographers*, 56, 549-568.

Dacey, M. F. (1971). Regularity in spatial distributions: a stochastic model of the imperfect central place plane. In *Statistical Ecology, Vol. 1*, G. P. Patil, E. C. Pielou, and W. E. Waters, eds. The Pennsylvania State University Press, University Park. 287-309.

David, F. N. (1970). Measurement of diversity I. *Sixth Berkeley Symposium in Mathematical Statistics and Probability*, 1, 631-648.

Delfiner, P. (1976). A test for spatial association. *Advances in Applied Probability*, 8, 653.

De Wit, C. T. (1960). *On Competition*. Verslagen van landbouwkundige onderzoekingen, Wageningen.

Diggle, P. J. (1975). Robust density estimation using distance methods. *Biometrika*, 62, 39-48.

Diggle, P. J. (1976). A spatial stochastic model of inter-plant competition. *Journal of Applied Probability*, 13, 662-671.

Diggle, P. J. (1977). *Spatial stochastic processes with applications in ecology*. Ph.D. thesis, University of Newcastle-upon-Tyne.

Diggle, P. J. (1979). Statistical methods for spatial point patterns in ecology. In *Spatial and Temporal Analysis in Ecology*, R. M. Cormack and J. K. Ord, eds. Satellite Program in Statistical Ecology, International Co-operative Publishing House, Fairland, Maryland.

Diggle, P. J., Besag, J. E., and Gleaves, J. T. (1976). Statistical analysis of spatial patterns by means of distance methods. *Biometrics*, 32, 659-667.

Donald, C. M. (1963). Competition among crop and pasture plants. *Advances in Agronomy*, 15, 1-118.

Eisenthal, R. and Cornish-Bowden, A. (1974). The direct linear plot: a new graphical procedure for estimating enzyme kinetic parameters. *Biochemical Journal*, 139, 715-720.

Fery, R. L. and Janick, J. (1971). Response of corn (*Zea mays* L.) to population pressure. *Crop Science*, 11, 220-224.

Fischer, R. A. and Miles, R. E. (1973). The role of spatial pattern in the competition between crop plants and weeds: a theoretical analysis. *Mathematical Biosciences*, 18, 335-350.

Fitter, A. H. (1976). Effects of nutrient supply and competition from other species on root growth of *Lolium perenne* in soil. *Plant and Soil*, 45, 177-189.

Ford, E. D. (1975). Competition and stand structure in some even-aged plant monocultures. *Journal of Ecology*, 63, 311-333.

Fries, J., ed. (1974). *Growth Models for Tree and Stand Simulation.* Royal College of Forestry, Stockholm.

Gates, D. J. (1978). Bimodality in even-aged plant monocultures. *Journal of Theoretical Biology.*

Gates, D. J. and Westcott, M. (1978). Zone of influence models for competition in plantations. *Journal of Applied Probability.*

Geary, R. C. (1954). The contiguity ratio and statistical mapping. *The Incorporated Statistician*, 5, 115-145.

Gerrard, D. J. (1969). Competition quotient: a new measure of the competition affecting individual forest trees. Michigan State University Agricultural Experimental Research Station Research Bulletin 20.

Gilbert, E. N. (1962). Random subdivision of space into crystals. *Annals of Mathematical Statistics*, 33, 958-972.

Glass, L. and Tobler, W. R. (1971). Uniform distribution of objects in a homogeneous field, cities on a plain. *Nature*, 233, 67-68.

Gleeson, A. C. (1978). *The statistical analysis of plant competition experiments.* Ph.D. thesis, University of New South Wales.

Green, P. J. and Sibson, R. (1978). Computing Dirichlet tessellations in the plane. *Computer Journal*, 21, 168-173.

Greig-Smith, P. (1952). The use of random and contiguous quadrats in the study of the structure of plant communities. *Annals of Botany, London*, 16, 293-316.

Harper, J. L. (1967). A Darwinian approach to plant ecology. *Journal of Ecology*, 55, 247-270.

Haynes, J. L. and Sayre, J. D. (1956). Response of corn to within-row competition. *Agronomy Journal*, 48, 362-364.

Hegyi, F. (1973). A simulation model for managing jack-pine stands. In J. Fries, ed., (*op. cit.*). 74-90.

Hill, M. O. (1973). The intensity of spatial pattern in plant communities. *Journal of Ecology*, 61, 225-235.

Hoel, P. G. (1943). The accuracy of sampling methods in ecology. *Annals of Mathematical Statistics*, 14, 289-300.

Holgate, P. (1965a). Some new tests of randomness. *Journal of Ecology*, 53, 261-266.

Holgate, P. (1965b). Tests of randomness based on distance methods. *Biometrika*, 52, 345-353.

Holgate, P. (1972). The use of distance methods for the analysis of spatial distribution of points. In *Stochastic Point Processes*, P. A. W. Lewis, ed. Wiley, New York. 122-135.

Hopkins, B. (1954). A new method of determining the type of distribution of plant individuals (appendix by J. G. Skellam). *Annals of Botany*, 18, 213-226.

Iwao, S. (1970a). Problems of spatial distribution in animal population. In *Random Counts in Scientific Work, Vol. 2*, G. P. Patil, ed. The Pennsylvania State University Press, University Park. 117-149.

Iwao, S. (1970b). Analysis of spatial patterns in animal populations: progress of research in Japan. *Review of Plant Protection Research*, 3, 41-54.

Iwao, S. (1972). Application of the $\overset{*}{m}$-m method to the analysis of spatial patterns by changing the quadrat size. *Researches on Population Ecology*, 14, 97-128.

Johnson, W. A. and Mehl, R. F. (1939). Reaction kinetics in processes of nucleation and growth. *Transactions of the American Institute of Mechanical Engineers*, 135, 416-458.

Keister, T. D. and Tidwell, G. R. (1975). Competition ratio dynamics for improved mortality estimates in simulated growth of forest stands. *Forest Science*, 21, 46-51.

Kelly, F. P. and Ripley, B. D. (1976). A note on Strauss's model for clustering. *Biometrika*, 63, 357-360.

Kendall, M. G. and Moran, P. A. P. (1963). *Geometrical Probability*. Griffin, London.

Kershaw, K. A. (1957). The use of cover and frequency in the detection of pattern in plant communities. *Ecology*, 38, 291-299.

Kira, T., Ogawa, H., and Sakazaki, N. (1953). Intraspecific competition among higher plants. I. Competition-yield-density interrelationship in regularly dispersed populations. *Journal Inst. Polytech., Osaka City University, Series D*, 4, 1-16.

Koyama, H. and Kira, T. (1956). Intraspecific competition among higher plants. VIII. Frequency distribution of individual plant weight as affected by the interaction between plants. *Journal Inst. Polytech., Osaka City University, Series D*, 73-94.

Krishna Iyer, P. V. A. (1949). The first and second moments of some probability distributions arising from points on a lattice, and their applications. *Biometrika*, 36, 135-141.

Lewis, J. E. and Rogers, T. (1974). Some remarks on random sets mosaics. In *Mathematical Problems in Biology*, P. Van de Driessche, ed. Springer-Verlag, New York. 146-151.

Lin, C. S. and Morse, P. M. (1975). A compact design for spacing experiments. *Biometrics*, 31, 661-671.

Little, D. V. (1974). A third note on recent research in geometrical probability. *Advances in Applied Probability*, 6, 103-130.

Ludwig, J. (1979). A new quadrat variance methods for the analysis of spatial pattern. In *Spatial and Temporal Analysis in Ecology*, R. M. Cormack and J. K. Ord, eds. Satellite Program in Statistical Ecology, International Co-operative Publishing House, Fairland, Maryland.

McGilchrist, C. A. and Trenbath, B. R. (1971). A revised analysis of plant competition experiments. *Biometrics*, 27, 659-671.

Machin, D. and Sanderson, B. (1977). Computing maximum-likelihood estimates for the parameters of the De Wit competition model. *Applied Statistics*, 26, 1-8.

Martin, F. B. (1973). Beehive designs for observing variety competition. *Biometrics*, 29, 397-402.

Matern, B. (1960). *Spatial Variation.* Meddelanden fran Statens Skogsforskningsinstitut, 49, 1-144.

Mead, R. (1966). A relationship between individual plant spacing and yield. *Annals of Botany (London)*, 118, 301-309.

Mead, R. (1967). A mathematical model for the estimation of inter-plant competition. *Biometrics*, 23, 189-206.

Mead, R. (1968). Measurement of competition between individual plants in population. *Journal of Ecology*, 56, 35-45.

Mead, R. (1970). Plant density and crop yield. *Journal of the Royal Statistical Society, Series C*, 19, 64-81.

Mead, R. (1971). Models for interplant competition in irregularly distributed populations. In *Statistical Ecology, Vol. 2*, G. P. Patil, E. C. Pielou, and W. E. Waters, eds. The Pennsylvania State University Press, University Park. 13-22.

Mead, R. (1974). A test for spatial pattern at several scales using data from a grid of contiguous quadrats. *Biometrics*, 30, 295-307.

Milne, A. (1961). Definition of competition among animals. *Symposium of the Society for Experimental Biology*, 15, 40-61.

Mollison, D. (1977). Spatial contact models for ecological and epidemic spread. *Journal of the Royal Statistical Society, Series B*, 39, 283-326.

Moran, P. A. P. (1948). The interpretation of statistical maps. *Journal of the Royal Statistical Society, Series B*, 10, 243-251.

Moran, P. A. P. (1950). Notes on continuous stochastic phenomena. *Biometrika*, 37, 17-23.

Moran, P. A. P. (1966). A note on recent research in geometrical probability. *Journal of Applied Probability*, 3, 453-463.

Moran, P. A. P. (1969). A second note on recent research in geometrical probability. *Advances in Applied Probability*, 1, 73-89.

Moran, P. A. P. (1973). A Gaussian Markovian process on a square lattice. *Journal of Applied Probability*, 10, 54-62.

Mountford, M. D. (1961). On E. C. Pielou's index of non-randomness. *Journal of Ecology*, 49, 271-275.

Murray, J. D. (1976). Spatial structures in predator-prey communities - a nonlinear time delay diffusional model. *Mathematical Biosciences*, 31, 73-86.

Nelder, J. A. (1962). New kinds of systematic designs for spacing experiments. *Biometrics*, 18, 283-307.

Nelder, J. A. (1966). Inverse polynomials, a useful group of multifactor response functions. *Biometrics*, 22, 128-141.

Opie, J. E. (1968). Predictability of individual tree growth using various definitions of competing basal area. *Forest Science*, 14, 314-323.

Ord, J. K. (1975a). Estimation methods for models for spatial interaction. *Journal of the American Statistical Association*, 70, 120-126.

Ord, J. K. (1975b). Some alternative approaches in spatial modelling. *Advances in Applied Probability*, 7, 452-453.

Ord, J. K. (1979). Time series and spatial patterns in ecology. In *Spatial and Temporal Analysis in Ecology*, R. M. Cormack and J. K. Ord, eds. Satellite Program in Statistical Ecology, International Co-operative Publishing House, Fairland, Maryland.

Paloheimo, J. E. (1971). On a theory of search. *Biometrika*, 58, 61-75.

Pielou, E. C. (1959). The use of point-to-plant distances in the study of the pattern of plant populations. *Journal of Ecology*, 47, 607-613.

Pielou, E. C. (1969). *An Introduction to Mathematical Ecology*. Wiley, New York.

Pielou, E. C. (1974). *Population and Community Ecology*. Gordon & Breach, New York.

Porter, W. R. and Trager, W. T. (1977). Improved non-parametric statistical methods for the estimation of Michaelis-Menten kinetic parameters by the direct linear plot. *Biochemical Journal*, 161, 293-302.

Ripley, B. D. (1976). The second-order analysis of stationary point processes. *Journal of Applied Probability*, 13, 255-266.

Ripley, B. D. (1977). Modelling spatial patterns. *Journal of the Royal Statistical Society, Series B*, 39, 172-212.

Roach, S. A. (1968). *The Theory of Random Clumping*. Methuen, London.

Ross-Parker, H. (1975). Inter-plant competition models and their sampling distributions. *Advances in Applied Probability*, 7, 453-454.

Sakai, K. I. (1955). Competition in plants and its relation to selection. *Cold Spring Harbour Symposium on Quantitative Biology*, 20, 137-157.

Sakai, K. I. (1957). Study on competition in plants. VII. Effect of competing and non-competing individuals. *Journal of Genetics*, 55, 227-234.

Schutz, W. M. and Brim, C. A. (1967). Inter-genotypic competition in soybeans. I. Evaluation of effects and proposed fieldplot designs. *Crop Science*, 7, 371-376.

Shinozaki, K. and Kira, T. (1956). Intraspecific competition among higher plants. VII. Logistic theory of the C-D effect. *Journal Inst. Polytech., Osaka City University*, 27, 35-72.

Smith-Gill, S. J. (1975). Cytophysiological basis of disruptive pigmentary patterns in the leopard frog *Rana pipiens*. II. *Journal of Morphology*, 146, 35-54.

Stiteler, W. M. and Patil, G. P. (1971). Variance to mean and Morisita's index as measures of spatial patterns in ecological populations. In *Statistical Ecology, Vol. 1*, G. P. Patil, E. C. Pielou, and W. E. Waters, eds. The Pennsylvania State University, University Park. 287-309.

Strauss, D. J. (1975a). Analyzing binary lattice data with the nearest-neighbor property. *Journal of Applied Probability*, 12, 702-712.

Strauss, D. J. (1975b). A model for clustering. *Biometrika*, 62, 467-475.

Strauss, D. J. (1977). Clustering on colored lattices. *Journal of Applied Probability*, 14, 135-143.

Street, A. P. and Macdonald, S. O. (1979). Balanced binary arrays. I. The square grid. In *Proceedings of the Sixth Australian Conference on Combinatorial Mathematics* (Armidale, 1978), A. F. Horadam and W. D. Wallis, eds. Springer-Verlag Lecture Notes in Mathematics.

Temperley, H. N. V., Rowlinson, J. S., and Rushbrooke, G. S., eds. (1968). *Physics of Simple Liquids*. North Holland, Amsterdam.

Theil, H. (1950). A rank-invariant method of linear and polynomial regression analysis. *Proc. Koninklijke Nederlandse Akad. van Wetenschappen, Series A*, 53, 386-392; 521-525; 1397-1412.

Thompson, H. R. (1958). The statistical study of plant distribution patterns using a grid of quadrats. *Australian Journal of Botany*, 6, 322-343.

Usher, M. B. (1975). Analysis of pattern in real and artificial plant populations. *Journal of Ecology*, 63, 569-586.

Veevers, A. and Boffey, T. B. (1975). On the existence of levelled beehive designs. *Biometrics*, 31, 963-967.

Veevers, A. and Boffey, T. B. (in preparation). Symmetric designs for observing variety competition.

Verhagen, A. M. W. (1971). Clustering of attributes on regular point lattices. *Journal of Applied Ecology*, 8, 665-682.

Whittle, P. (1954). On stationary processes in the plane. *Biometrika*, 41, 434-449.

Willey, R. W. and Heath, S. B. (1969). The quantitative relationships between plant population and crop yield. *Advances in Agronomy*, 21, 281-321.

Williamson, M. H. (1972). *The Analysis of Biological Populations*. Edward Arnold, London.

[*Received:* *July* 1977. *Revised:* *July* 1978]

R. M. Cormack and J. K. Ord, (eds).,
*Spatial and Temporal Analysis in Ecology*, pp. 213-246. 

# THE STATISTICAL PREDICTION OF THE FLUCTUATIONS IN ABUNDANCE IN NICHOLSON'S SHEEP BLOWFLY EXPERIMENTS

ROBERT W. POOLE

Section of Population Biology and Genetics
Division of Biology and Medicine
Brown University
Providence, Rhode Island 02912 USA

SUMMARY. This paper is a short introduction to the philosophy and methodology of statistical prediction in ecology and an analysis of a specific example. The first section presents forecasting as a general stochastic problem and discusses the pragmatic limitations on prediction in the field. The claim is made that forecasting, say, population abundance at some time $t+L$ in the future consists of finding the conditional probability distribution of the abundance given the past and present observed abundances and the past and present values of any other causally related variables available to us. The use of statistical linear predictors to compute this distribution is then presented.

The series of data analyzed is one of Nicholson's classic experiments on the Australian sheep blowfly. The first linear predictor of future adult sheep blowfly abundance created is a general 'seasonal' form of the ARIMA(p,d,q) model of Box and Jenkins (1970) based on the past and present values of the adult series only. The second linear predictor is a modification of the Box and Jenkins transfer function model and utilizes the past and present values of both egg and adult abundance. The egg data collected by Nicholson are included in the predictor because changes in the egg series foreshadow later fluctuations in the numbers of adults.

Both the ARIMA and transfer function models are good forecasters of the future fluctuations in adult abundance. However, the transfer function model's forecasts are superior to those of the ARIMA model. For thirty observations not included in either the estimation or identification procedures, variance reductions of 85 to 99 percent were observed both for the one-step ahead predictions and for the set of predictions from one to thirty steps ahead. The paper concludes with a short discussion of the limitations of statistical prediction and offers some suggested future modifications and uses.

KEY WORDS. statistical prediction, Box-Jenkins models, autoregressive-moving average models, transfer function models, population fluctuations.

## 1. INTRODUCTION

The prediction of future abundance fluctuations or changes in energy or nutrient levels are important ecological problems. By prediction I mean forecasting as defined in Poole (1979) rather than hypothesis generation or simulation. Although a multitude of factors are involved, forecasting is almost always plagued by certain pragmatic characteristics some of which are:

1. No ecological process is perfectly predictable. Important biological or physical environmental variables affecting the behavior of the forecasted process will be missed, no matter how well studied a population or ecosystem is, either through pure chance, mistake, unmeasurability, or lack of time and money.

2. The data available for forecasting are almost always limited to the past and present values of the monitored variables. For example, suppose we are studying an insect pest population and have made measurements on the population's abundance, temperature, humidity, and the density of a parasitoid population up to the present time t. The problem is to forecast the pest's abundance at some future time $t+L$. The prediction must be made solely on the basis of the observed values of the measured variables up to time $t$ since weather and parasitoid density are also unknown quantities from $t$ to $t+L$. This quality of forecasting is true of almost all ecological variables except those that can be strictly controlled. We can, of course, forecast temperature, humidity, and parasitoid abundance and use the predicted values and the observed pest fluctuations up to time $t$ to forecast the pest population abundance at time $t+L$. However, any discrepancies between the predicted and actual future values of the 'causal' variables increases the variability of the predictions of the pest's abundance.

3. The parameters of any model used in forecasting must be estimable from the type of data it is feasible to gather in the field.

4. A forecasting procedure must conform with the operational definition of a population as used in the field. While many representational models are based on theoretical zero dimensional populations, pragmatic forecasting methods must apply to populations containing a spatial component sometimes arbitrarily designated by time, money, or political jurisdiction.

Given these practical limitations on forecasting, the best a predictor can hope to do is determine the conditional probability distribution of possible abundances at time t+L. Prediction can be approached either by creating a stochastic representational model based on the biological and physical dynamics of the problem or by exploiting the statistical characteristics of the time series to predict the future from the present and past.

The purpose of this paper is to discuss the theoretical and practical bases of the statistical forecasting approach and to illustrate it by applying the methods to predict the fluctuations in one of Nicholson's classical sheep blowfly experiments. The blowfly example was chosen not because of any intrinsic practical value, but because it is one of the few long time series available in ecology. The real value of statistical prediction is in forecasting field populations and the blowfly example is used merely for illustrative purposes. Only a broad overview can be given, and the reader may wish to consult Box and Jenkins (1970) or Kashyap and Rao (1976) for more detail about the specific methodologies used.

## 2. THE THEORETICAL BASIS

The forecasting problem is illustrated in Figure 1. Suppose population abundance at time $t$ is denoted by $Y_t$ and temperature by $X_t$. The measured values of $Y_t$ and $X_t$ are available up to the present time $t$ and we wish to predict $Y_{t+L}$. Although abundance is discrete, it is usually treated as a continuous random variable in the domain $0 \leq Y_{t+L} < \infty$ with an unconditional distribution function $F(Y_{t+L})$ and density $f(Y_{t+L})$. Even if we knew nothing about the present and past history of the two variables, certain values of $Y_{t+L}$ are still more likely to happen than others. The greater the variability of possible abundances, the wider the spread of the probability density. If, however, the present and past values of $Y_t$ and $X_t$, $D_t = (Y_t, Y_{t-1}, \cdots; X_t, X_{t-1}, \cdots)$, are known, the conditional density of $Y_{t+L}$, $f(Y_{t+L} \mid D_t)$, is more concentrated because the past and present values of $Y_t$ and $X_t$ determine to some degree the future value of $Y_{t+L}$. Curve (b) in Figure 1 might be the conditional density of $Y_{t+L}$ given the present and past values of $X_t$ and $Y_t$. The stronger the dependence of $Y_{t+L}$ on $D_t$, the greater the concentration exhibited by the conditional density. On the other hand, the stronger the influence of non-included variables and chance factors and the longer the lead time of the prediction, the less concentrated the conditional density becomes.

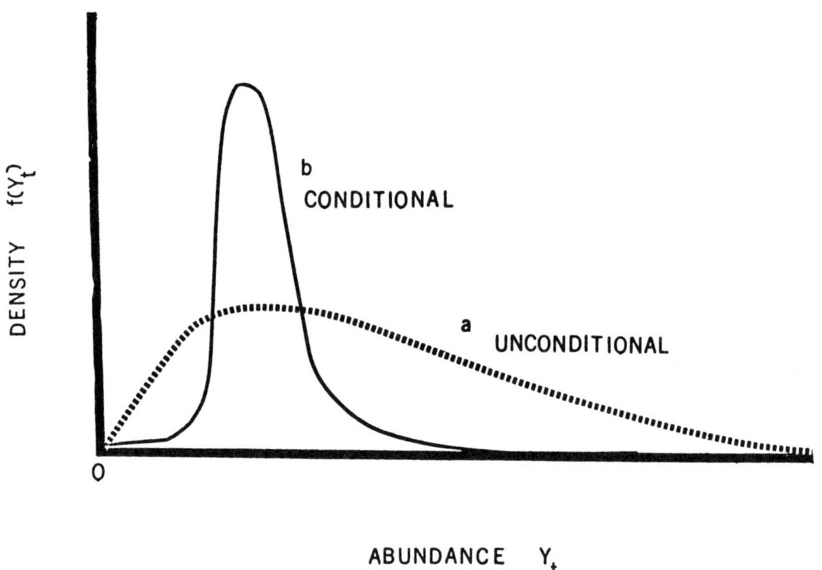


*FIG. 1: A diagrammatic comparison of the unconditional probability density of population abundance at time* t *with the conditional density that would exist if we knew the past and present histories of abundance and other causally related variables.*

With the available information $D_t$ the forecasting problem is "Find the conditional probability distribution $F(Y_{t+L} \mid D_t)$." If the conditional probability distribution is known, the forecasting problem is solved and within the practical constraints listed above is the best possible forecast given the information available. In applied problems, the conditional distribution is not known, of course, and the problem must be approached in a more roundabout manner.

Consider the related problem of finding an optimal point predictor of $Y_{t+L}$, $Y_t(L)$, given $D_t$. As a point predictor of $Y_{t+L}$ choose some function of available information $G(D_t) = Y_t(L)$. The usual optimality condition is to find the function $G$ minimizing the mean square error of the prediction, i.e.

$$\text{MSE} = E[\{Y_{t+L} - Y_t(L)\}^2] = E[\{Y_{t+L} - G(D_t)\}^2]. \tag{1}$$

If the conditional distribution of $Y_{t+L}$ is $F(Y_{t+L} \mid D_t)$, then the function $G$ minimizing MSE is the conditional expectation $E(Y_{t+L} \mid D_t)$ and the variance of the point prediction is the conditional variance $\text{Var}(Y_{t+L} \mid D_t)$ (Doob, 1953). The conditional mean and variance of $Y_{t+L}$ are generally not known either, although they are usually fairly complicated nonlinear functions. Even if they were known, the conditional mean and variance do not determine the distribution function $F(Y_{t+L} \mid D_t)$ in the general case.

There is, however, one important exception, the Gaussian or normal distribution. If $Y_{t+L}$ has a Gaussian distribution $F(Y_{t+L})$, then the conditional distribution $F(Y_{t+L} \mid D_t)$ is also Gaussian. The conditional expectation of a Gaussian distribution is a linear function and hence so is the predictor function $G$. Also, the mean and variance completely determine a Gaussian distribution. Therefore, if random variables $Y_t$ and $X_t$ are jointly normal, then the forecasting problem is: 1) find the linear function $G$ minimizing MSE, i.e., $G$ is the conditional expectation of $Y_{t+L}$; 2) compute the conditional variance from equation (1); and 3) use the mean and variance to calculate the conditional distribution $F(Y_{t+L} \mid D_t)$. The distribution of a population's abundance is, of course, not Gaussian because of the non-linearities implicit in population dynamics. However, if transformations could be found approximately normalizing the stochastic processes $(Y_t;\ t\varepsilon T)$ and $(X_t;\ t\varepsilon T)$, then the procedure would be applicable on the transformed scale.

Let $y_t$ be the transformed values of $Y_t$ and $x_t$ the transformed values of $X_t$, i.e., $y_t = h_1(Y_t)$ and $x_t = h_2(X_t)$. The forecasting problem viewed as a statistical procedure can be summarized by the following steps:

1. Find transformations approximately normalizing the stochastic processes $Y_t$ and $X_t$.

2. Find the optimal linear predictor of $y_{t+L}$ on the transformed scale.

3. Calculate the prediction $y_t(L)$ (the conditional expectation of $y_{t+L}$) and the variance of the prediction, the conditional variance of $y_{t+L}$.

4. Given the conditional mean and variance of $y_{t+L}$, calculate the conditional distribution of $y_{t+L}$, $F(y_{t+L} \mid y_t, y_{t-1}, \ldots; x_t, x_{t-1})$.

5. Finally, backtransform the point prediction, its variance, and the probability distribution of $y_{t+L}$ to the original scale.

In theory the predictions should be as good as those of a 'perfectly specified' representational model based on the species biology utilizing the same available information. There is, however, one important condition which must be met; weak stationarity. Weak stationarity implies that the stochastic processes $Y_t$ and $X_t$ have certain weak consistent statistical properties. Specifically, the unconditional expectation of the random variables $Y_t$ and $X_t$ are assumed to be constants,

$$E(Y_t) = \mu_Y(t) = \mu_Y,$$

and the unconditional covariance between $Y_{t+s}$ and $Y_t$ must depend only on $s$ and not on $t$, i.e.,

$$E[(Y_t - \mu_Y)(Y_{t+s} - \mu_Y)] = \text{cov}_Y(t, t+s).$$

For a more thorough discussion of covariance structure of a time series see Ord (1979). The mean function condition is analogous to the assumption that there exists a stable equilibrium point in a deterministic model. The restriction on the covariance function implies that there are consistent second order statistical properties to the series. The weak stationarity condition insures that the parameters of the linear predictor function depend only on the lead time of the prediction $L$ and, therefore, can be estimated from the data. Several forms of nonstationarity can be corrected. As we shall see, time series with periodic trends caused by seasonal cycles or stable limit cycle behavior can be approximately transformed to stationary processes as can series with randomly wandering means or increasing or decreasing trends. Weak stationarity, in a sense, is equivalent to the assumption that a representational model based on the biology of a species once constructed can be used to represent the population's future behavior.

## 3. LINEAR PREDICTORS

For the moment we will consider predictors of $y_{t+L}$ based on the present and past history of $y_t$ alone. The inclusion of $x_t$ in the predictor is discussed later. After correction for a stationary mean, the linear predictor of the transformed series $y_t$ given $y_{t-1}, y_{t-2}, \cdots, y_{t-p}$ is

$$(y_t - \mu_y) = \phi_1(y_{t-1} - \mu_y) + \phi_2(y_{t-2} - \mu_y) + \cdots + \phi_p(y_{t-p} - \mu_y) + a_t, \tag{2}$$

or if we let $\tilde{y}_t$ stand for $(y_t - \mu_y)$,

$$\tilde{y}_t = \phi_1\tilde{y}_{t-1} + \phi_2\tilde{y}_{t-2} + \cdots + \phi_p\tilde{y}_{t-p} + a_t,$$

where the $a_t$ are assumed to be independently and identically distributed random variables with zero mean and variance $\sigma_a^2$. If $y_t$ is a Gaussian process, then the $a_t$ are also normally distributed. Equation (2) is usually called an autoregressive model of order p. Denote the backward operator by $B$ so $By_t = y_{t-1}$ and $B^d y_t = y_{t-d}$. Then equation (2) can be written in operator notation as

$$(1 - \phi_1 B - \phi_2 B^2 - \cdots - \phi_p B^p)\tilde{y}_t = a_t,$$

or

$$\phi(B)\tilde{y}_t = a_t. \tag{3}$$

This expression may be inverted to

$$\tilde{y}_t = \phi^{-1}(B)a_t = \psi(B)a_t = (1 + \psi_1 B + \psi_2 B^2 + \psi_3 B^3 + \cdots)a_t, \tag{4}$$

where $\psi(B)a_t$ is an infinite weighted sum of the present and past error terms $(a_t, a_{t-1}, a_{t-2}, \cdots)$. If $\phi(B)$ is factored, $(1 - G_1B)(1 - G_2B)\cdots(1 - G_pB) = \phi(B)$, the condition for convergence of $\psi(B)a_t$ and hence stationarity of $y_t$ is that all of the roots $G_i$ must be less than one in absolute value or modulus (Box and Jenkins, 1970). According to equation (4), a finite autoregression can be written as an infinite moving average. Similarly, a finite moving average can be written as an infinite autoregression (see Box and Jenkins, 1970). Consequently, it is sometimes useful to use a linear predictor combining autoregressive and moving average terms, the so-called autoregressive moving average process ARMA(p,q) to insure a parsimonious representation. This has the form

$$\tilde{y}_t = \phi_1\tilde{y}_{t-1} + \phi_2\tilde{y}_{t-2} + \cdots + \phi_p\tilde{y}_{t-p} + a_t - \theta_1 a_{t-1} - \theta_2 a_{t-2} - \cdots - \theta_q a_{t-q}, \tag{5}$$

where p and q are the orders of the autoregressive and moving average components.

The condition that the roots $G_i$ are less than one in absolute value or modulus is equivalent to stability of the critical point $\tilde{y}=0$ in a deterministic p*th* order difference equation. If any of the $G_i$ lie outside the unit circle, the deterministic solution moves off to plus or minus infinity. Deterministically, a root $G_i = 1.0$ corresponds to neutral stability. If, therefore, one of the roots $G_i$ is set to one, the series exhibits a randomly wandering mean. If a constant term $\theta_0$ is added, the series possesses an increasing or decreasing trend. Forcing one root to one is equivalent to the difference transformation $(1-B)y_t = \nabla y_t = y_t - y_{t-1} = w_t$ where $w_t$ is stationary if $y_t$ is non-stationary with a wandering mean. The difference transformation $(1-B)(1-B)y_t = \nabla^2 y_t = w_t$ reduces a series with a wandering mean and slope to a stationary series $w_t$. Box and Jenkins refer to models with one or two differencings as autoregressive-integrated-moving average models ARIMA(p,d,q) where d is the number of differencings.

One final difference transformation is particularly important to this paper. Suppose a time series possesses a periodic trend due to either a seasonal cycle or a stable limit cycle phenomenon with a period of $d$ time units. The transformation $w_t = (1-B^d)y_t = y_t - y_{t-d}$ reduces the periodic series $y_t$ to the stationary series $w_t$. In all of these differencing transformations, the new variable $w_t$ should have a zero mean even if the $y_t$ series mean is non-zero, unless an increasing or decreasing trend is present (see Box and Jenkins, 1970). Models with the differencing transformation $(1-B^d)y_t = y_t - y_{t-d}$ are called 'seasonal models' by Box and Jenkins although they are used in this paper for preliminary identification of an appropriate model for time series with stable limit cycles. In a seasonal model, the values of $y_t$ are related to the values of $y_{t-d}$, $y_{t-2d}$ and so forth by the ARMA model

$$\Phi(B^d)(1 - B^d)y_t = \theta(B^d)\ \alpha_t. \tag{6}$$

The error components $\alpha_t, \alpha_{t-1}$, and so forth are not in general uncorrelated as assumed by the ARMA model. Therefore, they are related by their own ARMA model to uncorrelated error terms $a_t$

$$\phi(B)\alpha_t = \theta(B)a_t,$$

or

$$\alpha_t = \phi^{-1}(B)\theta(B)a_t.$$

Substituting in equation (6)

$$\phi(B)\Phi(B^d)(1 - B^d)y_t = \theta(B)\Theta(B^d)a_t.$$

Making the following redefinitions

$$\phi^*(B) = \phi(B)\Phi(B^d);\theta^*(B) = \theta(B)\Theta(B^d)$$

produces the seasonal model

$$\phi^*(B)(1 - B^d)y_t = \theta^*(B)a_t.$$

## 4. THE TIME SERIES

The time series studied in this paper are from Nicholson (1954). The first 106 observations are graphed in Figure 2. The solid line is the number of adults of the Australian sheep blowfly, *Lucilia cuprina*, censused at two day intervals. The dashed line is the number of eggs laid every other day. The actual egg data are given daily and plotted two days before they were actually laid. The egg data have been used as plotted with values taken every other day to conform with the adult observations. The culture was reared at 25°C with abundant water and sugar for the adults and excess larval food. The control variable is the amount of ground liver available to the adults; 0.5 grams were added each day. The amount of ground liver available to the adults has a strong effect on adult fertility.

The outstanding feature of both time series is the intense fluctuations in the abundance of both eggs and adults. Biologically, the fluctuations are easily explained. Significant egg laying occurs only when the adult population is very low. At high adult densities, competition among the adults for the ground liver is so severe that few individuals secure sufficient protein to develop eggs. Normal adult mortality causes the population to decrease until competition is relaxed enough to permit some adults to secure adequate liver for egg development. These eggs later give rise to new adults and the process is repeated. The biology of the species, consequently, contains a strong time lag. A representational model of this species has been constructed by Oster *et al.* (1976) based on a Von Foerster continuous time, continuous age distribution model. Their model exhibits strong limit cycle behavior which is apparent in Figure 2 along with a suggestion of a higher frequency component, the notches in the top of most of the cycles.

In fitting and identifying the linear predictors of this paper, the first 106 observations (212 days) were used to identify the model and estimate the parameters. The last 30 observations (60 days) plotted in Figure 5 were excluded from the analysis and are used to test the predictive ability of the fitted predictors.

## 5. TRANSFORMATION TO NORMALITY

The first step in the statistical prediction of the blowfly population is approximate normalization of the egg and adult series. Let $X_t$ be the untransformed egg series and $Y_t$ the untransformed adult series. The transformation used is a modification of the power transformation of Box and Cox (1964)

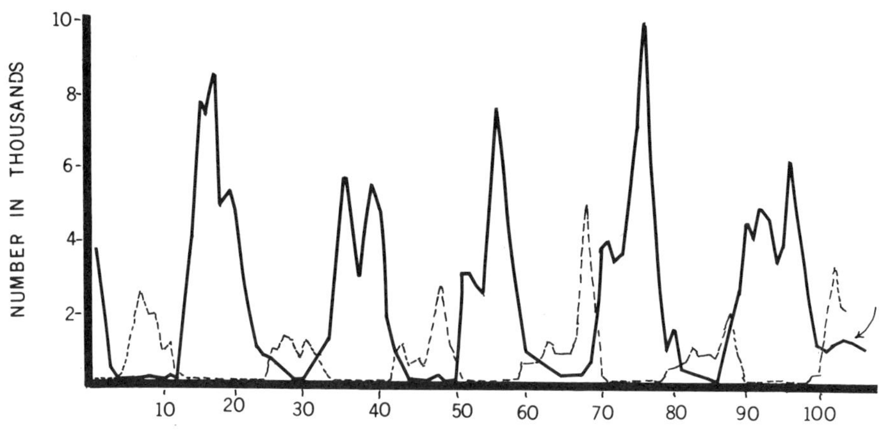


*FIG. 2: The first* 106 *observations on the egg and adult abundance in one of Nicholson's sheep blowfly experiments. The observations are at two day intervals. The solid line is the adult series and the dotted line is the egg series. (After Nicholson,* 1954*)*

$$y_t = \begin{cases} (Y_t + m)^\lambda & (\lambda \neq 0), \\ \log_e(Y_t + m) & (\lambda = 0). \end{cases}$$

The normality criterion is to vary $\lambda$ until a value is found minimizing the correlation between the present value $y_t$ and the next absolute deviation $|y_t - y_{t+1}|$. This criterion is based on the independence of the mean and variance of a normally distributed random variable. One of the most characteristic features of non-normally distributed abundance data is the strong dependence of the variance of the fluctuations on the population's present level. If abundance is high, the range of possible increases or decreases is much greater than if the population is low. The statistical properties of the reverse transformation $Y_t = h^{-1}(y_t)$ where $y_t$ is normally distributed are discussed in Granger and Newbold (1976).

The constant $m$ is an arbitrary parameter chosen to make the normalizing $\lambda$ fall at either 1/2, 1/3, or 0.0. In calculating the variance of the point predictions after backtransformation, it is very convenient to have the normalizing $\lambda$ fall at one of these three values.

The transformations produced by this procedure for the adult and egg series are both natural log implying an underlying log-normal distribution of the original series.

$$y_t = \log_e(Y_t + 300.0), \quad \text{and}$$

$$x_t = \log_e(X_t + 25.0).$$

## 6. UNIVARIATE MODEL IDENTIFICATION AND PARAMETER ESTIMATION

The methodology involved in the preliminary identification of an ARIMA(p,d,q) model and the estimation of its parameters is fairly involved and cannot be discussed in any detail. An excellent presentation of the methods may be found in Box and Jenkins (1970); see also Ord (1979). Preliminary identification is based on the characteristics of the autocorrelation and partial autocorrelation functions of the series (see Ord, 1979). Usually a tentative ARIMA(p,d,q) model can be identified and rough estimates of the parameters found by examining these two estimated functions. Given the identified model and initial parameter estimates, final parameter estimates were calculated with the nonlinear least square procedure in Box and Jenkins (1970). The identification and parameter estimation procedure used, in summary, is:

1. After transformation, the autocorrelation and partial autocorrelation functions of the series are estimated.

2. If the autocorrelation function or biological intuition indicates the need for a differencing transformation, the differencing is performed and the autocorrelation and partial autocorrelation functions of the differenced data calculated.

3. The autocorrelation and partial autocorrelation functions of the differenced data suggest a preliminary model and rough estimates of the parameters.

4. Given the preliminary model and parameter estimates, the least squares procedure is used to compute final estimates of the parameters.

5. The autocorrelation and partial autocorrelation functions of the residuals of the fitted model are computed to check for

independence. The covariance matrix of the parameter estimates is examined for strong correlations between estimates or large standard errors. The existence of a strong correlation between two parameters is a possible indication of parameter redundancy and over-parameterization. Diagnostic checking is also discussed in Ord (1979).

6. If the analysis of the residuals indicates a modification of the original model, the newly identified model is fit by the least squares procedure and the residuals again analyzed. For example, if the original model is

$$\phi(B)(1 - B^d)y_t = \theta(B)\alpha_t$$

and the autocorrelation function of the residuals indicates that the error terms could be modeled by $\phi_\alpha(B)\alpha_t = \theta_\alpha(B)a_t$, then the newly identified model is

$$\phi_\alpha(B)\phi(B)(1 - B^d)y_t = \theta_\alpha(B)\theta(B)a_t.$$

7. This iterative procedure is followed until: a) the autocorrelation and partial autocorrelation functions of the residuals indicate residual independence, and b) the standard errors of the estimates and the correlations between estimates indicate no serious instability in the estimation procedure.

Parsimony is of paramount importance in identifying a model as emphasized by Box and Jenkins (1970). If too many parameters are included in the model, serious instability can and usually does occur during parameter estimation, particularly for models with both autoregressive and moving average components. Even if the parameter estimates do converge for the over-identified model producing an estimated error variance less than the error variance of a simpler model, it is my experience that the model's predictions may be significantly worse than those of the more parsimonious model (see Kashyap and Rao (1976) for an example).

The autocorrelation function of the transformed adult data is shown in Figure 3 for lags of zero to 30 (0 to 60 days). The failure of the autocorrelation function to damp to zero is indicative of a periodic component of period 19 (38 days), undoubtedly the stable limit cycle behavior implicit in the fly's biology. Consequently, a 'seasonal' difference transformation was applied $w_t = y_t - y_{t-19} = (1 - B^{19})y_t$. Since a difference transformation is used, means are not subtracted. The autocorrelation function of the differenced data is characteristic of a first or possibly

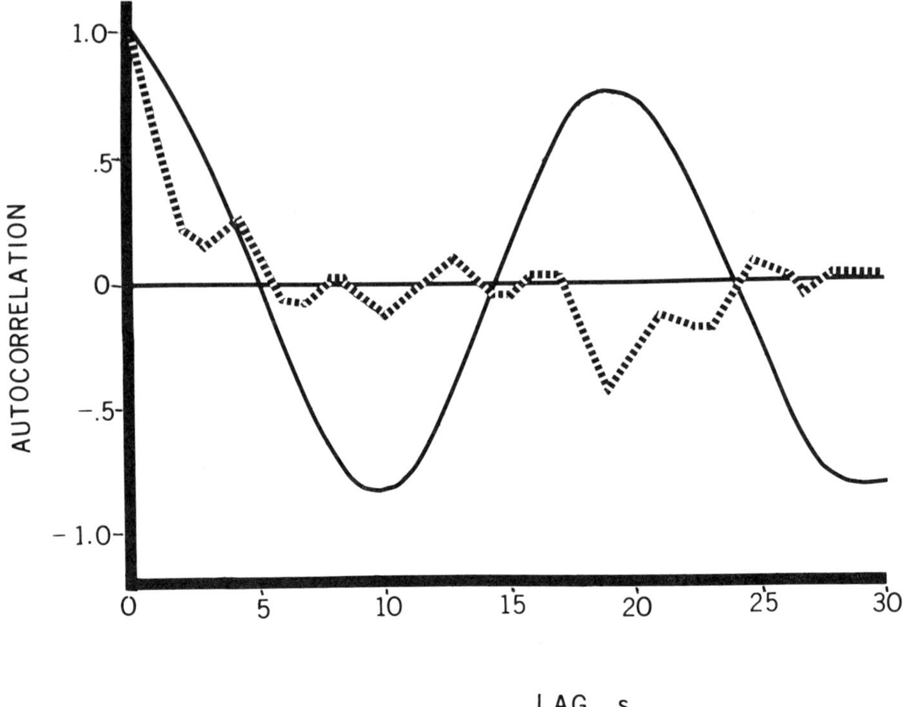


*FIG. 3: The first 30 autocorrelations of the normalized adult data. The solid line is the function without a seasonal differencing of the data and the dotted line is the autocorrelation function after a seasonal differencing of the data of the form* $(1 - B^{19})y_t$.

second order autoregression. The significant autocorrelation at lag 19 suggests a moving average term of lag 19. As a first step, however, the tentatively identified model is

$$(1 - \phi_1 B - \phi_2 B^2)(1 - B^{19})y_t = \alpha_t.$$

After fitting the model the autocorrelation and partial autocorrelations of the residuals indicated that the noise terms $\alpha_t$ should be represented by a first order moving average $\alpha_t = (1 - \theta_1 B^{19})a_t$. Therefore, the identified model is

$$(1 - \phi_1 B - \phi_2 B^2)(1 - B^{19})y_t = (1 - \theta_1 B^{19})a_t.$$

The fitted model is

$$(1 - 0.81B + 0.26B^2)(1 - B^{19})y_t = (1 - 0.78B^{19})a_t$$

$$y_t = 0.81y_{t-1} - 0.26y_{t-2} + 1.00y_{t-19} - 0.81y_{t-20} + 0.26y_{t-21} + a_t - 0.78a_{t-19}, \quad (7)$$

with an estimated residual variance of $s_a^2 = 0.1353$. The autocorrelation and partial autocorrelation functions of the residuals and the correlation matrix of the parameter estimates indicated no reason to reject the model.

The 'seasonal differencing' scheme was used for preliminary identification of the model. Given the fitted 'seasonal' model equation (7), a more general predictor was created by using the parameters in equation (7) as initial estimates in the 'general seasonal model'

$$y_t = \phi_1 y_{t-1} + \phi_2 y_{t-2} + \phi_3 y_{t-19} + \phi_4 y_{t-20} + \phi_5 y_{t-21} + a_t - \theta_1 a_{t-19}.$$

A modification of the computation of the unconditional residuals as described in Box and Jenkins (1970) was necessary. The modification and the reason for it are described in the section on the identification and estimation of transfer function models later in the paper. The fitted general model is

$$y_t = 0.82y_{t-1} - 0.21y_{t-2} + 0.97y_{t-19} - 0.79y_{t-20} + 0.27y_{t-21} + a_t - 0.78a_{t-19},$$

with an estimated residual variance of $s_2^2 = 0.1104$. The estimated residual variance is the one-step prediction variance on the transformed scale (see below). The unconditional variance of the original transformed series is $s_y^2 = 0.9693$. Therefore, the fitted general seasonal model accounts for about 89 percent of the variance in the fluctuations of the transformed series for prediction lead times of one time unit. No similar statement can be made about the variance reductions on the original scale since the conditional variance depends on the conditional mean. The estimated residual variance is probably a slight underestimate of the true variance. As described in the next section the one-step predictions were calculated for the 30 observations not included in the estimation of the parameters. For these one-step predictions the sample variance is $s_a^2 = 0.1361$.

Using the same identification and estimation procedures, the 'general seasonal' model for the egg series is

$$x_t = 0.58x_{t-1} - 0.26x_{t-2} + 0.95x_{t-19} - 0.51x_{t-20} + 0.24x_{t-21}$$

$$+ a_t - 0.86a_{t-19}$$

with an estimated residual variance of $s_a^2 = 0.3827$. The unconditional variance of the transformed series is $s_x^2 = 2.2788$. The egg ARMA model is used in the identification estimation, and forecasting of the transfer function model of the adult series later in the paper.

## 7. FORECASTING WITH THE ARIMA PREDICTOR

For a lead time of $L$ time units, the ARIMA(p,d,q) model is

$$y_{t+L} = \phi_1 y_{t+L-(1)} + \phi_2 y_{t+L-(2)} + \cdots + \phi_p y_{t+L-(p)}$$

$$+ a_{t+L} - \theta_1 a_{t+L-(1)} - \cdots - \theta_q a_{t+L-(q)}. \qquad (8)$$

The parentheses $(1),\cdots,(p);(1),\cdots,(q)$ indicate that the lags are not necessarily always in steps of one time unit. Since the minimum mean square error predictor of $y_{t+L}$ is the conditional expectation of $y_{t+L}$ given $y_{t-(1)},\cdots,y_{t-(p)};a_{t-(1)},\cdots,a_{t-(q)}$, take conditional expectations in equation (8) and denote them by $[y_{t+L}]$ and so forth.

$$[y_{t+L}] = y_t(L) = \phi_1[y_{t+L-(1)}] + \cdots + \phi_p[y_{t+L-(p)}]$$

$$+ [a_{t+L}] - \theta_1[a_{t+L-(1)}] - \cdots - \theta_q[a_{t+L-(q)}]. \qquad (9)$$

Suppose $y_t, y_{t-1},\cdots,y_{t-(p)};a_t,a_{t-1},\cdots,a_{t-(q)}$ are the present and past known values of the adult abundance and the residuals. Then the conditional expectation of $y_{t+L-(j)}$ is either its point prediction $y_t(L-j)$ if $t+L-(j)$ is greater than $t$ or its observed value if $t+L-(j) \leq t$. Because of the independence of the error terms, the conditional expectation of $a_{t+L-(j)}$ is zero if $t+L-(j)$ is greater than $t$ or equal to its observed value if $t+L-(j) \leq t$. Therefore, the point predictions $y_t(L)$ of $y_{t+1},\cdots,y_{t+L}$ are calculated recursively from equation (9).

The variance of the point prediction can be calculated from equation (4) by writing $y_{t+L}$ as

$$y_{t+L} = (a_{t+L} + \psi_1 a_{t+L-1} + \cdots + \psi_{L-1} a_{t+1}) + (\psi_L a_t + \psi_{L+1} a_{t-1} + \cdots ) = e_t(L) + y_t(L).$$

By taking conditional expectations in this equation it is apparent that $y_t(L)$ is the conditional expectation of $y_{t+L}$ and $e_t(L)$ is the error of the $L$ step prediction. Consequently, the variance of the point prediction, the conditional variance of $y_{t+L}$ is

$$\mathrm{Var}[y_{t+L} \mid y_t, y_{t-(1)},\cdots,y_{t-(p)}; a_t, a_{t-(1)},\cdots,a_{t-(q)}]$$
$$= \mathrm{Var}[e_t(L)] = V(L) = (1 + \psi_1^2 + \psi_2^2 + \cdots + \psi_{L-1}^2)\, \sigma_a^2.$$

Since the variable $y_{t+L}$ is assumed to be normally distributed, the conditional distribution given $y_t, y_{t-(1)},\cdots,y_{t-(p)}$; $a_t, a_{t-(1)},\cdots,a_{t-(q)}$ is normally distributed with mean $y_t(L)$ and variance $V(L)$.

After predictions are made on the basis of the transformed series, they are backtransformed to the original scale. Since the backtransformation from $y_{t+L}$ to $Y_{t+L}$ is one to one of the form $Y_{t+L} = h^{-1}(y_{t+L})$, the conditional distribution of $Y_{t+L}$ is

$$F(Y_{t+L}) = \Pr[Y_{t+L} \leq a] = \Pr[h^{-1}(y_{t+L}) \leq h^{-1}(a)].$$

The conditional density of $Y_{t+L}$ can be computed numerically from the distribution function. In the case of the adult and egg series, the distribution is lognormal and the density can be calculated directly.

The conditional distribution of $Y_{t+L}$ is, of course, the whole point of forecasting. It is also useful, however, to backtransform the point predictions $y_t(L)$. The point prediction can be thought of intuitively as the average future behavior of $Y_{t+L}$. Unfortunately, there is some ambiguity in the term 'average behavior.' The mean, median, and mode of a normally distributed variable such as $y_t$ are all identical. They do not coincide, however, for a general non-normal distribution. The lognormal distribution of the adult and egg time series is particularly bad in this regard. In Figure 4 a lognormal density with median 1 and variance 1 has been drawn and the mean, median, and mode indicated. The mode of a distribution is its most likely value, the median is the point with 50 percent of the probability to the left and 50 percent to the right, and the mean is the expectation of the distribution. All three characteristics of the distribution of $Y_{t+L}$ could be used as a measure of the 'average behavior' of the random variable.

My personal experience is that the median produces the most intuitively satisfying point estimates with the mode running a close second. The mode point predictions often look better but are more difficult to relate directly to the probabilistic basis of the procedure. The mean does not produce good intuitive estimates because it is strongly affected by rare but large possible outcomes, particularly for small $Y_t$ and large $V(L)$. The mean point prediction appears to be consistently larger than the abundance actually observed. On the other hand, the variance of the median and model point predictions are slightly larger than the variance of the mean point forecast. However, if the point predictions are subsidiary to the conditional distribution forecasts, this fact is not particularly important. Consequently, all point predictions in this paper are the median of the backtransformed distributions. The reader is reminded that mode point predictions usually look somewhat better than the median predictions.

The median prediction on the original scale of $Y_{t+L}$ is simply the backtransformed value of the median prediction on the normalized scale, i.e.,

$$y_t(L) = h^{-1}[y_t(L)].$$

The mode may be found by inspecting the probability density of $Y_t(L)$. The mean is difficult to find unless $\lambda = 1/2, 1/3$, or 0.0. The properties of the median and mean point predictions are

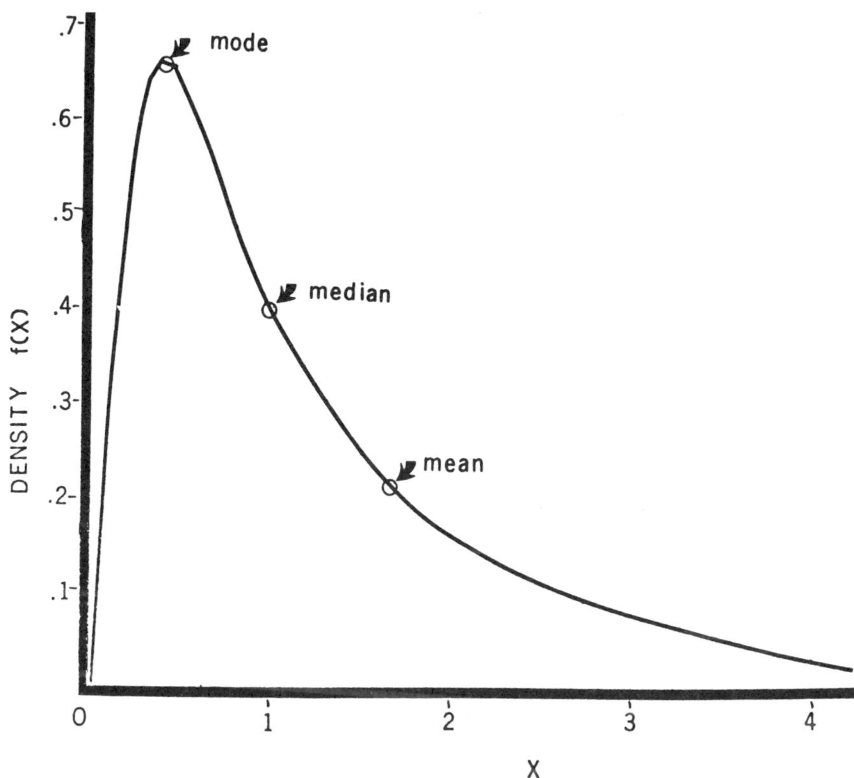


*FIG. 4: A lognormal probability density with a median and variance of one illustrating the differences in the mode, median, and mean. The mode is* .37, *the median is* 1.0, *and the mean is* 1.65.

listed in the Appendix. A general analysis of the backtransformation problem is given by Granger and Newbold (1976).

The one-step predictions (L=1) of the 30 observations not used in the predictor's identification and estimation are shown in Figure 5 along with the observed adult abundances. The general shape of the two cycles is generally good, but much of the detail is not reproduced. Forecasts for lead times of L=1 to L=30 beginning at observation t=107 are shown in Figure 6. As the lead time increases past L=1,2, and 3 the specific detail is rapidly lost and the predictions are essentially just the periodic component as induced by the previous 19 data values up to time t=106. Certainly it should be possible to do better than this.

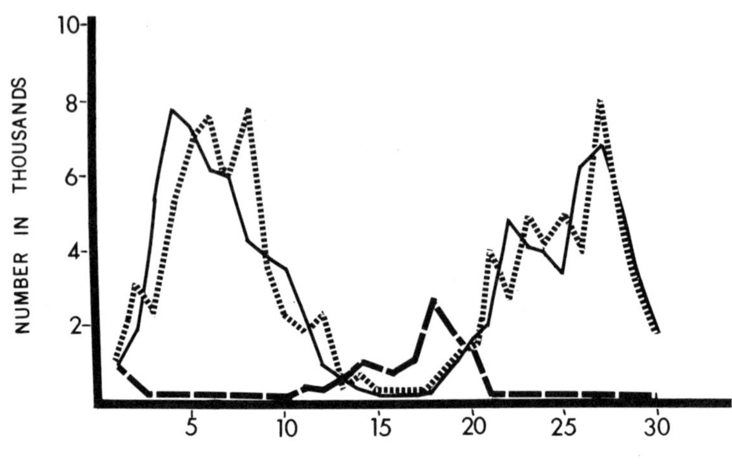


*FIG. 5: The one-step median point predictions of the* 30 *reserved observations based on the general seasonal* ARIMA *model of the adult series. The solid line is the observed and the dotted line the predicted series. The dashed line is the observed egg abundance.*

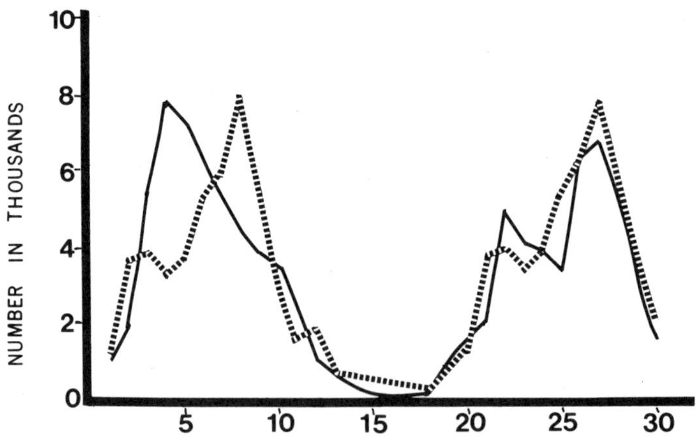


*FIG. 6: The* 1 *to* 30 *step predictions of the* 30 *reserved observations based on the general seasonal* ARIMA *model. The solid line is the observed and the dotted line the predicted series for lead times of* 1 *to* 30.

The conditional probability distribution function of $Y_t(1)$ is shown in Figure 7 and the conditional density in Figure 8. For comparison the unconditional density of $Y_{t+L}$ based on the sample mean and variance of the adult series is also shown in Figure 8.

The data in Figure 2 seem to indicate that much of the detail in the fluctuations of the adults can be anticipated by changes in the number of eggs laid 8 to 9 time periods previous (16 to 18 days). The predictions, perhaps, can be improved if the distribution of $Y_{t+L}$ is conditioned on both past adult and egg abundance. The next few sections examine this possibility.

## 8. THE TRANSFER FUNCTION MODEL

The development of eggs into adults can be broadly viewed as an input-output relationship. We can try to find a linear predictor conditioning the distribution of $Y_{t+L}$ on the past values of both adult and egg abundance. Since the future adult fluctuations are somewhat foreshadowed by the egg series, the egg series is a 'leading indicator' of future adult abundance. In a more specific setting Box and Jenkins (1970) refer to such a predictor as a transfer function model. To my knowledge the first use of the transfer function predictor in ecology is by Hacker *et al.* (1975). Since a 'seasonal' difference transformation will be applied to both the adult and egg series, means are not subtracted. On the normalized scale the linear relationship between the input and output takes the form

$$(1 - \delta_1 B - \delta_2 B^2 - \cdots - \delta_r B^r)(1 - B^d)y_t = (\omega_0 - \omega_1 B - \omega_2 B^2 - \cdots - \omega_s B^s)(1 - B^d)x_{t-b}$$

which may be written as

$$\delta(B)(1 - B^d)y_t = \omega(B)(1 - B^d)x_{t-b} \qquad (10)$$

or as

$$(1 - B^d)y_t = \delta^{-1}(B)\omega(B)(1 - B^d)B^b x_t,$$

where $(1 - B^d)$ is the seasonal operator, in this case as before $d=19$ time intervals, and $b$ is the time lag between the laying of the eggs and the emergence of the adults, approximately 8 to 9

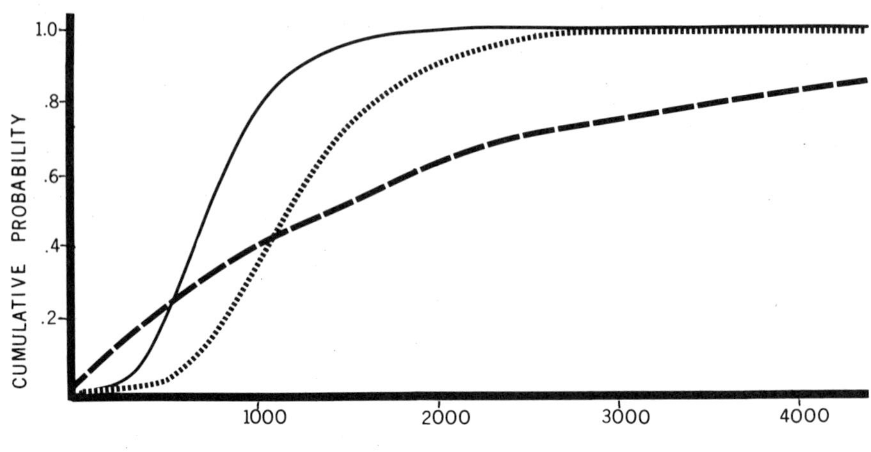


*FIG. 7: The probability distributions of observation* 107 *(the first of the reserved observations). The dashed line is the unconditional distribution, the dotted line is the conditional distribution generated by the one-step prediction based on the* ARIMA *model, and the solid line is the conditional distribution of the one-step prediction based on the general transfer function model.*

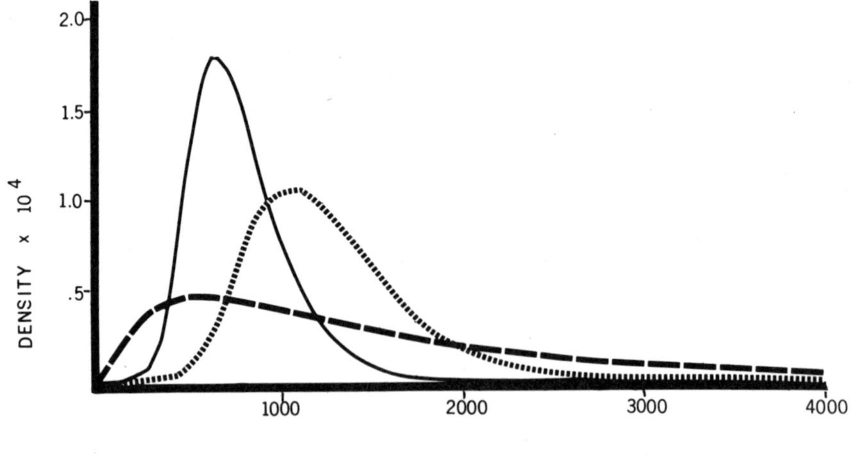


*FIG. 8: The densities corresponding to the distributions in Figure 7.*

time intervals. The constants $r$ and $s$ are the orders of the polynomials of the output and input series respectively. If we let $w_t = (1 - B^d)y_t = y_t - y_{t-d}$ and $z_t = (1 - B^d)x_t$ be the seasonally differenced values of $y_t$ and $x_t$, then

$$w_t = \delta^{-1}(B)\omega(B)B^d z_t = v(B)z_t = v_0 z_t + v_1 z_{t-1} + v_2 z_{t-2} + \cdots,$$

where $v(B) = \delta^{-1}(B)\omega(B)B^b$ is called the impulse response function which predicts $w_t$ as an infinite weighted sum of all of the past input values $z_t$, hence the term transfer function.

The discrepancy between the actual value of $y_t$ and the point prediction generated by equation (1) is denoted by $N_t$. These 'noise' terms are generally not independent. However, they may be represented by an ARIMA(p,d,q) model

$$\phi(B)N_t = \theta(B)a_t.$$

Adding the noise terms to the transfer function model, equation (10) produces

$$(1 - B^d)y_t = \delta^{-1}(B)\omega(B)(1 - B^d)B^b x_t + \phi^{-1}(B)\theta(B)a_t. \qquad (11)$$

We can then multiply through by $\delta(B)$ and $\phi(B)$ to produce the transfer function model

$$\phi(B)\delta(B)(1 - B^d)y_t = \phi(B)\omega(B)(1 - B^d)B^b x_t + \delta(B)\theta(B)a_t.$$

Once the Box and Jenkins transfer function model has been identified, the definition of $y_t$ as the output from the input series $x_t$ can be dropped. The identified transfer function model is then used as the basis of a more general predictor by making the substitutions

$$\delta^*(B) = \phi(B)\delta(B)(1 - B^d),$$

$$\omega^*(B) = \phi(B)\omega(B)(1 - B^d),$$

$$\theta^*(B) = \delta(B)\theta(B).$$

The generalized transfer function predictor is

$$\delta^*(B)y_t = \omega^*(B)x_{t-b} + \theta^*(B)a_t. \tag{12}$$

## 9. IDENTIFICATION AND PARAMETER ESTIMATION

The initial step in identifying and estimating the general relationship in equation (12) is to identify the transfer relationship $w_t = v(B)z_t$ in equation (10). Substituting $w_t = v(B)z_t$ into equation (10) produces the relationship

$$(1 - \delta_1 B - \delta_2 B^2 - \cdots - \delta_r B^r)(v_0 + v_1 B + v_2 B^2 + \cdots)$$
$$= (\omega_0 - \omega_1 B - \cdots - \omega_s B^s)B^b . \tag{13}$$

Given $\delta(B)$ and $\omega(B)$ the impulse response function $v(B)$ can be found by equating coefficients in equation (13).

The identification procedure is extensively discussed in Box and Jenkins (1970) and is too complex to repeat here. In essence it takes the form:

1. Roughly estimate the impulse response function $v(B)$ using the general ARIMA model applied to the input series (eggs).

2. Given the impulse response function, use it to identify the orders $r$ and $s$ of the output and input polynomials.

3. Given $r$, $s$, and $v(B)$ use relationship (13) to arrive at crude estimates of the parameters $\delta_1,\cdots,\delta_r;\ \omega_0,\omega_1,\cdots,\omega_s$.

4. Calculate the noise terms $n_t$ as $n_t = w_t - v(B)z_t$.

5. Identify an ARIMA(p,d,q) model for the noise terms and find initial estimates in $n_t = \phi^{-1}(B)\theta(B)a_t$ where the $a_t$ are assumed to be independently and identically distributed error terms.

The nonlinear least squares procedure described in Box and Jenkins (1970) is used to compute final parameter estimates. The estimation procedure for the Box-Jenkins transfer function model is based on minimizing the residuals generated in the following manner:

1. Given the series of values of $x_t$ from $t=1$ to $t=N$, the predicted values of the differenced output are calculated as

$$\hat{w}_t = \delta^{-1}(B)\omega(B)z_{t-b},$$

or

$$\hat{w}_t = \delta_1\hat{w}_{t-1} + \cdots + \delta_r\hat{w}_{t-r} + \omega_0 z_{t-b} - \omega_1 z_{t-b-1} - \cdots - \omega_s z_{t-b-s}$$

starting at $t=b$ and continuing up to $t=N$. Zeroes are used for any unknown values of $\hat{w}_t$ before $t=b$ and their predicted values from $t=b+1$ and onward.

2. The noise terms are computed as

$$n_t = w_t - \hat{w}_t,$$

and the residuals calculated as

$$\hat{a}_t = \theta^{-1}(B)\phi(B)n_t.$$

Notice that the predicted $w_t$ are based solely on the input series. One reason for using the more general model equation (12) is to eliminate this restriction and to construct a model less dependent on the input-output relationship.

A critical assumption in forecasting with the estimated model is that the error terms $\alpha_t$ of the ARIMA model fitted to the egg or input series and the $a_t$ of the transfer function model are independent. Possible model misidentification of the fitted transfer function model is diagnosed by: 1) checking the covariance matrix of the parameter estimates for strong correlations and large standard errors, 2) calculating the autocorrelation and partial autocorrelation functions of the residuals $a_t$ and 3) calculating the cross-correlation function between $a_t$ and $\alpha_t$. If strong correlations between two or more of the parameters exist or if the standard error of one of the parameters is relatively large, overparameterization is a possibility. If significant autocorrelations exist in the residuals, the noise model is reidentified. Finally, significant cross-correlations between the $\alpha_t$ and $a_t$ indicate misidentification of the transfer function relationship between the input and the output. Based on these three tests and others, the model is possibly modified and the procedure continued until all three diagnostic checks indicate model adequacy. The identified transfer function model and its parameters is then used as the preliminary form of the general model, equation (12).

This procedure was applied to the adult fly series. The approximate impulse response function and the autocorrelation and partial autocorrelation functions of the noise series suggested a transfer function model with $r=0$, $s+1=2$, $p=0$, and $q=3$ of the form

$$(1 - B^{19})y_t = (\omega_0 - \omega_1 B)(1 - B^{19})B^8 x_t + (1 - \theta_1 B - \theta_2 B^{19} - \theta_3 B^{20})a_t.$$

The estimated transfer function model finally produced was

$$(1 - B^{19})y_t = (0.28 + 0.15B)(1 - B^{19})B^8 x_t + (1 - 0.55B - 0.80B^{19} - 0.53B^{20})a_t.$$

The general model based on the transfer function model is

$$y_t = \delta_1 y_{t-19} + \omega_0 x_{t-8} - \omega_1 x_{t-9} - \omega_2 x_{t-27} - \omega_3 x_{t-28} + a_t - \theta_1 a_{t-1} - \theta_2 a_{t-19} - \theta_3 a_{t-20}.$$

In estimating the parameters of the general model, the parameters of the transfer function model are used as preliminary estimates. The residuals were calculated in the following way:

1. Define $h$ as the longest lag in the output and input polynomials and $q$ as the longest lag in the noise polynomial.

2. Given the parameter values at each iteration and the ARIMA model fit to the input series, the $y_t$ and $x_t$ series are backforecast from $t=0$ to a negative value of $t$ equal to the longest lag in either the output or input segment plus twice the length of the longest lag in the noise component, i.e., from $t=0$ to $t=-h-2q$.

3. Given the backforecasted values of the input and output values, the forward predictions of $y_t$, $\hat{y}_t$, are computed from equation (12) from $t=-2q$ to $t=N$ setting any unknown residuals to zero. The residuals from $t=-2q$ to $t=N$ are calculated as $\hat{a}_t = y_t - \hat{y}_t$ where $y_t$ is the backforecast value from $t=-2q$ to $t=0$ and the actual value from $t=1$ to $t=N$ where $N$ is the number of observations.

4. The sum of the squared residuals from $t=1$ to $t=N$ is the sum of squares.

The unknown residuals which must be set to zero to begin the computations introduce a transient if the procedure is started with $a_t = 0.0$ from $t=1$ to $t=h$. The reason for backforecasting beyond the end of the series is to eliminate as much as possible this transient behavior by allowing it to damp out somewhat before $t=1$.

The fitted general model is

$$y_t = 1.01y_{t-19} + 0.18x_{t-8} + 0.15x_{t-9} - 0.21x_{t-27} - 0.14x_{t-28}$$

$$+ a_t + 0.63a_{t-1} - 0.77a_{t-19} - 0.56a_{t-20}$$

with an estimated residual variance of $s_a^2 = 0.1034$. There are good reasons for suspecting that the estimated residual variance is an overestimate. The sample variance of the one-step prediction residuals of the 30 reserved observations is $s_a^2 = 0.0660$. There are three reasons for believing that the residual sum of squares of the estimated model is inflated: 1) the transient behavior induced by setting the first 20 residuals to zero may not have been entirely eliminated; 2) the small abundances are difficult to read accurately from Nicholson's figure; on a log scale relatively small errors in reading these low numbers represent fairly large errors in the logarithms; many of the larger residuals appear to be associated with these small values of $Y_t$ and $X_t$; 3) the data values marked by an arrow in Figure 2 are atypical of the series and contributed over 15 percent of the residual sum of squares.

If we use the sample variance of the one-step prediction residuals as an estimate of the true error variance of the model, the error variance of the transfer function model is approximately half of the variance of the ARIMA model fit to the adult data alone, i.e., 0.0660 for the general transfer function model versus 0.1361 for the ARIMA model. This halving of the variance seems borne out by a subjective comparison of the one-step and the one to 30 step predictions of the two models (compare Figures 5 and 7 and 6 and 8).

## 10. FORECASTING WITH THE GENERAL TRANSFER FUNCTION MODEL

The general transfer function model is

$$\delta^*(B)y_t = \omega^*(B)x_{t-b} + \theta^*(B)a_t \tag{14}$$

with orders $r^*$, $s^*$, and $q^*$ for the output, input, and moving components respectively. The point prediction for a lead time L is calculated recursively as in the ARIMA model from equation (14)

$$\begin{aligned} y_t(L) = [y_{t+L}] &= \delta_1^*[y_{t+L-(1)}] + \delta_2^*[y_{t+L-(2)}] + \cdots \\ &+ \delta_r^*[y_{t+L-(r)}] + \omega_0^*[x_{t+L-b}] - \omega_1^*[x_{t+L-b-(1)}] - \cdots \\ &- \omega_s^*[x_{t+L-b-(s)}] + [a_{t+L}] - \theta_1^*[a_{t+L-(1)}] - \cdots \\ &- \theta_q^*[a_{t+L-(q)}]. \end{aligned} \tag{15}$$

The brackets indicate conditional expectation. Therefore, values of $y_{t+L-(j)}$ are the known values if $t+L-(j) \leq t$ and are the predicted values if $t+L-(j) > t$. For the input values $x_{t+L-(j)}$ the known values are used if $t+L-(j) \leq t$ and their predicted values from the fitted ARIMA model are used if $t+L-(j) > t$. The conditional expectations of the residuals are the known values if $t+L-(j) \leq t$ and zero if $t+L-(j) > t$.

Appropriate equations for computing the variance of $y_t(L)$ may be found in Box and Jenkins (1970). The variance of the prediction for a lead time L is a combination of two components; the variance of the residuals of the transfer function model and the variance of the forecasted values of the input series. The predicted value of $y_{t+L}$, $y_t(L)$, is normally distributed and the conditional mean (15) and conditional variance (see Box and Jenkins, 1970) specify the conditional distribution of $y_{t+L}$. The transformed point predictions and the probability distributions are then backtransformed to the original scale as in the ARIMA model.

The one-step median point predictions of the 30 reserved observations are plotted in Figure 9 along with the observed values. On the whole the median point predictions match the actual observations quite well and are far superior to the median point predictions of the ARIMA model of the adult series only. The shapes of the two cycles are particularly well matched. On the other hand, the predictions seem to lag slightly behind the actual values by about one time period, and the dip in the second cycle exhibited by the data is not as closely matched as one would want.

The median point predictions for lead times of $L=1$ to $L=30$ (2 to 60 days) are shown in Figure 10. Again the general shapes of the cycles are well matched. However, in the second cycle the predicted dip in the abundance lags two time intervals (4 days) behind the actual data and does not reproduce the properties of the dip all that well. On the other hand, given the lead time (46 to 60 days) the discrepancy is not unexpected.

The conditional density of $Y_{t+1}$ (observation 107) is shown in Figure 8 based on the conservative error variance estimate of 0.1034 for the one-step predictions on the transformed scale. This density is far more concentrated than the corresponding density of the ARIMA model. If 0.0660 had been used as the one-step error variance, the degree of concentration of the conditional density would, of course, have been even greater. The corresponding one-step conditional distribution is shown in Figure 7 along with the conditional distribution of the same observation for the ARIMA model and the unconditional distribution.

## 11. DISCUSSION AND CONCLUSIONS

The unconditional variance of any value of $Y_t$ on the original scale is 2,320,080 and the standard deviation is 1523. It is interesting to compare the unconditional value with the variance and standard error of the one-step conditional distribution of observation 107 (the first of the reserved values) based on the general transfer function model. The conditional variance of $Y_{107}$ is 39,294 and the standard deviation is 198. The variance reduction from the unconditional to the conditional distribution generated by the predictor is 98.3 percent for the variance and the percent reduction in the standard deviation is 86.9 percent. Similar comparisons could be made for every one-step or $L$ step prediction. However, the conditional variance depends on the conditional expectation at time $t+L$ as well as the lead time $L$. Therefore, no general statements can be made about the degree of variance reduction. However, the variance reductions

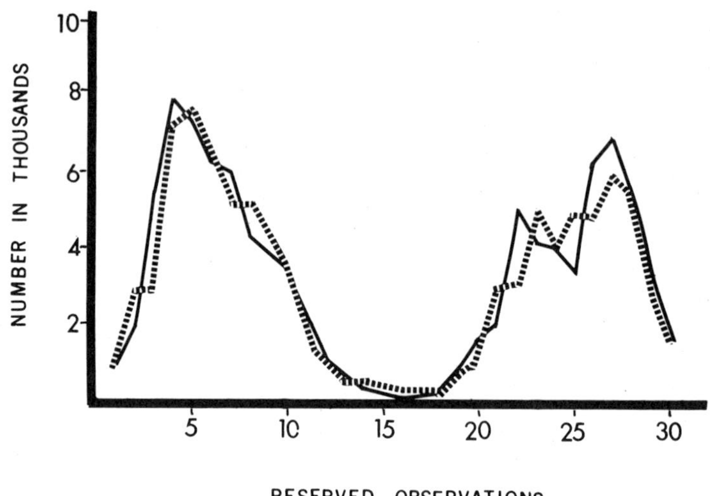


*FIG. 9: The one-step median point predictions of the* 30 *reserved observations based on the general transfer function model. The solid line is the observed series and the dotted line is the predicted series.*

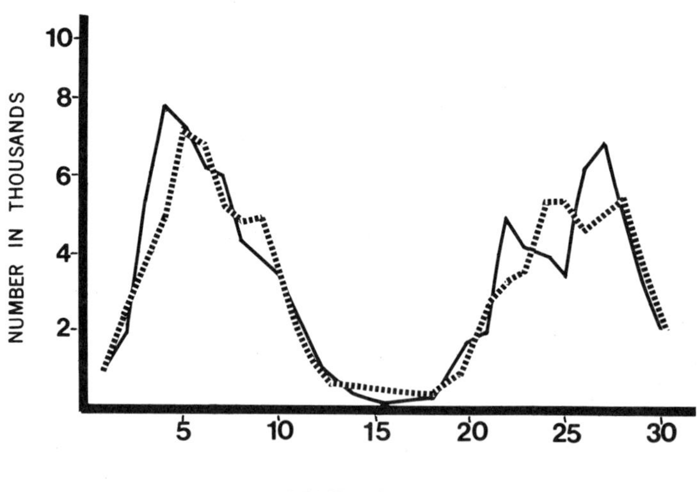


*FIG. 10: The* 1 *to* 30 *step median point predictions of the* 30 *reserved observations based on the general transfer function model. The solid line is the observed data and the dotted line the predicted series.*

of the 1 and 1 to 30 step predictions generally fell in the 85 to 99 percent range on the original scale.

The high predictability of this experimental population is probably due in part to the relatively constant conditions of the laboratory. In field populations, other factors enter such as fluctuations in weather and food supply. However, we can create a multiple leading indicator model with eggs, temperature, humidity, and food supply as leading indicators provided again that the variables of the predictor possess consistent statistical properties. The identification and estimation of these models is more difficult, but the basic principles are the same. For groups of interacting species or ecosystem components, multivariate models can be used. The possibility also exists that statistical predictors can be used for optimization and control purposes. However, this has yet to be tried.

## REFERENCES

Anderson, T. W. (1971). *The Statistical Analysis of Time Series.* John Wiley and Sons, New York.

Box, G. E. P. and Cox, D. R. (1964). An analysis of transformations. *Journal of the Royal Statistical Society, Series B,* 26, 211-243.

Box, G. E. P. and Jenkins, G. M. (1970). *Time Series Analysis: Forecasting and Control.* Holden-Day, San Francisco.

Doob, J. L. (1953). *Stochastic Processes.* John Wiley and Sons, New York.

Granger, C. W. J. and Newbold, P. (1976). Forecasting transformed series. *Journal of the Royal Statistical Society, Series B,* 38, 189-203.

Hacker, C. S., Scott, D. W., and Thompson, J. R. (1975). A transfer function forecasting model for mosquito populations. *Canadian Entomologist,* 107, 243-249.

Kashyap, R. L. and Rao, A. R. (1976). *Dynamical Stochastic Models from Empirical Data.* Academic Press, New York.

Nicholson, A. J. (1954). An outline of the dynamics of animal populations. *Australian Journal of Zoology,* 2, 9-65.

Nicholson, A. J. (1957). The self-adjustment of populations to change. *Cold Spring Harbor Symposium on Quantitative Biology,* 22, 153-172.

Ord, J. K. (1979). Time-series and spatial patterns in ecology. In *Spatial and Temporal Analysis in Ecology*, R. M. Cormack and J. K. Ord, eds. Satellite Program in Statistical Ecology, International Co-operative Publishing House, Fairland,Maryland.

Oster, G. F., Auslander, D. M., and Allen, T. T. (1976). Deterministic and stochastic effects in population dynamics. *Transactions American Society Mechanical Engineers, Dynamic Systems, Measurement Control,* 98, 44-48.

Poole, R. W. (1979). Ecological models and the stochastic-deterministic question. In *Contemporary Quantitative Ecology and Related Ecometrics,* G. P. Patil and M. Rosenzweig, eds. International Co-operative Publishing House, Fairland, Maryland.

## APPENDIX

This appendix lists the properties of the median and mean point predictions after backtransformation to the original scale. For notational purposes $y_t(L)$ is the conditional expectation of $y_{t+L}$ and $\sigma_y^2(L)$ is the conditional variance of the point prediction on the transformed scale. For a more thorough presentation see Granger and Newbold (1976).

*Median point predictions*

$\lambda=1/2$ $\quad Y_t(L) = [y_t(L)]^2 - m; \quad E[Y_{t+L} - Y_t(L)] = \sigma_y^2(L)$

$$E\{[Y_{t+L} - Y_t(L)]^2\} = \sigma_y^2(L)[3\sigma_y^2(L) + 4y_t^2(L)]$$

$\lambda=1/3$ $\quad Y_t(L) = [y_t(L)]^3 - m; \quad E[Y_{t+L} - Y_t(L)] = 3y_t(L)\sigma_y^2(L)$

$$E\{[Y_{t+L} - Y_t(L)]^2\} = 3\sigma_y^2(L)\{3y_t^4(L) + 15y_t^2(L)\sigma_y^2(L) + 5[\sigma_y^2(L)]^2\}$$

$\lambda=0.0$ $\quad Y_t(L) = \exp[y_t(L)] - m;$

$$E[Y_{t+L} - Y_t(L)] = \exp[y_t(L)]\{\exp[\sigma_y^2(L)] - 1\}$$

$$E\{[Y_{t+L} - Y_t(L)]^2\} = \exp[2y_t(L) + \sigma_y^2(L)]\{\exp[\sigma_y^2(L)] + \exp-[\sigma_y^2(L)] - 2\exp-[\sigma_y^2(L)/2]\}$$

*Mean point predictions*

$\lambda=1/2$ $\quad Y_t(L) = y_t^2(L) + \sigma_y^2(L) - m; \quad E[Y_{t+L} - Y_t(L)] = 0$

$$E\{[Y_{t+L} - Y_t(L)]^2\} = \sigma_y^2(L)[4y_t^2(L) + 2\sigma_y^2(L)]$$

$\lambda=1/3$ $\quad Y_t(L) = y_t^3(L) + 3\sigma_y^2(L)y_t(L) - m; \quad E[Y_{t+L} - Y_t(L)] = 0$

$$E\{[Y_{t+L} - Y_t(L)]^2\} = 3\sigma_y^2(L)\{3y_t^4(L) + 14y_t^2(L)\sigma_y^2(L) + 5[\sigma_y^2(L)]^2\}$$

$\lambda=0.0$ $\quad Y_t(L) = \exp[y_t(L) + \sigma_y^2(L)/2] - m;$

$$E[Y_{t+L} - Y_t(L)] = 0$$

$$E\{[Y_{t+L} - Y_t(L)]^2\} = \exp[2y_t(L) + \sigma_y^2(L)] \times \{\exp[\sigma_y^2(L)] - 1\} .$$

Although the median point predictions are intuitively superior to the means, note that the mean point predictions are unbiased and have a smaller variance.

[*Received: July* 1978. *Revised: December* 1978]

R. M. Cormack and J. K. Ord, (eds).,
*Spatial and Temporal Analysis in Ecology*, pp. 247-270. 

# THE DENSITY OF SPATIAL PATTERNS: ROBUST ESTIMATION THROUGH DISTANCE METHODS

W. G. WARREN

Environment Canada
Forestry Directorate
Western Forest Products Laboratory
6620 N. W. Marine Drive
Vancouver, British Columbia Canada V6T 1X2

C. L. BATCHELER

Forest and Range Experiment Station
New Zealand Forest Service
Christchurch, New Zealand

SUMMARY. In the first part of this paper the work of one of us (Batcheler) on the development of robust methods, for estimating the density of non-randomly dispersed populations through distance methods, is summarized. This covers the essentially empirical evolution of the methods and their successful application as reported in a series of papers from 1970 through 1975. The second part of the paper presents a summary of closely allied theory, particularly as derived by the senior author, and gives previously unpublished details on the analytical development of the nearest neighbor distribution for three special cases of clustered populations. These three examples illustrate 'building blocks' from which results for a fairly wide spectrum of situations could be constructed.

KEY WORDS. non-random point processes, clustered processes, nearest neighbor distances, distance distribution, plant populations, animal populations.

## 1. INTRODUCTION

Formulae for estimating the density of individuals dispersed over an area, based on distances either from randomly chosen points to the nearest individuals, or from individuals to their nearest neighbors, are readily obtained if the individuals can be regarded as a realization of a two-dimensional Poisson process. It is well known, however, that the estimates given by these formulae may be seriously biased if the individuals follow some non-random pattern, that is, are not a realization of a Poisson process (see e.g., Persson 1971, Batcheler 1973). Pielou (1969) noted: Since randomness of a population "cannot be assumed without a test based on additional observations, and those observations entail estimating, or determining, the density ... we must also carry out quadrat sampling or a complete count of the population. Thus ... there is no longer any need to use the distance measurements for density estimation".

Since the search for robust distance methods has proved to be relatively intractable, it is worth re-examining the grounds for any work on them, particularly since there is abundant theoretical evidence that populations can readily be estimated with quantifiable precision from accurate counts within random bounded plots.

There are several reasons, one or more of which are often encountered as practical problems in sampling. Firstly, there are many situations in which demarcating and searching plots (with the consequent possibility of missing individuals) is more time consuming and often prone to serious boundary effect biases than is implicit in measuring distances to the nearest individuals. Secondly, there are frequently problems relating to defining the entities being counted which lead to density estimates being a function of plot size (see Batcheler 1975b). Thirdly, as is well known, the variance/mean ratio of a sample of plots is an ambiguous function of the relationship between plot size and density (i.e., related to the mean number per plot). Thus plot variance is not a dimensionless parameter and it is spurious to use it as a comparable parameter of different populations in which sample means per plot differ. (This is always true except when the tedious approach of contiguous sampling by a series of plots of different sizes is adopted). Conversely, when distances to nominated individuals are measured, estimates obtained by distance sampling, whatever other limitations they may possess, are at least free from such factors of scale. Fourthly, there are many kinds of populations (e.g., rabbit fecal pellets on a range of hills, or thistles or other weeds on range lands) which occur in almost unstratifiable and usually intricately interspersed density phases (e.g., Morisita 1957, Pielou 1969). Such populations can be surveyed reliably only by use of a very large number of samples, regardless of the size or configuration of the sample units employed, and regardless of the "face value" of estimates of precision. In such cases, the real practical problem is to determine the extent of a density within each of the density phases. Consequently, on practical grounds, the sample units have to be made small, and this entails a high chance of many plots containing no members within the lower density phases. Conversely, by distance sampling, a nominated number of members will be sampled at each point regardless of local density.

There are, in short, many inter-related reasons why researchers have continued to use alternative sampling systems, and a scan of the literature of the past two decades is an adequate reminder that, despite the known theoretical problems, they continue to show a lively interest in the use of distance methods. The essence of them is that, if the difficulties can be solved, distance sampling is always unambiguous in its relationship to density and distribution of the organisms being sampled. A precisely predictable number of specimens will be encountered in a chosen number of point-samples (e.g. 1 per plot from shortest distance sampling (Clark and Evans 1954, four per point from point centred quarter (Cottam and Curtis 1956)). This advantage is particularly relevant when it is desired to estimate other attributes of the organisms (e.g., weight, height, crown area, etc.) in addition to merely estimating their numbers.

Accordingly, after a decade of quiescence, various authors have continued to search for estimators which are robust against departures from randomness. Here we recall in particular the early work of Morisita (1957), and cite the more recent work of Diggle (1975, 1977), Cox (1976) and the series of papers by Batcheler (1971, 1973, 1975a, 1975b, 1975c) and Batcheler and Bell (1970). Those by Morisita, Diggle and Cox contain developments of theory and methodology relating to what they call angle-order, T-square, and a conditioned distance method. However, the papers by Batcheler emphasize the development of a distinctive distance sampling called joint point distance-nearest neighbor distance (see also Cox and Lewis 1976) which was derived empirically from extensive sampling studies of both natural and simulated populations. It had in mind also the severe practical condition that a method had to be capable of application to populations in which some members may be hundreds of metres from a sample point, i.e. to be generally applicable, a distance method has to be useable with missing values. The objective of the present paper is to summarize the empirical work, and to provide a possible starting point for further development of the method of presenting the distribution theory of distances for a particular class of aggregated populations. This theory will be illustrated by a few special cases.

## 2. DEVELOPMENT OF A ROBUST ESTIMATOR

Batcheler's first attempts at the problem are embodied in two papers (Batcheler and Bell, 1970; Batcheler, 1971). The starting point is the maximum likelihood estimate (MLE) for the Poisson process of the density $\lambda$ (the number per unit area) from distances $r_i$ from random points to the nearest individuals. Suppose that p observations $r_1, r_2, \ldots, r_p$ are less than R, and (N-p) observations lie outside the area searched; then the MLE is:

$$d = p/\pi[r_1^2 + r_2^2 + \ldots + r_p^2 + (N-p)R^2].$$

In examining the behavior of this estimator for a variety of dispersal patterns (regular, random, aggregated) as a function of R, or rather as a function of p/N (p/N = 1 for R sufficiently large), Batcheler (1971) observed that the ratio of the estimated density to the true population density increased with p/N for the regular populations, decreased for aggregated populations, and, of course, was essentially constant ($\doteq 1$) for random populations. Not uncommonly the ratio passed through unity from below for regular populations and from above for aggregated populations. It was conjectured, therefore, that an intermediate value of R would give the most stable estimate and, in fact, that the most appropriate value of R would be that under which distances would be measured at 50% of the points (i.e. p/N = 0.5). These were called the 50% PDE's (Point Density Estimators).

These were, of course, not entirely satisfactory. Batcheler observed that this estimate tended to be low for aggregated populations, for which the distance from a random point to the nearest individual ($r_p$) tends to be larger than the distance between nearest neighbors ($r_n$). He reasoned, therefore, that some function of $\Sigma r_p/\Sigma r_n$ could be used as a correction factor for the 50% PDE. (To obtain the $r_n$ Batcheler measured the distance between the individual nearest the sampling point and its nearest neighbor). In the 1970 paper the 50% PDE was simply multiplied by the correction factor, $b_o = \Sigma r_p/\Sigma r_n$; that is,

$$b_o\ p/\left\{\pi[r_1^2 + r_2^2 + \ldots + r_p^2 + (N-p)\ R^2]\right\}.$$

In the 1971 paper this was amended by regressing $\log_{10}(d/\lambda)$ on $b_o$ where $\lambda$ denotes the true population density. From this he obtained

$$\hat{\lambda} = 0.772 \times 1.450^{b_o} \times d \qquad (1)$$

where d is the 50% PDE.

Batcheler and Bell (1970) consider a difficulty omitted from the 1971 paper, namely that the correction factor for non-randomness, $\Sigma r_p/\Sigma r_n$, requires measurement of the point- and nearest neighbor distances from all sample points. Therefore, since the 50% PDE uses only half the measurements, whereas the correction factor uses all, the field technique is a hybrid that is not applicable to the problem originally studied. To allow for this Batcheler and Bell consider the estimator $\hat{\lambda} = b_1 d$ where the correction factor is

$$b_1 = n^2 N\ \Sigma r_p/p^3 \Sigma r_n$$

and n is the number of nearest neighbor distances less than or equal to R. They found that this correction apparently failed in the case of clusters within clusters. Accordingly, they considered the measurement of distances, not only from point to nearest individual and from the individual to its nearest neighbor, but also from this nearest neighbor to its nearest neighbor ($r_m$), exclusive of the first, there being m such distances less than or equal to R. the correction factor for the level of clustering within clusters was therefore taken as $m^2 p \Sigma r_n/n^3 \Sigma r_m$, and this was cross multiplied with $b_1$ to give a second index of non-randomness:

$$b_2 = m^2 N \Sigma r_p/p^2 n \Sigma r_m.$$

In short, the idea is to compare the ratios of $\Sigma r_p/\Sigma r_n$ and $\Sigma r_p/\Sigma r_m$ in conjunction with the frequencies with which measurements might be expected to occur within R (assuming that $p/N \doteq (n/N)^2$ and $\doteq (m/N)^3$), and to use these composite factors to correct the biased estimate of density obtained from the maximum likelihood estimator. As noted by Batcheler and Bell, in an addendum to their paper, the effect of such corrections, calculated from measurements within a common R, with respect to the improved estimator of the 1971 paper, namely

$$\hat{\lambda} = 0.722 \times 1.450\ \Sigma r_p/\Sigma r_n \times d$$

has not been evaluated.

The experience of the next year or so, particularly based on the extensive simulation studies of James (1971), showed clearly that correction factors based on the $r_p$-$r_n$-$r_m$ ratios, though suitable for uniform to random populations, failed to give reasonable estimates for moderately to severely aggregated cases. James' data led Batcheler to a new estimator which incorporated the coefficient of variation of the point distances into the correction factors. Let $f = p/N$ and

$$c = \sqrt{p[\Sigma r_p^2 - (\Sigma r_p)^2/p]}/\Sigma r_p$$

(c is the coefficient of variation of the point distance, i.e. the distance from a random point to the nearest individual). Let $A_1 = \{c/E(c)\}\sqrt{b_1}$, $A_2 = \{c/E(c)\}\sqrt{b_1 b_2}$ where E(c) is the expected value of c for a random population. Batcheler (1973) gives a table of E(c) based on the approximating formula

$$\ln E(c) = -1.03107 + 0.48924f^2 - 0.71817f^4 + 0.60946f^6.$$

His final formula for estimating density is then

(2.1) if $A_1 = A_2$ (> 0.5), $\lambda = \dfrac{d(1 + 2.717f)^{A_1}}{(1 + 2.473f)}$ ;

(2.2) if $A_1 > 0.5$ ($\neq A_2$),

$$\hat{\lambda} = \frac{d}{2(1 + 2.473f)}\left[(1+2.717f)^{A_1} + (1+2.717f)^{A_2}\right] ;$$

(2.3) if $A_1 < 0.5$, $\hat{\lambda} = d/(1.0065 + 0.3401f + 0.4235f^2)$.

Note that (2,2) is incorrectly expressed in Batcheler (1973); the correction is noted in Batcheler (1975a).

It will be readily recognized from the literature that Batcheler's series of papers make novel but elementary use of some well established lines of argument to develop a possible new approach to robust distance sampling. They obviously start with the ideas presented by Hopkins (1954) for estimation of non-randomness from measurements from random points to nearest plants and random plants to nearest neighbors (Batcheler's innovation here is to couple the measurements (see also Persson 1964) in the interests of practical convenience) and extends the sampling to the neighbor's neighbor; the method then incorporates the well-known coefficient of variation from random points to nearest plants (Morisita 1954, Eberhardt 1967). Conceptually, its elements appear to be that the coefficient of variation senses non-randomness of individuals and clusters of individuals with respect to the distribution of (nominally random) sample points, while the neighbor measurements estimate the nature and degree of compaction of clusters and clusters within clusters.

The empirical nature of these formulae has been well recognized; see for example, the discussion in Batcheler's 1973 paper. Application to some very different populations has, however, yielded very satisfactory results - these in addition to those simulated and real populations from which the formulae were derived. Specifically one can cite Laycock and Batcheler (1975) - sampling tussock grassland species; Batcheler (1975b) - deer census from pellet groups; and Batcheler and Hodder (1975) - volume of pine plantations. This last mentioned, however, might have been better accomplished, for the forest conditions described, by a version of line intersect sampling, specifically horizontal line sampling as described by Strand (1958).

## 3. THEORETICAL CONSIDERATIONS

The apparent success of Batcheler's formulae as robust estimators of density provides a challenge to the theoretician. Is it possible to determine, theoretically, the properties of this method for a broad class, or classes of spatial pattern? This would not only give greater credibility to the method, but also may lead to a means for its extension and/or improvement.

A possible starting point for one such class is what is commonly known as the Neyman-Scott process (see e.g. Fisher, 1972), but also known as the Poisson cluster process (Diggle, 1978) and the center-satellite process (Warren, 1962, 1971). While permitting only a single level of clustering (although this could be extended) this model has considerable intuitive appeal and, by adjustment of the 'parameters,' can be made to cover a wide range of conditions, including the difficult case of overlapping clusters.

Following Bartlett (1974), Diggle (1975, 1978) has given the cumulative distribution function (CDF) of distance from a random point to the nearest individual as:

$$F(x) = 1- \exp[-2\pi\rho\int_0^\infty \{1-P(r,x)\}\, rdr] \tag{2a}$$

for the case of offspring only, and

$$F(x) = 1- \exp(-\pi\rho[x^2 + 2\int_x^\infty \{1-P(r,x)\} rdr]) \tag{2b}$$

if parents are included. The density of the parent events is denoted by $\rho$, and $P(r,x)$ is the probability that no offspring from a parent a distance $r$ from the origin will be found in the circle, center at the origin and radius $x$.

The distance from an arbitrary individual to the nearest neighbor has CDF

$$G(y) = 1 - [1-F(y)]\, P_2(y) \quad , \tag{3}$$

where $P_2(y)$ denotes the probability that no individual from the same cluster as the arbitrary individual will be found within a distance $y$ of the arbitrary individual.

These CDF's have been given in a somewhat more explicit form by Warren (1962, 1971†), namely

$$F(x) = 1-\exp\ [-\rho\int \{1-g[1-p(\underset{\sim}{u})]\}d\underset{\sim}{u}] \tag{4}$$

$$G(y) = 1-[1-F(y)] \int f(\underset{\sim}{u})\ g^*[1-p(\underset{\sim}{u})]d\underset{\sim}{u} \quad , \tag{5}$$

where $g(t)$ is the probability generating function (PGF) of the number of individuals in a cluster, and $p(u)$ the probability that an individual of a cluster with center at $\underset{\sim}{u}$ is located in a specific region $w$; i.e. $p(\underset{\sim}{u}) = \int h(\underset{\sim}{x}-\underset{\sim}{u})\ d\underset{\sim}{x}$ where $h(.)$ is the conditional probability density function of the distance of an individual from its associated center. In the case of $F(x)$, $w$ is a circular region about the random point, while for $G(y)$, $w$ is a circular region about a randomly selected individual. With

---

†Note that in the derivation of equation (21) in Warren (1971), $\lambda_w$ was accidently substituted for $\lambda$ and consequently the factor $\pi r^2$ erroneously included. The formulation in Warren (1962) is, however correct.

no loss of generality these points can be taken as the origin of the coordinate system. Finally $g^*(t) = [g(t) - g(0)]/t[1-g(0)]$ .

The equivalence of (2a) and (4) can be seen as follows. Let $q(\underset{\sim}{u})$ denote the probability that an individual of a cluster with center at $\underset{\sim}{u}$ be located outside the region $w$ . Then

$$g[1-p(\underset{\sim}{u})] = g[q(\underset{\sim}{u})] = p_o + p_1\, q(\underset{\sim}{u}) + p_2\, [q(\underset{\sim}{u})]^2 + \cdots$$

$$= \sum_k \Pr(N=k)[q(\underset{\sim}{u})]^k \quad ,$$

where $N$ is the number of individuals in a cluster. Thus $\Pr(N=k)[q(\underset{\sim}{u})]^k$ is the probability that there are $k$ individuals in a cluster with center at $\underset{\sim}{u}$ and that all $k$ lie outside the region $w$ . Hence $g[q(\underset{\sim}{u})]$ is the probability that $w$ is empty of all individuals of that cluster. Thus

$$\int [1-g(1-p(\underset{\sim}{u}))]d\underset{\sim}{u} = \int [1-g(q(\underset{\sim}{u}))]\, d\underset{\sim}{u}$$

$$= \int_0^{2\pi} \int_0^{\infty} [1-g(q(\underset{\sim}{u}))] \quad r dr d\theta$$

$$= 2\pi \int_0^{\infty} [1-P(r,x)] \quad r dr \quad .$$

Note also that (2b) readily follows from (2a). Firstly observe that $\exp(-2\pi\rho \int_0^{\infty} \{1-P(r,x)\}\, r dr)$ is the probability that the circular region with center at the random point is empty of offspring only. We now require that the parent be included, i.e. there is an individual at the cluster center. Now

Pr(Region contains no parent and no offspring)

= Pr(Region contains no offspring|Region contains no parent) $\times$ Pr(Region contains no parent) .

Since the parents (centers) follow a Poisson process with density $\rho$

$$\Pr(\text{Region contains no parent}) = \exp(-\pi\rho x^2) \quad .$$

Also Pr(Region contains no offspring|Region contains no parent) is

$$\exp(-\rho\int\{1-g[1-p(\underset{\sim}{u})]\}\, d\underset{\sim}{u}) \quad ,$$

where the range of integration is over all $\underset{\sim}{u}$ such that $|\underset{\sim}{u}| \geq x$ , since there can be no center with $|\underset{\sim}{u}| < x$ . The expression therefore becomes

$$\exp\{-\rho\int_0^{2\pi}\int_x^{\infty} [1-g(q(\underset{\sim}{u}))]\ r dr d\theta\}$$

$$= \exp\{-2\pi\rho\int_x^{\infty} [1-P(r,x)]\ r dr\} \quad .$$

The equivalence between $P_2(y)$ and $\int f(\underset{\sim}{u}) g^*[1-p(\underset{\sim}{u})]\, d\underset{\sim}{u}$ can be readily demonstrated. Since we are measuring from an individual to its nearest neighbor the cluster with center at $\underset{\sim}{u}$ must contain at least one member. Thus $g^*(t)$ represents the PGF of the number of individuals remaining in the cluster. Given the cluster center $\underset{\sim}{u}$ , it follows from the argument given above that $g^*[1-p(\underset{\sim}{u})]$ is the probability that the region does not contain any member of that cluster. The unconditional probability is then obtained by the above integral.

Although not obvious from the form of expressions (4) and (5), the general theory presented in 1962, which underlies them, permits, if required, the inclusion of an individual at the center.

Evaluation is formidable but has been carried out for certain special cases (Warren, 1962; Diggle, 1975) which are permissive of extension. Examples are given in the following section. With the advent of the modern computer the possibility of evaluation for more general cases, by multiple quadrature methods, now exists. Nevertheless, even for the Neyman-Scott process we are still far from developing the tools necessary for exploring Batcheler's formulae. Batcheler uses the distance from a random point to the nearest individual, and the distance from that individual to its nearest neighbor (and the distance from the latter to its nearest neighbor exclusive of the first). The truncation of distances presents a further complication. Thus, although for the Neyman-Scott process the CDF of $r_p$ is as given above, the CDF for $r_n$ is not that which has been presented. This can be readily seen from Figure 1.

If 0 be the random point and X the nearest individual, then there can be no other individual in the circle, center at 0 radius $r_p$ . Thus when integrating over a circle with center X , the

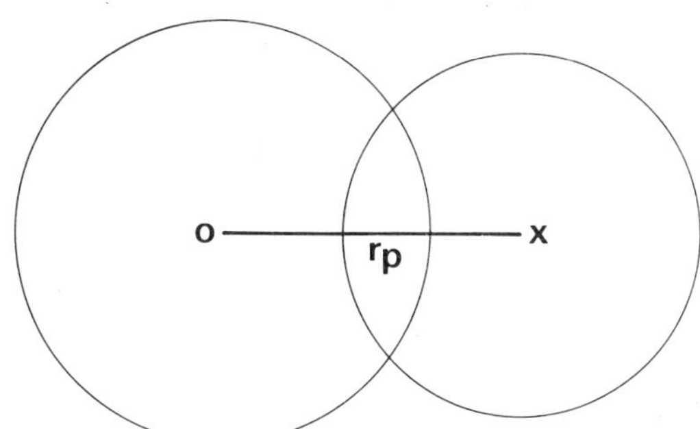


*FIG. 1; Relationship between random point, 0, and nearest individual,* X.

intersection of this circle and the circle with center 0 and radius $r_p$ must be excluded. The CDF of $r_n$ for the case of a simple Poisson process is given by Kendall and Moran (1963). Their result is not strictly correct; the complete version is given by Cox and Lewis (1976). It appears possible that the same approach could be applied to the Neyman-Scott process. The problems are, perhaps, not as intractable as they might at first appear.

Note that in Batcheler's first revision his empirical correction factor is unity if $0.1416 - 0.1613\Sigma r_p/\Sigma r_n = 0$ , i.e. if $\Sigma r_p/\Sigma r_n = 0.878$ which differs little from Cottam and Curtis's empirical value of 0.833 (Cottam and Curtis, 1956).

## 4. EXAMPLES

We now develop the CDF of the distance from an individual to its nearest neighbor in three very special cases. Note that the distribution of the distance from a random point to the nearest individual is obtained simultaneously. Firstly let us write

$$P(C^*) = \exp(-\int \rho \ \{1-g[1-p(\underset{\sim}{u})]\} \ d\underset{\sim}{u})$$

$$P(C) = \int f(\underset{\sim}{u}) \ g^* \ [1-p(\underset{\sim}{u})] \ d\underset{\sim}{u}$$

then $F(x) = 1-P(C^*)$ and $G(y) = 1-P(C^*)P(C)$ .

*Case 1.* To each cluster there belong exactly two individuals independently located on the circumference of a circle of radius

$a$ ; thus $g(t) = t^2$ . Recall that $p(\underset{\sim}{u})$ denotes the probability that a member of a cluster with center at $|\underset{\sim}{u}|$ is contained in the region $w$ , which we now take as a circle with center at the origin and radius $x$ . Let us write the distance between the origin and the cluster center as $r = |\underset{\sim}{u}|$ . As can be seen from Figure 2a, if $x < a$

$$p(\underset{\sim}{u}) = \begin{cases} 0 & , r \leq a - x \\ \phi/\pi & , a - x \leq r \leq a + x \\ 0 & , a + x \leq r \end{cases}$$

where $\phi = \cos^{-1}[(a^2 + r^2 - x^2)/2ar]$ . Hence

$$P(C^*) = \exp[-4\rho\int_{a-x}^{a+x} \phi r dr + (2\rho/\pi) \int_{a-x}^{a+x} \phi^2 r dr] \ .$$

From Figure 2b, if $x \geq a$, we have

$$p(\underset{\sim}{u}) = \begin{cases} 1 & , r \leq x-a \\ \phi/\pi & , x-a \leq r \leq x+a \\ 0 & , x+a \leq r \ . \end{cases}$$

whence

$$P(C^*) = \exp[-4\rho\int_{x-a}^{x+a} \phi r dr + (2\rho/\pi)\int_{x-a}^{x+a} \phi^2 r dr] \exp[-\rho\pi(x-a)^2] \ .$$

It can be shown that

$$\int_{a-x}^{a+x} \phi r dr = \pi x^2/2 \quad \text{and} \quad \int_{x-a}^{x+a} \phi r dr = \pi(ax-a^2/2) \ .$$

It is also possible to obtain $\int_{a-x}^{a+x} \phi^2 r dr$ and $\int_{x-a}^{x+a} \phi^2 r dr$ as convergent infinite series; direct quadrature nevertheless appears to be a more efficient means of evaluation.

$P(C)$ is most easily obtained by recalling that the origin is now located at one member of the cluster and observing that $\theta$ , the angle subtended at $\underset{\sim}{u}$ by the chord joining the two cluster members, is uniformly distributed. The PDF of the distance between individuals is then

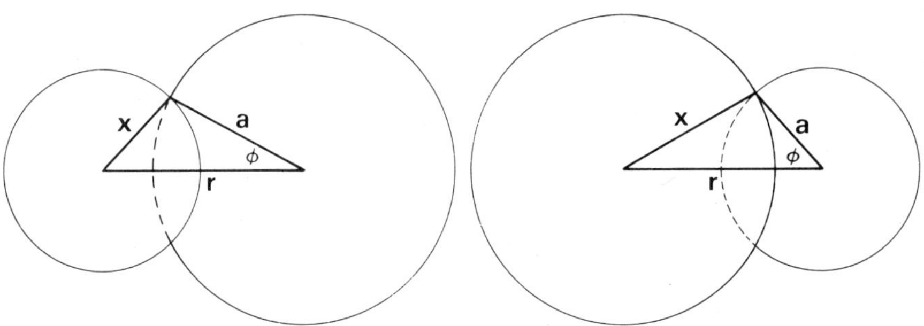


*FIG. 2: Diagram to assist in the derivation of* P(C*) *for Example 1.*

$$f(y) = \begin{cases} 2/\left(\pi\sqrt{4a^2 - y^2}\right) & , 0 \le y \le 2a \\ 0 & , \text{otherwise} \end{cases}$$

(see Figure 3). Now $g^*(t) = t$ , so that $g^*(1-p(\underset{\sim}{u})) = 1-p(\underset{\sim}{u})$ and $p(\underset{\sim}{u}) = Pr(y \le x)$ , and it follows that $P(C) = 1 - \int_0^x f(y)dy$ is given by

$$P(C) = \begin{cases} 1-(2/\pi)\sin^{-1}(x/2a), & 0 \le x \le 2a \\ 0 & , \text{otherwise} \end{cases} .$$

Bringing together these components gives

$$G(x) = \begin{cases} 1 & , x \ge 2a \\ 1-[1-(2/\pi)\sin^{-1}(x/2a)] \exp\{-\rho\pi(x^2+2ax-a^2\} \\ \qquad \times \exp\{2\rho/\pi)\int_{x-a}^{x+a} \phi^2 r dr\} & , a \le x \le 2a \\ 1-[1-(2/\pi)\sin^{-1}(x/2a)] \exp\{-2\rho\pi x^2\} \\ \qquad \times \exp\ \{(2\rho/\pi)\int_{a-x}^{a+x} \phi^2 r dr, & 0 \le x \le a \end{cases} .$$

*Case 2.* To each cluster there can be one, two, or three members but, as in case (1), these are independently located on the circumference of a circle of radius a ; thus

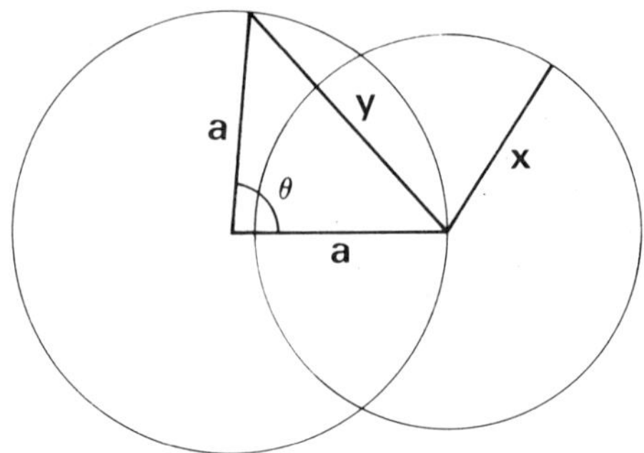


*FIG. 3: Diagram to assist in the derivation of* P(C) *for Example 1.*

$$g(t) = p_1 t + p_2 t^2 + p_3 t^3 \quad , \quad p_1 + p_2 + p_3 = 1 \quad .$$

The values of $p(\underset{\sim}{u})$ are, of course, unchanged from case (1). Hence for $x \leq a$ ,

$$g[1-p(\underset{\sim}{u})] = \begin{cases} p_1(1-0) + p_2(1-0)^2 + p_3(1-0)^3 = 1 & , \; r \leq a-x \\ p_1(1-\phi/\pi) + p_2(1-\phi/\pi)^2 + p_3(1-\phi/\pi)^3 & , \; a-x \leq r \leq a+x \\ 1 & , \; a+x \leq r \end{cases}$$

while for $x \geq a$

$$g[1-p(\underset{\sim}{u})] = \begin{cases} 0 & , \; r \leq x-a \\ p_1(1-\phi/\pi) + p_2(1-\phi/\pi)^2 + p_3(1-\phi/\pi)^3 & , \; x-a \leq r \leq x+a \\ 1 & , \; x+a \leq r \quad . \end{cases}$$

Let $m_j = g^j(1)/j!$ and $I_j(p,q) = \int_{p-q}^{p+q} (\phi/\pi)^j \, rdr$ . Then for

$x < a$

$$P(C^*) = \exp\{-2\pi\rho \, [m_1 I_1(a,x) - m_2 I_2(a,x) + m_3 I_3(a,x)]\}$$

and for $x \geq a$

$$P(C^*) = \exp[-2\pi\rho\{m_1I_1(x,a) - m_2I_2(a,x) + m_3I_3(x,a)\}]\exp[-\pi\rho(x-a)^2] \quad .$$

To determine $P(C)$ we firstly note that for $x \geq 2a$, $p(\underset{\sim}{u}) = 1$ so that

$$g^*[1-p(\underset{\sim}{u})] = p_1 + p_2(1-p(\underset{\sim}{u})) + p_3(1-p(\underset{\sim}{u}))^2 = p_1 \quad ,$$

Thus, in a manner parallel to case (1)

$$P(C) = \begin{cases} p_1 & , \quad x \geq 2a \\ p_1 + p_2\ [1-(2/\pi)\ \sin^{-1}(x/2a)] & \\ \quad + p_3\ [1-(2/\pi)\sin^{-1}(x/2a)]^2, & 0 \leq x \leq 2a \quad . \end{cases}$$

Let us write $s = 1-(2/\pi)\sin^{-1}(x/2a)$ then

$$G(x) = \begin{cases} 1-p_1\ P(C^*) & , \quad x \geq 2a \\ 1-(p_1 + p_2s + p_3s^2)P(C^*), & a \leq x \leq 2a \\ 1-(p_1 + p_2s + p_3s^2)P(C^*), & 0 \leq x \leq a \quad , \end{cases}$$

recalling that $P(C^*)$ differs in form as $x \leq a$ or $x \geq a$ . Note that case (1) can be treated as a special case of case (2) $[p_1 = p_3 = 0\ ,\ p_2 = 1]$. The extension to general $g(t)$ is clear.

Basically what is required is $\int_{x-a}^{x+a} (\phi/\pi)^j r dr$ for $x \geq a$ and $\int_{a-x}^{a+x} (\phi/\pi)^j r dr$ for $x \leq a$ , for $j = 1, 2, \cdots$. As noted above, these are perhaps best obtained by quadrature.

Diggle (1975) considered $g(t) = \exp(-\rho(1-t))$ , $a = 1$, and includes, in addition, an individual at the cluster center.

*Case 3.* There are exactly two individuals in each cluster, i.e. $g(t) = t^2$ , but these are independently located on circumferences

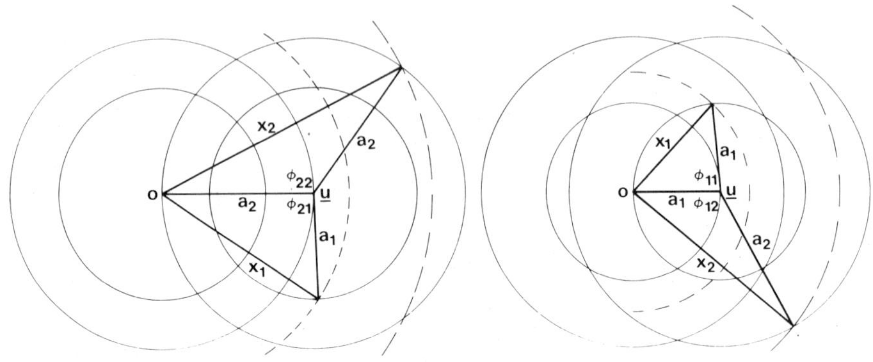


*FIG. 4: Diagram to assist in the derivation of* P(C) *for Example 3.*

of circles with common center, $\underset{\sim}{u}$ , and radii, $a_1$ and $a_2$ , $(a_2 > a_1)$ with probabilities $\pi_1$ and $\pi_2$ , respectively $(\pi_1 + \pi_2 = 1)$.

The derivation of P(C) is fairly straightforward. By reference to Figure 4, we see, analogously to case (1), that conditional on the individual from which the distance to the nearest neighbor is being determined being on circumference $a_1$ , (Figure 4a) we have, with probability $\pi_1$ ,

$$p(\underset{\sim}{u}) = \begin{cases} 1 & , \ 2a_1 \leq x \\ \phi_{11}/\pi, & x \leq 2\, a_1 \end{cases}$$

and with probability $\pi_2$

$$p(\underset{\sim}{u}) = \begin{cases} 1 & , \ a_1 + a_2 \leq x \\ \phi_{12}/\pi, & a_2 - a_1 \leq x \leq a_1 + a_2 \\ 0 & , \ x \leq a_2 - a_1 \end{cases}$$

Also, conditional on the individual being on circumference $a_2$ (Figure 4b) we have, with probability $\pi_1$

$$p(\underset{\sim}{u}) = \begin{cases} 1 & , a_1 + a_2 \leq x \\ \phi_{21}/\pi , & a_2 - a_1 \leq x \leq a_2 + a_1 \\ 0 & , x \leq a_2 - a_1 \end{cases}$$

and with probability $\pi_2$

$$p(\underset{\sim}{u}) = \begin{cases} 1 & , 2a_2 \leq x \\ \phi_{22}/\pi , & x \leq 2a_2 \end{cases} ,$$

where $\phi_{11} = 2 \sin^{-1}(x/2a_1)$, $\phi_{22} = 2 \sin^{-1}(x/2a_2)$, and $\phi_{12} = \phi_{21}$ $= \cos^{-1}[(a_1{}^2 + a_2{}^2 - x^2)/2a_1a_2]$ . For simplicity we impose the condition that $a_2 < 3a_1$ . It then follows that

$$P(C) = \begin{cases} 1-\pi_1^2\, \phi_{11}/\pi - \pi_2^2\, \phi_{22}/\pi & , x \leq a_2 - a_1 \\ 1-\pi_1^2\, \phi_{11}/\pi - 2\pi_1\pi_2\, \phi_{12}/\pi - \pi_2^2\, \phi_{22}/\pi , & a_2 - a_1 \leq x \leq 2a_1 \\ 1-\pi_1^2 - 2\pi_1\pi_2\, \phi_{12}/\pi - \pi_2^2\, \phi_{22}/\pi & , 2a_1 \leq x \leq a_1 + a_2 \\ 1-\pi_1^2 - 2\pi_1\pi_2 - \pi_2^2\, \phi_{22}/\pi & , a_1 + a_2 \leq x \leq 2a_2 \\ 1-\pi_1^2 - 2\pi_1\pi_2 - \pi_2^2 = 0 & , 2a_2 \leq x \end{cases} .$$

The determination of $P(C^*)$ is somewhat more complex. Let $p_j(\underset{\sim}{u})$ denote the probability that, given an individual on the circumference with radius $a_j$ , it will be in a circle with center at the random point and radius $x$ . Then $p(\underset{\sim}{u}) = \Sigma\pi_j p_j(\underset{\sim}{u})$ and

$$P(C^*) = \exp\{-\rho\int [2\pi_1 p_1(\underset{\sim}{u}) + 2\pi_2 p_2(\underset{\sim}{u}) - \pi_1^2 p_1^2(\underset{\sim}{u}) - 2\pi_1\pi_2 p_1(\underset{\sim}{u}) p_2(\underset{\sim}{u}) - \pi_2^2 p_2^2(\underset{\sim}{u})]\, d\underset{\sim}{u}\} .$$

Now $\int p_j(\underset{\sim}{u})\, d\underset{\sim}{u}$, $\int p_j^2(\underset{\sim}{u})\, d\underset{\sim}{u}$ $(j=1,2)$ are obtainable as in case (1). The only new element is $\int p_1(\underset{\sim}{u}) p_2(\underset{\sim}{u})\, d\underset{\sim}{u}$ . Various cases arise;

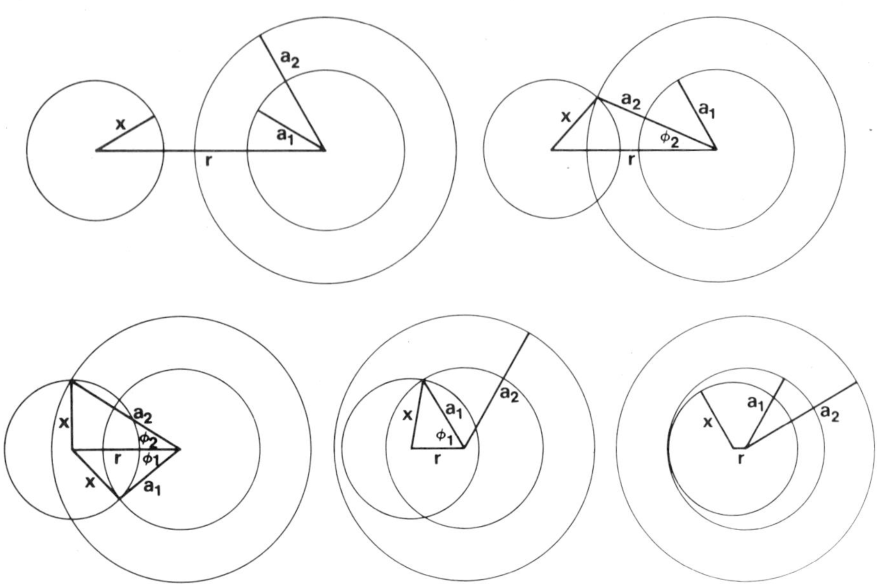


*FIG. 5; Diagram to assist in the derivation of* P(C*) *for Example 3.*

we illustrate solely for $a_2 - a_1 < x < a_1$, $a_2 > a_1$. Then from Figure 5 we see

(a) if $r \geq a_2 + x$, $p_1(\underset{\sim}{u}) = p_2(\underset{\sim}{u}) = 0$

(b) if $a_2 + x \geq r \geq a_1 + x$, $p_1(\underset{\sim}{u}) = 0$, $p_2(\underset{\sim}{u}) = \phi_2/\pi$

(c) if $a_1 + x \geq r \geq a_1$, $p_1(\underset{\sim}{u}) = \phi_1/\pi$, $p_2(\underset{\sim}{u}) = \phi_2/\pi$

(d) if $a_1 \leq r \leq a_1 - x$, $p_1(\underset{\sim}{u}) = \phi_1/\pi$, $p_2(\underset{\sim}{u}) = 0$

(e) if $a_1 - x \leq r$, $p_1(\underset{\sim}{u}) = p_2(\underset{\sim}{u}) = 0$ .

In this situation we require only $\int_{a_1}^{a_1+x} \phi_1\phi_2/\pi^2 r\,dr$ where $\phi_i = \cos^{-1}[(a_i^2 + r^2 - x^2)/2a_i r]$ $(i=1,2)$ . Again quadrature appears to be the most suitable method of evaluation. It should be clear that other values of $x$ relative to $a_1$ and $a_2$ can be handled in a parallel fashion.

## 5. DISCUSSION

These three examples have illustrated what can be called the "building blocks" from which results for a fairly wide spectrum of situations can be constructed.

We expect that once the theoretical problems relating to the joint neighbor distance method are solved for a single species, it could then be extended to simultaneous sampling of any number of species. By way of a suggestion, we conclude with a brief outline of a possible approach which empirical studies suggests some promise, and which is both elegant and simple to use in the field.

Assume that a sample of joint $r_p$, $r_n$ and $r_m$ distances are measured from random points in a mixed population of several species, and that each plant is labelled as it occurs. Data for a particular point will then be of the form: ($r_p$) (species name); ($r_n$) (species name); ($r_m$) (species name). We have found by computer simulation that if one species in the mixed population (say species A) is less clumped than another (say C), then more A's are recorded in the total sample than would be expected from its proportion of the total population. Conversely, C is under-represented. That is, the bias of proportional sampling by species in mixed populations is related to their relative departures from randomness: Relatively clumped species are under-represented, and *vice versa*.

However, we have also found from trial data, and it is intuitively obvious, that the distances between members of a relatively clumped species (i.e., $r_n$ and $r_m$) are usually smaller than are the corresponding distances between members of diffusely or uniformly distributed species.

We are essentially interested in making use of this relationship, as follows.

Assume for example that there are three species A, B and C in a mixed population and that over N points, a plants of species A are encountered at $k_a$ points, b of species B occur at $k_b$ and c of C occur at $k_c$ points. Altogether, if R is unconstrained, a + b + c = 3N. If a large number of samples are taken, the data list can conveniently be partitioned into those points which included at least one plant of A (i.e., $k_a$), then those with at least one B (e.g., $k_b$) etc. (obviously, some points will be included in up to three partitions in a

3-species system). We find that if we then calculate separate estimates of density for these independent sub-systems by formula 2.2 (page 252) then, because of the relationship between interplant distances and relative clumping of the species, estimates for relatively uniform species are lowered, while those for clumped species are raised (i.e., $A_1$ and $A_2$ tend to be lower for systems dominated by uniform species, and high for systems dominated by clumped species). We label these estimates for the three sub-systems as $d_A$, $d_B$ and $d_C$. We then make some allowance by proportions for the degree of "contamination" of each sub-system by other species as: -

$$\hat{\lambda}_a \text{ (density of species A)} \doteq d_a \cdot a/3k_a,$$
$$\hat{\lambda}_b \text{ (density of species B)} \doteq d_b \cdot a/3k_b,$$
$$\hat{\lambda}_c \text{ (density of species C)} \doteq d_c \cdot a/3k_c.$$

Clearly, $\hat{\lambda}_t$ should equal $\hat{\lambda}_a + \hat{\lambda}_b + \hat{\lambda}_c$, where $\hat{\lambda}_t$ is the combined density of all species. However, because of sampling errors and the partial duplication of the summation and calculating procedures, $\hat{\lambda}_t$ will not precisely equal $\hat{\lambda}_T$, the estimate made from all the distances together. Therefore, an obvious reconciliation procedure is to scale the estimate for each species into the global $\hat{\lambda}_T$ by $\hat{\lambda}_T(\hat{\lambda}_a/\hat{\lambda}_t) + \hat{\lambda}_T(\hat{\lambda}_b/\hat{\lambda}_t) + \hat{\lambda}_T(\hat{\lambda}_c/\hat{\lambda}_t)$ (where the three summed terms represent A, B and C respectively), as the final estimates of density and proportions of the three species.

This suggested approach has been developed from empirical study of five populations, containing 2-14 species, for which we have total counts of the plants or plot samples (Batcheler, unpubl. data), and it now seems worth suggesting the approach for consideration by a wider mathematical fraternity.

Apart from the academic challenge, we consider this somewhat premature recommendation is justified by the belief that such an approach to field sampling, if properly developed, will be an extremely useful addition to the repertoire of biologists concerned with the everyday problem of sampling complex biological communities.

## REFERENCES

Bartlett, M.S. (1974). The statistical analysis of spatial pattern. *Advances in Applied Probability*, 6, 336-358.

Batcheler, C.L. (1971). Estimation of density from a sample of joint point and nearest neighbor distances. *Ecology*, 51, 703-709.

Batcheler, C.L. (1973). Estimating density and dispersion from truncated or unrestricted joint point-distances nearest neighbor distances. *Proceedings of the New Zealand Ecological Society*, 20, 131-147.

Batcheler, C.L. (1975a). Probable limit of error of the point distance-neighbor distance of density. *Proceedings of the New Zealand Ecological Society*, 22, 28-33.

Batcheler, C.L. (1975b). Development of a distance method for deer census from pellet counts. *Journal of Wildlife Management*, 39, 641-652.

Batcheler, C.L. and Bell, D.J. (1970). Experiments in estimating density from point- and nearest neighbor distance samples. *Proceedings of the New Zealand Ecological Society*, 17, 111-117.

Batcheler, C.L. and Hodder, R.A.C. (1975). Tests of a distance technique for inventory of pine plantations. *New Zealand Journal of Forestry Science*, 5, 3-17.

Clark, P.J. and F.C. Evans. (1954). Distance to nearest neighbor as a measure of spatial relationships in populations. *Ecology* 35, 445-453.

Cottam, J.T. and Curtis, J.T. (1956). The use of distances in phytosociological sampling. *Ecology*, 37, 451-460.

Cox, T.F. (1976). The robust estimation of the density of a forest stand using a new conditioned distance method. *Biometrika* 63, 493-499.

Cox, T.F. and Lewis, T. (1976). A conditioned distance ratio method for analyzing spatial patterns. *Biometrika*, 63, 483-491.

Diggle, P.J. (1975). Robust density estimation using distance methods. *Biometrika*, 62, 39-48.

Diggle, P.J. (1977). A note on robust estimation for spatial point patterns. *Biometrika,* 64, 91-95.

Diggle, P.J. (1979). Statistical methods for spatial point patterns in ecology. In *Spatial and Temporal Processes in Ecology,* R.M. Cormack and J.K. Ord, eds. Satellite Program in Statistical Ecology, International Co-operative Publishing House, Fairland, Maryland.

Eberhardt, L.L. (1967). Some developments in 'Distance Sampling'. *Biometrics* 23, 207-216.

Fisher, L. (1972). A survey of the mathematical theory of multi-dimensional point processes. In *Stochastic Point Processes,* P.A.W. Lewis, ed. Wiley, New York. 462-513.

Hopkins, B. (1954). A new method for determining the type of distribution of plant individuals. *Annals of Botany, New Series* 18, 213-227.

James, I.L. (1971). A computer study of corrected density estimators for distance sampling of non-random populations. Dipl. of Agr. Sci. Thesis, Massey University, New Zealand.

Kendall, M.G. and Moran, P.A.P. (1963). *Geometrical Probability.* Griffin, London.

Laycock, W.A. and Batcheler, C.L. (1975). Comparison of distance measurement techniques for sampling tussock grassland species in New Zealand. *Journal of Range Management,* 28, 235-239.

Morisita, M. (1954). Estimation of population density by spacing method. *Memoirs of the Faculty of Science, Kyushu University.* Series E1, 187-197.

Morisita, M. (1957). A new method for the estimation of density by the spacing method applicable to non-randomly distributed populations. *Physiology and Ecology (Kyoto),* 7, 134-143.

Persson, O. (1964). The use of distance measurements in the estimation of seedling density and open space frequency. *Studia Forestalia Suecica.* No. 15.

Persson, O. (1971). The robustness of estimating density by distance measurements. In *Statistical Ecology, Vol. 2,* G.P. Patil, E.C. Pielou, and W.E. Waters, eds. The

Pennsylvania State University Press, University Park, Pennsylvania. 175-187.

Strand, L. (1958). Sampling for volume along a line. *Norske Skogsførsksvesen*, No. 51.

Warren, W.G. (1962). Contributions to the theory of spatial point processes. University of North Carolina, Institute of Statistics Mimeo Series, No. 337.

Warren, W.G. (1971). The center-satellite concept as a basis for ecological sampling. In *Statistical Ecology, Vol. 2*, G.P. Patil, E.C. Pielou, and W.E. Waters, eds. The Pennsylvania State University Press, University Park, Pennsylvania. 87-116.

[*Received: August* 1978. *Revised: February* 1979]

R. M. Cormack and J. K. Ord, (eds).,
*Spatial and Temporal Analysis in Ecology*, pp. 271-288. 

# THE ANALYSIS OF ECOLOGICAL MAPS AS MOSAICS

B. MATÉRN

Department of Forest Biometry
Swedish University of Agricultural Sciences
S-75007 Uppsala, Sweden

SUMMARY. Mosaics can be visualized as partitions of the plane into subsets marked by different colors. Stochastic models of mosaics are obtained by starting from a (possibly random) subdivision of the plane into cells. Deciding the color of the cells by random experiments, we get various types of stochastic mosaics. Other types are obtained from models of random clumpings. Methods of analyzing observed mosaics are reviewed. Special methods are possible when the underlying cell structure is observable.

KEY WORDS. geometric probabilities, Poisson process, random sets, tessellations.

## 1. INTRODUCTION

By a *mosaic* we understand a subdivision of a planar region into subsets of different kinds. For simplicity, we imagine a map of the region in which the different kinds of subsets are marked by different *colors*. In the simplest case, we have a *two-phase mosaic* with only two colors, black and white (say). We shall here concentrate on this mosaic and examine some methods and concepts for describing and analyzing the pattern formed by the white and black portions of such a map. We shall, however, very briefly consider also the *n phase mosaic*.

This presentation is influenced by the approach in Pielou (1969) and the work of Switzer (1971). In consequence, the emphasis is rather much on *stochastic models of mosaics*. In the present stage of the theory it appears that a study of different mechanisms producing random spatial patterns will be useful for the development of methods for analyzing actual ecological maps.

## 2. RANDOM DIVISION OF THE PLANE INTO CELLS

As a preliminary to the study of random mosaics, we consider stochastic methods of dividing the plane $(R_2)$ into contiguous subsets called *cells*. When the subsets are convex polygons we speak of a *tessellation* in the plane.

We start with a fixed tessellation consisting of *congruent* convex polygons. By randomly translating and rotating the system we get a random division of $R_2$ into cells. Two examples are shown in Figure 1 (A and B). Let the area and the perimeter of a constituent cell be A and P respectively. Consider two points in $R_2$ at given distance $v$ apart. The probability that they shall be in the same cell is a function, $\psi(v)$, of the distance $v$. Simple use of Bayes' Theorem shows that

$$\psi(v) = \frac{A\ a(v)}{2\pi v}\ , \tag{2.1}$$

where $a(v)$ is the probability density of the distance between two points chosen independently with uniform distribution in the constituent polygon. For small values of $v$, we can derive the approximation

$$a(v) = 2v\{\pi A - Pv + o(v)\}/A^2 \tag{2.2}$$

by arguing that, except for a band of width $v$ around the perimeter, two points are at distance $v$ if one lies in an annulus of length $2\pi v$ around the other; in the band this area is reduced. This gives

$$\psi(v) = 1 - Pv/(A\pi) + \cdots \tag{2.3}$$

Exact expressions for $a(v)$ in the case of rectangles, equilateral triangles, and regular hexagons, are listed in Matérn (1960, p. 24-25).

We now pass to a rather different network of cells. Consider points, in the sequel called centers, located in the plane according to a Poisson process with density $\lambda$ (see Diggle, 1979).

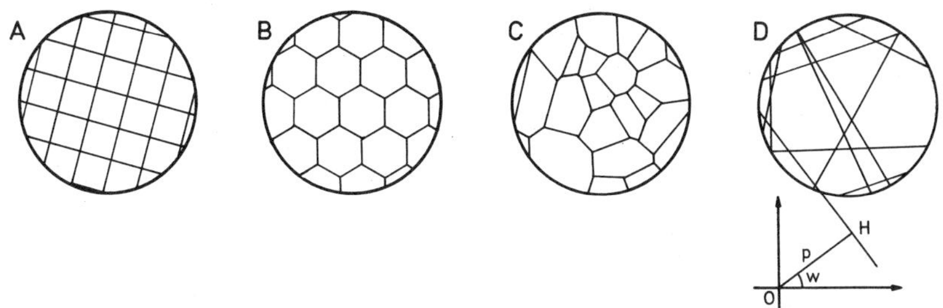


*FIG. 1. Subdivision of a plane region into cells.* (A) *and* (B)*: Cells of congruent convex polygons.* (C)*: Dirichlet cells of a Poisson process.* (D)*: Cells formed by random lines.*

Attach to each center its *Dirichlet cell* (also called Voronoi polygon) consisting of those points that are closer to this center than to any other. The Dirichlet cells of a realization of a Poisson process constitute a random division of the plane into polygons varying in size and shape, see Figure 1(C). The probability that two points, at mutual distance $v$, shall belong to the same cell is in this case

$$\psi(v) = 4\lambda \int_0^\infty \int_0^\infty \exp\{-\lambda U(\sqrt{(x-\tfrac{v}{2})^2+y^2},\ \sqrt{(x+\tfrac{v}{2})^2+y^2};\ v)\}dxdy \tag{2.4}$$

where $U(a,b;c)$ denotes the area of the union of two circles with radii $a$ and $b$ respectively, and with centers at distance $c$ apart. By consideration of geometric probabilities it can be seen that several properties of the mosaic are functionals of $\psi(v)$. The expected length per unit area of $R_2$ of borders between cells is, by the same argument as gave (2.2),

$$-\frac{\pi}{2}\cdot\psi'(0). \tag{2.5}$$

The expected area of the cell to which a given point belongs is

$$2\pi\int_0^\infty v\ \psi(v)\ dv. \tag{2.6}$$

This quantity must be distinguished from the expected area of a randomly chosen cell, which is

$$E(A) = 1/\lambda. \tag{2.7}$$

The mean value in (2.6) is seen to equal

$$E(A^2)/E(A), \tag{2.8}$$

where, as in (2.7), expectation is taken over all cells. The formulae (2.5) and (2.6) are valid for any isotropic system of cells. It is easy to check that they give correct results in the previous case of congruent polygons. For the Dirichlet cells of a Poisson process we find

$$-\frac{\pi}{2}\psi'(0) = 2\sqrt{\lambda}\ .$$

This result, and (2.7), show that the average perimeter per cell is $4/\sqrt{\lambda}$. Inserting (2.4) into (2.6), we get

$$\frac{E(A^2)}{E(A)} = \frac{4\pi}{\lambda}\int_0^\infty\int_0^\infty \{U(\sqrt{(x-\tfrac{1}{2})^2+y^2},\ \sqrt{(x+\tfrac{1}{2})^2+y^2};\ 1)\}^{-2}\,dxdy \tag{2.9}$$

Numerical integration of (2.9) gives the value $1.28018/\lambda$. Using also (2.7), we conclude that the variance of the size of a cell is

$$0.28018/\lambda^2. \tag{2.10}$$

In an analogous way, Dirichlet cells can be attached also to other random arrangements of centers. With the exception of strictly regular arrangements, the corresponding pattern of cells has the property that the probability is zero that four or more cells shall meet in the same corner. From Euler's polyhedral formula, it then follows that the expected number of corners of a cell is 6. (In the case of a mosaic of congruent regular polygons, the number of corners is 3, 4, or 6, corresponding to the triangular, square, and hexagonal network respectively.)

A third type of network is obtained by the cells formed by *intersecting random curves* in $R_2$. In the simplest case, the curves are straight lines, see Figure 1(D). In this case, we assume that the probability is zero that a point is common to more than two lines. Applying Euler's theorem once more, we find that the expected number of corners in a cell is 4. To specify the model we assume that the lines represent a realization of a *Poisson process of lines in* $R_2$. This process is defined in the following way. In Figure 1(D), let the parameter $p$ denote the length of the perpendicular OH from the origin to the line, $w$ the angle that OH makes with an arbitrary axis. Then, in the region $0 < p < \infty$ and $0 < w < 2\pi$ of the pw-plane, points are chosen in accordance with a planar Poisson process with density $\mu$. We attach the line

$$y\cdot\cos w + x\cdot\sin w - p = 0 \tag{2.11}$$

in the xy-plane to a selected pair $(p,w)$. If we define

$$\lambda = \pi\mu^2 \tag{2.12}$$

then the average area per cell is $1/\lambda$. Further, we find that the probability that two points belong to the same cell is

$$\psi(v) = \exp(-2\mu v), \qquad (2.13)$$

where, as before, $v$ is the distance between the two points. This simple expression for $\psi(v)$ can be deduced as a consequence of the fact that the points of intersection between a given straight line $L$ (say) in the xy-plane and the lines (2.11) form a one-dimensional Poisson process along $L$ with density $2\mu$. Inserting (2.13) into (2.5) and (2.6) we find that the expected perimeter of a cell is $2/\mu = \sqrt{4\pi/\lambda}$, and that the variance of the size of a cell is $(\pi^2 - 2)/(2\lambda^2) = 3.9348/\lambda^2$. The variance is considerably higher than in the case of Dirichlet cells of a Poisson process in $R_2$ with intensity $\lambda$.

It should be observed that we have only considered *isotropic* mechanisms for the production of networks of cells. In the case of non-isotropic, but spatially homogeneous, mechanisms, the function $\psi(v)$ should be replaced by a function $\psi(x,y)$, giving the probability that two points with the coordinates $(x_0, y_0)$ and $(x_0 + x,\ y_0 + y)$ belong to the same cell.

## 3. RANDOM MOSAICS

As mentioned already, we shall chiefly deal with two-phase mosaics, where the two phases are called black and white, respectively. We shall also concentrate on mosaics derived from the cell models of the previous section.

The obvious way to construct a random two-phase mosaic from a network of cells, is to make a series of random experiments for deciding which color to give to each cell. We first introduce a certain symmetry among the cells by assuming that the experiments are identical. Hence, the probability that a cell is black is always equal to $p$ (say), and the probability that the cell is white is $1-p$. If we make the additional assumptions that the cell model is *isotropic*, and that all the experiments are *independent*, we shall speak of a *purely random mosaic*. (This definition is much wider than the one used by Pielou, 1969, p. 149). Alternatives of possible interest are those that involve a *correlation between neighboring cells*. The correlation may be positive, when neighboring cells have a comparatively high probability of being colored alike, or negative, when neighboring cells have a tendency to be of different colors. The detailed definition of such alternatives can be made in different ways. For the models based on congruent polygons, we may use autoregressive schemes. They will however be more complex than the corresponding models for time series, see Whittle (1954) and Bartlett (1971).

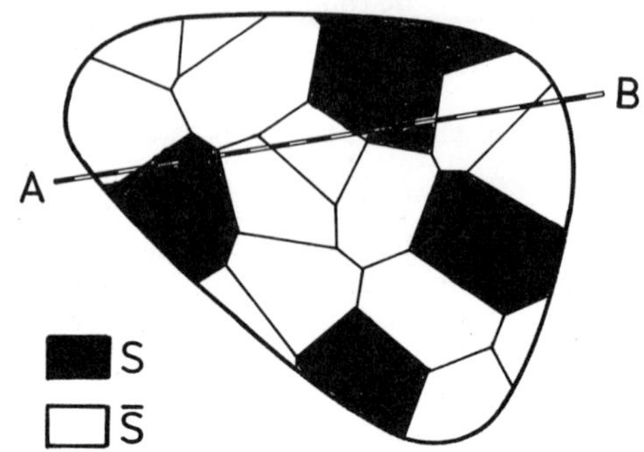


*FIG. 2. Mosaic formed by allocating at random two phases (black, and white) to the Dirichlet cells of points in* $R_2$. *(See text).*

When each cell has got its color, black or white, $R_2$ is divided into two sets: S, the union of all black cells, and its complement $\bar{S}$, the union of all white cells. S may, or may not, consist of distinct clumps, that appear as 'islands' in $\bar{S}$. Figure 2 shows an example based on the Dirichlet cells of points in $R_2$. A realization of a two-phase mosaic in $R_2$ is completely defined by a stochastic function $z(x,y)$ (say) which takes the value 1 if $(x,y)$ belongs to S, and 0 if $(x,y)$ is in $\bar{S}$. For the spatially homogeneous case we have the expectation

$$E[z(x,y)] = p \,. \tag{3.1}$$

For a purely random mosaic, we have

$$c(v) = \mathrm{Cov}[z(x_1,y_1),\ z(x_2,y_2)] = p(1-p)\psi(v)\,, \tag{3.2}$$

where $\psi$ is the function defined in Section 2, and $v$ is the distance between $(x_1,y_1)$ and $(x_2,y_2)$. From (3.1) and (3.2) it is seen that the values of $z(x,y)$ along any line in $R_2$ (as the line AB in Figure 2) represent a stationary one-dimensional stochastic process with mean $p$, variance $p(1-p)$, and auto-correlation function $\psi(v)$. Hence $\psi(v)$, which is an expression of the properties of the individual cell, also characterizes an essential property of the purely random mosaic based on the net-work of cells. We note that the expected length of border lines between S and $\bar{S}$ is $-\pi c'(0)$ per unit area, cf. (2.5).

In the patterns considered so far, there is a certain symmetry between the two phases. We shall now consider a simple example of a type of model where the two sets, S and $\bar{S}$, are not interchangeable. Consider again a Poisson process of centers

*FIG. 3. A 'mosaic' formed by randomly located circles.*

in $R_2$ with density $\lambda$. Define S as the union of all circles around these centers with the fixed radius r, and $\bar{S}$ as the complement $R_2-S$, see Figure 3. We find for this model:

$$E[z(x,y)] = 1 - \exp(-\lambda\pi r^2), \tag{3.3}$$

$$\mathrm{Cov}[z(x_1,y_1),\ z(x_2,y_2)] = \exp[-\lambda U(r,r;v)] - \exp(-2\lambda\pi r^2), \tag{3.4}$$

where v is the distance between $(x_1,y_1)$ and $(x_2,y_2)$. This model can obviously be modified in various ways. We can, for example, assign to each center a figure at random from a given set of figures, varying in size and shape. Also, the location of centers can follow a random scheme other than the Poisson process. On these models, see Kendall and Moran (1963, esp. p. 112-117), and Roach (1968). In the literature, the interest concentrates on the formation of clumps by overlapping among the figures dropped on the plane. Although the problems concerning the probability of formation of clumps seem "tantalizingly simple" (Roach, 1968, p. 26), few exact solutions have been found. However several approximations have been worked out. To quote one example, let us again consider the simple case of circles with constant radius, r. Then the expected number of isolated clumps per unit area formed by n circles is approximately

$$\lambda a(1-a)^{n-1}/n, \tag{3.5}$$

with $a = \exp(-4\lambda\pi r^2)$. The formula is exact for $n=1$. The approximate number of isolated clumps of all sizes is found by summing (3.5) for $n=1,2,\cdots$. The sum is

$$4\lambda^2 \pi r^2 / [\exp(4\lambda\pi r^2) - 1]. \tag{3.6}$$

We shall now comment briefly on models of n-phase mosaics. The method of producing a two-phase mosaic from a network of cells can be extended in a straightforward way to the formation of n-phase networks, at least in the purely random case. We only have to allocate the $n$ colors to each cell with probabilities $p_1, p_2, \cdots, p_n$ $(p_1 + p_2 + \cdots + p_n = 1)$, independently of the coloring of other cells. There is no obvious way of generalizing the 2-phase mosaic formed by dropping figures at random over $R_2$. One might give each figure a color by a random experiment, and have a rule for ranking the colors in the case of overlapping. It should then always be clear which one of two colors dominates. A realization of an n-phase mosaic can be completely described by $n$ indicator functions $z_1(x,y), \cdots, z_n(x,y)$, where $z_j(x,y)$ is 1 if $(x,y)$ is situated in a region with the *jth* color, 0 otherwise. In the purely random case, the formulae (3.1) and (3.2) can be applied to $z_j(x,y)$ if we substitute $p_j$ for $p$. A general covariance-formula, involving both the auto-covariances, considered previously, and cross-covariances is

$$\mathrm{Cov}[z_j(x_1,y_1),\ z_k(x_2,y_2)] = p_j(\delta_{jk} - p_k)\psi(v), \tag{3.7}$$

where $\delta_{jk}$ is Kronecker's delta (1 if $j = k$, 0 if $j \neq k$).

To get alternatives to the purely random model we should introduce a correlation, positive or negative, between pairs of colors. Thus two colors should be given a tendency to appear in contiguous cells or to avoid this kind of contact. How such a procedure should be carried out in detail remains to be investigated. When we consider the complex structure of natural communities, the shortcomings of the simple mechanical devices considered here are obvious. We may hope to get more realistic models by including a *dynamic element*. A mathematical theory intended to show how an ecological mosaic comes into existence and develops should presumably lead to a more logical model of the mosaic existing at a particular moment. It can, however, be pointed out that at least two of the above models can be given a dynamic interpretation. The network of Dirichlet cells (see Figure 1(C) and Figure 2) can be considered as an equilibrium situation resulting from a process of growth: From seeds germinated in random spots, plants grow (e.g. by rhizomes) uniformly in all directions until running into the neighbors' territories. Such a model has been developed by Fischer and Miles (1973). Also the model shown in Figure 3 can similarly be completed by introducing the time dimension. By varying stochastically the speed of growth in different individuals and in varying directions, we can produce patterns of a less stereotyped appearance.

## 4. SOME COMMENTS ON INFERENCE PROBLEMS

An empirical investigation of ecological mosaics may serve different objectives. We may want to perform a detailed testing of a particular model and an estimation of its parameters. Our purpose may instead be to estimate some general characteristics of the mosaic, not related to any particular model, such as the total areal extent of the different phases. The methods of collecting the information may also vary. Our observations can emanate from sample points or from a sample of transects, quadrats etc. The observations may consist of measurement of areas, lengths, distances, etc., or they may be confined to classifying the sampling units by phases. All this means that we meet a great variety of statistical problems. Some of them are the common problems of sampling, estimation, hypothesis testing, etc. We shall not here take up these general statistical problems. We shall instead comment on some of the problems that seem to be specific to the study of mosaics.

In dealing with models based on *networks of cells*, we can distinguish between the case when the *cells are observable*, and the case when *only the resulting mosaic* - and not the borders between cells 'colored alike' - *can be observed.*

In the first case, we can collect data on the size and shape of individual cells. Among the questions encountered in such studies, mention may be made of 'stereological' problems. These concern cases in which our observations are made in a space of lower dimension than that of the geometric object under investigation. Suppose, for example, that we measure the length of chords cut out from the cells by a sample of transects and want to estimate the size distribution of the cells. If shape is independent of size, we can write down an integral equation connecting the distribution of size of cells and the distribution of length of chords. An explicit solution is known in the case of circular cells. More details and references are found in Kendall and Moran (1963, p. 67-70, 86-92) and Moran (1966, p. 460). The problem is illustrated in Figure 4. It appears that much remains to be done in this field, especially in regard to the accuracy of estimation.

If the cell network is isotropic, we may wish to estimate the function $\psi(v)$, the probability that two points at distance $v$ apart belong to the same cell. A simple way to collect data for this purpose is to select clusters of points, the points of

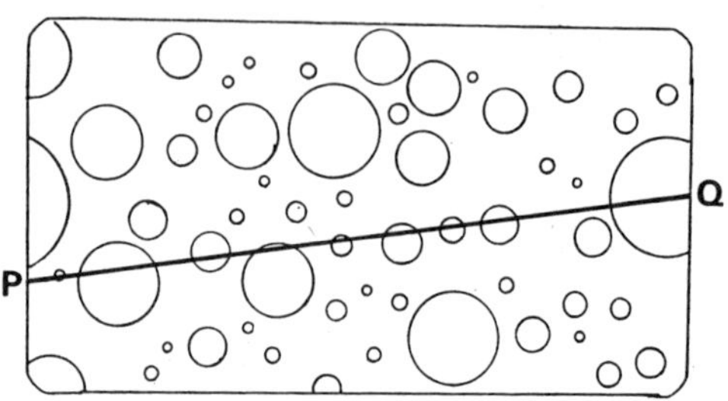


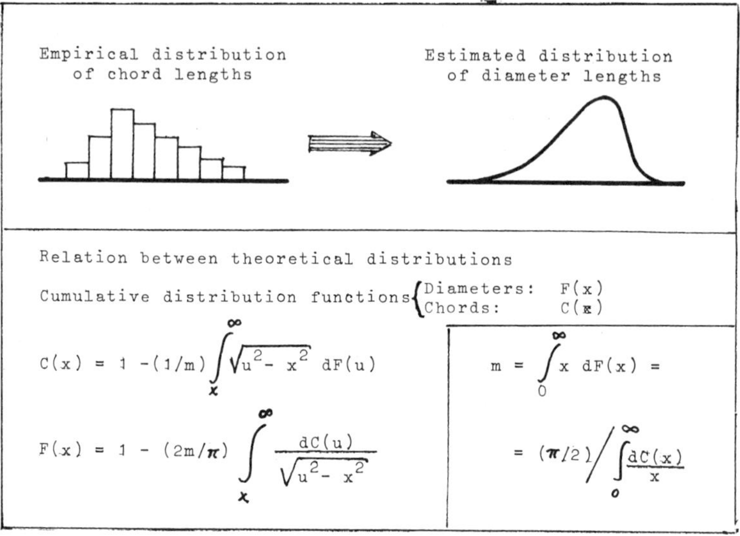


*FIG. 4: The relation between the distributions of diameters and chords in circles, when chords are measured along transects (see* PQ *in the topmost figure).*

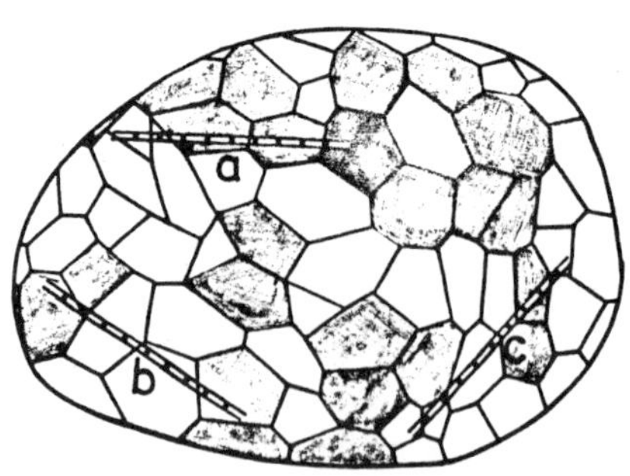


*FIG. 5: Three clusters (*a, b, *and* c*) of points for the estimation of* $\psi$(v) *(see text).*

a cluster being situated equidistantly along a line of fixed length (not smaller than the maximum diameter of a cell). See the three segments a, b, and c in Figure 5. For each point we record to which cell it belongs. If the clusters are chosen at random over a large area, standard sampling theory can be applied to the estimation of $\psi(v)$ for values of $v$ corresponding to distances between observation points on the segment.

If we have a 2-phase mosaic with a discernible cell structure, we may want to *test if the mosaic is purely random.* An easy way to collect information to elucidate the problem is to observe the succession of phases along transects, such as the line AB in Figure 2, where we find the sequence

$$W\ B\ W\ W\ B\ W\ W. \tag{4.1}$$

It seems intuitively clear that the relevant question is whether the $m$ W's and the $n$ B's in a sequence of type (4.1) occur in a purely random order. It should however be noted that we must assume that the cells are *convex.* Otherwise, the assumption of a purely random mosaic would not necessarily mean that the order of the letters in a sequence such as (4.1) is purely random. We now make the assumption of convexity. (It should perhaps have been included already in the definition of a purely random mosaic?) The most popular statistical tests of the randomness of a sequence of symbols, are based on *runs.* In (4.1) we have the five runs

$$W,\quad B,\quad WW,\quad B,\quad WW.$$

In the *one-sample runs rest,* the test quantity is simply the number of runs, $r$ (say). For short series, the probability distribution of $r$ can be obtained simply by enumerating the possible cases, or from tables; (see, for example, Siegel (1956, Table F).) In (4.1) the observed number of runs (5) is the highest possible with $m = 5$, $n = 2$. We easily find, however, that the probability of getting $r = 5$ is in this case as high as $2/7$. If $m$ and $n$ are large, we can use the normal approximation with mean

$$E[r] = 1 + \frac{2mn}{m+n}\ , \tag{4.2}$$

and variance

$$\mathrm{Var}(r) = \frac{2mn(2mn-m-n)}{(m+n)^2(m+n-1)}\ . \tag{4.3}$$

Under the two alternative hypotheses, vaguely formulated above, $r$ has a tendency to be either lower (positive correlation between adjoining cells) or higher (negative correlation between adjoining cells) than (4.2). So the one-sample runs test has a certain intuitive appeal in the present situation. If we have transects sufficiently far from one another, the corresponding r's can be regarded as independent. We can then merge the sequences into one long series and apply (4.2) and (4.3) to this series.

If the cell network consists of *congruent convex polygons* (Types A and B in Figure 1), we can apply methods devised for testing the random mixture of different objects arranged in the lattice. For a network of two colors the methods, which can be regarded as generalizations of the runs test, are discussed by Ord (1979).

Similar methods can be applied when we have n phases (or colors). In the test of the randomness of the arrangement of the B's, it is immaterial whether W stands for one color or is a common designation for several colors. The problem of testing if two particular colors in an n-phase mosaic are randomly mixed without making any assumption about the arrangement of the other colors is slightly different. We take again as example a square network of cells, see Figure 6. We want here to test whether the 35 B's and 39 W's in the lattice are

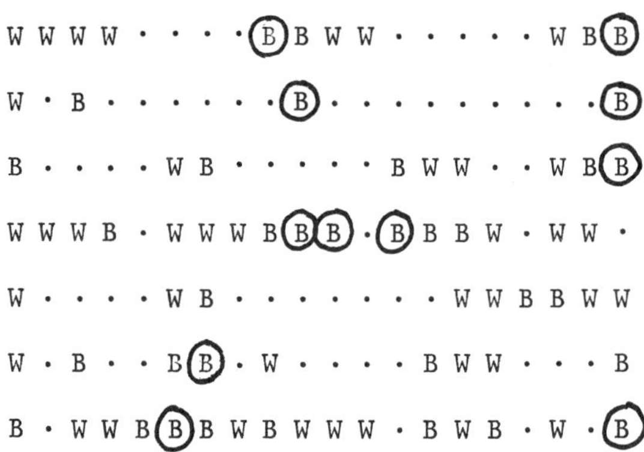


*FIG. 6: The arrangement of black* (B) *and white* (W) *cells in a square lattice. The dots represent other colors. The circle cells are "isolated* B*'s".*

distributed completely at random over the 74 cells occupied by a B or a W. We shall (cf. Verhagen, 1971) use as test quantity the number of *isolated black cells*, defined to mean that there is no W in any of the adjoining cells. If we consider two cells as adjoining when they have one side in common the number of isolated Bs in Figure 6 is 11. Verhagen (1971) considers also the case of diagonal joins, both in square and triangular grids. The expected number of isolated B's - under the hypothesis of random mingling of the two colors - is found by summing, for all the occupied cells, the probabilities that they are isolated B's. Using the notation $x^{(r)} = x(x-1)\cdots(x-r+1)$, the probability that an occupied cell having $k$ occupied neighbors will contain an isolated $B$ is

$$m^{(1+k)}/N^{(1+k)}, \tag{4.6}$$

where $m$ is the total number of B's and $(N-m)$ the total number of W's. In the example of Figure 6 the expected number of isolated B's is found to be 10.81. To find the expectation of the square of the number of isolated B's (and hence the variance) we sum for all pairs of occupied cells the probabilities that both cells are isolated B's. For two occupied cells the probability that both shall be isolated B's is

$$m^{(2+j)}/N^{(2+j)}, \tag{4.7}$$

where $j$ is the total number of additional occupied cells adjoining one or both of the two cells. There is no need to perform this calculation in the example, where the observed number is as close as possible to the expected. Verhagen (1971) points out that the computation of the mean and variance, by summing expressions of the type (4.6) and (4.7), can be readily automated with the aid of a digital computer. In the case of a regular area in which all sites are occupied, general formulae can be obtained for the mean and the variance. They can be substantially simplified by wrapping the lattice around a torus, so that all cells have the same number of neighbors. Verhagen (1971) gives such formulae in a number of cases.

An *irregular network of cells* can be examined in a similar way by counting (for example) the *number of contacts* between cells belonging to the same phase. We find in Figure 2 three contacts between the seven black cells. We can here perform a randomization test, consisting of selecting in all possible ways seven out of the 21 cells, and count the number of contacts between them. For a large number of cells, where a total enumeration of all cases is impractical, we may perform Monte Carlo experiments. Alternatively, the normal approximation may be applied. We then have to compute the mean and variance of the number of contacts. To do this we can (in analogy with the above tests, quoted from Pielou and Verhagen) write the test statistic as the quadratic form

$$\Sigma\ c(i,j)\ x_i x_j, \tag{4.8}$$

with summation over all pairs of different cells. Here $c(i,j)$ has the value 1 or 0, according as the *ith* and *jth* cells adjoin or not. The random variable $x_i$ is given the value 1 if the *ith* cell is black, 0 otherwise. To find moments of this sum, we use the formula

$$E[x_{i_1} x_{i_2} \cdots x_{i_k}] = m^{(s)}/N^{(s)},$$

where N is the total number of cells, m is the number of black cells, and s is the number of different integers among $i_1, i_2, \cdots, i_k$. It is worth noting that this test is most useful when the number of W cells is relatively small ($\leq$ 20% of total cells in lattice, say). Otherwise, the expected number of isolated B's becomes very small and the test lacks power.

We now pass to *methods which do not presuppose that the cell pattern is discernible.*

The fundamental characteristics of an isotropic mosaic are the proportions of the different phases (as p in formula 3.1), the autocovariance function (see 3.2 and 3.4), and the cross-covariance function (3.7) with $j \neq k$). The estimation of p in a practical case amounts of course to the estimation of areas. We shall not take up the many methods used for this purpose. The estimation of an autocovariance function is illustrated in Figure 7. The diagram shows correlograms, the graphic representation of the estimated correlation function. The sampling is analogous to the one indicated in Figure 5. As indicated by the correlograms taken in different directions we may often have occasion to investigate whether a mosaic is isotropic, or anisotropic.

If the areas with one particular color consist of patches of finite size, surrounded by areas of other color, we can estimate the properties of these patches by methods similar to those used for the estimation of cells (in mosaics with a visible cell-structure). If the cells of a *2-phase mosaic* cannot be observed, it is difficult to test whether the mosaic is purely random in the sense defined earlier. It may even seem meaningless to speak about randomness in a 2-phase mosaic without visible cells. One may argue that very similar patterns can arise from a purely random model with large cells as from a model with small cells, involving high correlation between neighboring cells.

When we pass to *n-phase mosaics* ($n > 2$), the situation is changed, as can be illustrated by the succession of phases along a transect. In a two-phase mosaic, we can only have patterns of the type

$$\cdots W\ B\ W\ B\ W\ B\ W\ B\ W\ B\ W\ B\ W\ B\ W\ B\ W\ B \cdots$$

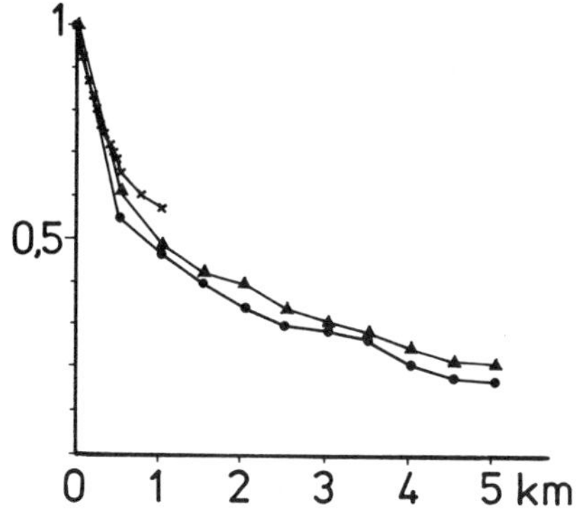


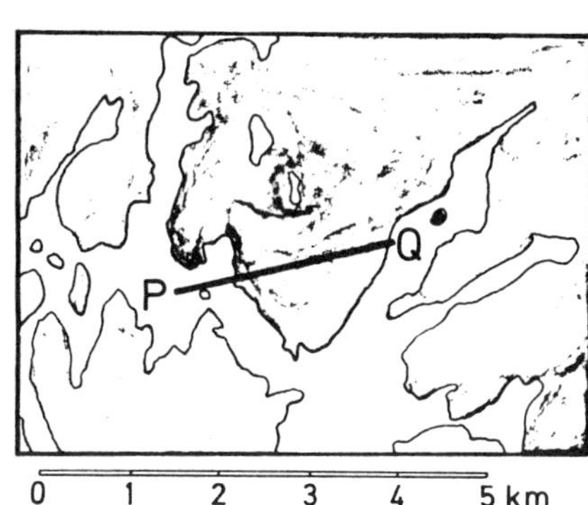


*FIG. 7: Correlogram of the distribution of land area according to a map of the Stockholm area (scale* 1:50000*)*

x *Observations from* 72 *rays of length* 50 *mm, in various directions (example* PQ *in the inserted portion of the map).*

· *Observations from transects in direction east-west.*

Δ *Observations from transects in direction north-south.*

which give no information. In a 4-phase mosaic, for example, with the phases B, G, W, and Y, we can observe a sequence such as

··· B W Y B Y G W G W G B Y B W G W G W G B Y B G B Y B Y W ···

Here there may appear to be a positive association between W and G, and likewise between B and Y. We can evidently test at least some aspects of the randomness in an n-phase mosaic. If we assume that the mosaic is purely random, based on a system of *convex* cells, then a sequence such as the last should be a completely random mixture of the phases, with the restriction that no two adjacent letters are the same. The sequence of letters is therefore a Markov chain with a matrix of transition probabilities that is of a special simple type. The statistical methods applicable to Markov chains can hence be used to test the hypothesis of a purely random mosaic. We can apply a $\chi^2$ test to the frequencies of transitions from one color to another, see e.g. Pielou (1969).

We shall mention finally the problem of mapping a mosaic by observations made in a sample, for example a (systematic) sample of points in a square grid, or a sample of equidistant lines. The problem has been studied by Switzer (1971).

ACKNOWLEDGEMENT

The author is indebted to the editors for many suggestions that have improved the text.

REFERENCES

Bartlett, M. S. (1971). Two-dimensional nearest-neighbour systems and their ecological applications. In *Statistical Ecology, Vol. 1*, G. P. Patil, E. C. Pielou, W. E. Waters, eds. The Pennsylvania State University Press, University Park. 179-194.

Diggle, P. (1979). Statistical methods for spatial point processes in ecology. In *Spatial and Temporal Analysis in Ecology*, R. M. Cormack and J. K. Ord, eds. Satellite Program in Statistical Ecology, International Co-operative Publishing House, Fairland, Maryland.

Fischer, R. A. and Miles, R. E. (1973). The role of spatial pattern in the competition between crop plants and weeds: a theoretical analysis. *Mathematical Biosciences*, 18, 335-350.

Kendall, M. G. and Moran, P. A. P. (1963). *Geometrical Probability*. Griffin, London.

Matérn, B. (1960). Spatial variation. *Meddelanden från statens skogsforskningsinstitut*, 49, 5.

Moran, P. A. P. (1966). A note on recent research in geometrical probability. *Journal of Applied Probability*, 3, 453-463.

Ord, J. K. (1979). Time-series and spatial patterns in ecology. In *Spatial and Temporal Analysis in Ecology*, R. M. Cormack and J. K. Ord, eds. Satellite Program in Statistical Ecology, International Co-operative Publishing House, Fairland, Maryland.

Pielou, E. C. (1969). *An Introduction to Mathematical Ecology*. Wiley, New York.

Roach, S. A. (1968). *The Theory of Random Clumping*. Methuen, London.

Switzer, P. (1971). Mapping a geographically correlated environment. In *Statistical Ecology, Vol. 1,* G. P. Patil, E. C. Pielou, W. E. Waters, eds. The Pennsylvania State University Press, University Park. 235-269.

Verhagen, A. M. W. (1971). Clustering of attributes on regular point lattices. *Journal of Applied Ecology,* 8, 665-682.

Whittle, P. (1954). On stationary processes in the plane. *Biometrika,* 41, 434-449.

[*Received: April* 1977. *Revised: February* 1979]

R. M. Cormack and J. K. Ord, (eds).,
*Spatial and Temporal Analysis in Ecology*, pp. 289-304. 

# A TEST OF DIFFERENT QUADRAT VARIANCE METHODS FOR THE ANALYSIS OF SPATIAL PATTERN

JOHN A LUDWIG

Biology Department
New Mexico State University
Las Cruces, New Mexico 88003 USA

SUMMARY. Methods based on Quadrat Variances have been tested against the known intensity and grain of artificial spatial distribution patterns. In 1973, Hill described a Two-Term Local-Quadrat Variance (TTLQV) method for the analysis of spatial pattern. It was found to show accurately the known scales of spatial distribution tested here. In 1974, Goodall introduced a new method, based on the Random Pairing of Quadrats for Variance (RPQV) estimates at different spacings. In theory this RPQV method has many clear advantages over earlier methods. However, in practice it is limited to relatively few spacings since random pairing without replacement leads to only a few degrees of freedom for each variance estimate (unless a very large number of quadrats have been sampled - not the usual case). Another method based on Paired-Quadrat Variance (PQV) is described here. It differs from Goodall's RPQV method in that all possible pairs of quadrats are used to estimate the variance at a given spacing. It is equivalent to the correlogram used in time-series analysis. It was found to reveal accurately an existing small-scale pattern, but a second larger-scale pattern may be obscured.

KEY WORDS. plant dispersion, spatial pattern methods, contiguous quadrats, blocked-quadrats, paired-quadrats, variances.

## 1. INTRODUCTION

Spatial pattern analysis is the quantitative description of the horizontal distribution of individuals of a species within a homogeneous plant community, where this distribution may be random, regular, or aggregated. Approaches to the analysis of spatial pattern originally centered on statistical tests of randomness, e.g., use of the Poisson Distribution. However, since many dispersion patterns are non-random, recent interest has centered on approaches that provide quantitative descriptions of the dispersion, i.e., the intensity and grain of spatial pattern (Pielou, 1977). As surveyed by Pielou (1977), these approaches are based either on distances between individuals (or random points) or on abundances in contiguous quadrats. Recent developments in the use of contiguous quadrats have sought to overcome several drawbacks in the original hierarchical analysis of variance methods described by Goodall (1954), Greig-Smith (1952), and Kershaw (1957). Even though Goodall (1974) and Pielou (1977) have listed and discussed these drawbacks, the major ones are repeated here along with the recent developments attempting to overcome them.

Perhaps the most severe drawback in the original methods is a very limited choice of spacings or quadrat distances, since the contiguous quadrats are blocked into successive powers of two for variance estimates. In other words, a set of quadrat spacings at block sizes 1, 2, 4, 8, 16, 32, 64, etc. are imposed by these methods, here referred to as the Blocked Quadrats Variance (BQV) methods. A graph of variance against block size is traditionally examined and the peaks and troughs interpreted as to the intensity and scale of any heterogeneity pattern.

Hill (1973) has proposed several alternative modifications of the original BQV methods, of which the most promising appears to be the Two-Term Local Quadrat Variance (TTLQV) method. This method emphasizes the intensity of spatial pattern at *any block size* up to half the number of contiguous quadrats. Using field data for *Acacia ehrenbergiana* (Greig-Smith and Chadwick, 1965), Hill compared TTLQV with BQV and also with Spectral Analysis (SA), a descriptive technique applicable to spatial data (Bartlett, 1964), as well as its usual time-series application.

Another serious drawback in the BQV methods is that, if a regular pattern exists, the starting point of Blocking can affect the results. Usher (1969) proposed a modification of the BQV methods involving a Stepped Blocking of the Quadrats for Variance (SBQV) estimation. Using artificial data, Errington (1973) and Usher (1975) have shown that peak variance drifts to the right or left, depending on the starting position of the blocking. The SBQV method emphasizes the interpretation of pattern from the shifts in the peak variance, with the suggestion that the average variances for block sizes 1, 2, 4, etc. over all analyses with different starting positions be used for a generalized interpretation. Using these artificial data and also field data for *Lotus corniculatus* (Burden, 1970), Usher (1975) compared his SBQV method with Hill's TTLQV method and with SA.

A limiting drawback in the BQV methods, and the SBQV and TTLQV modifications of it, is that the variance estimates at each of the block sizes or spacings are calculated from the same data and significance tests use the same residual sum of squares. A significantly different variance at one block size invalidates tests at other block sizes unless one invokes a trivial case, where the observations in each quadrat are all independent and normally distributed (Thompson, 1958). To avoid this, Mead (1974) applied a special '2 within 4 randomization test' and Goodall (1974) proposed a Random Pairing of Quadrats to obtain Variance (RPQV) estimates. Instead of the usual 'dependent' variance estimates from blocked quadrats, Goodall obtains 'independent' variance estimates by selecting and using the differences between random pairs of quadrats. The terms 'dependent' and 'independent' here refer to whether variances are calculated from the same quadrat data or from random subsets of the data respectively. Independent variances allow for the formation and testing of hypotheses about the significance of differences in variance over a wide range of selected quadrat spacings without invoking a trivial case.

Another quadrat variance method based on Paired-Quadrat Variances (PQV) has been described by Ludwig and Goodall (1978). The PQV method calculates variances from *all possible paired quadrats* at a given spacing or distance, rather than from *randomly paired quadrats* as in the RPQV method. The PQV method is equivalent to the Correlogram used in time-series analysis (see Chatfield, 1975), but is applied to spatial data. Of course, variances obtained by the PQV procedure will not be independent for *different* spacings since the same data will be used each time, unlike the RPQV method which uses the procedure of random pairing without replacement. Lack of independence is of less concern in the exploration for pattern in quadrat data than in the subsequent testing of hypotheses about any pattern found.

The purpose of this paper is to describe the different quadrat variance methods for the analysis of spatial pattern and to test further the performance of these different methods on simulated spatial patterns of different levels of complexity. These methods were compared on extensive field data and on some artificial data by Ludwig and Goodall (1978); however, the aim here will be to test these methods on more complex patterns of artificial data.

## 2. QUADRAT VARIANCE METHODS

Although detailed descriptions of the BQV, SBQV, TTLQV, RPQV, and PQV methods are in Usher (1975), Hill (1973), Goodall (1974), and Ludwig and Goodall (1978), respectively, a brief emphasis of the differences in these methods is provided by using Hill's mathematical formulations. If the observations in quadrats along a simple belt transect are given by

$$x_1, x_2, x_3, \ldots, x_n, x_{n+1}, x_{n+2}, \ldots, x_{n+m},$$

where $n$ and $m$ are integers representing total numbers of quadrats ($n$ equal to a power of 2), then the calculation of mean squares or variances for these quadrat variance methods can be compared for the two smallest block-sizes or spacings, with analogous calculations at higher block-sizes or spacings.

*2.1 Blocked-Quadrat Variance Methods.* For the BQV method, the variance at a block-size of one is given by

$$\mathrm{Var}(X)_1 = (2/n)[(1/2)(x_1-x_2)^2 + (1/2)(x_3-x_4)^2 + \cdots + (1/2)(x_{n-1}-x_n)^2];$$

at a block-size of two by

$$\mathrm{Var}(X)_2 = (4/n)[(1/4)(x_1+x_2-x_3-x_4)^2 + (1/4)(x_5+x_6-x_7-x_8)^2 + \cdots + (1/4)(x_{n-3}+x_{n-2}-x_{n-1}-x_n)^2];$$

with analogous calculations at spacings of 4, 8, 16, etc. Note that only the quadrats up to $n$ in number can be used due to the restriction of using blockings in powers of two with this method.

For the SBQV method, the variance at a block-size of one is given by

$$\mathrm{Var}(X)_1 = [1/(m+1)] \sum_{i=0}^{m} [(2/n)\{(1/2)(x_{1+i}-x_{2+i})^2 + (1/2)(x_{3+i}-x_{4+i})^2 + \cdots + (1/2)(x_{n-1+i}-x_{n+i})^2\}];$$

at a block-size of two by

$$\mathrm{Var}(X)_2 = [1/(m+1)] \sum_{i=0}^{m} [(4/n)\{(1/4)(x_{1+i}+x_{2+i}-x_{3+i}-x_{4+i})^2 + (1/4)(x_{5+i}+x_{6+i}-x_{7+i}-x_{8+i})^2 + \cdots + (1/4)(x_{n-3+i}+x_{n-2+i}-x_{n-1+i}-x_{n+i})^2\}];$$

and so forth. Note that Usher's SBQV procedure is to average m+1 BQV analyses, each at a different starting position by stepping along the transect as many times as there are m additional quadrats beyond n.

For the TTLQV method, the variance at a block size of one is given by

$$\mathrm{Var}(X)_1 = [1/(n+m-1)][(1/2)(x_1-x_2)^2 + (1/2)(x_2-x_3)^2 + \cdots + (1/2)(x_{n+m-1}-x_{n+m})^2];$$

at a block-size of two by

$$\mathrm{Var}(X)_2 = [1/(n+m-3)][(1/4)(x_1+x_2-x_3-x_4)^2+(1/4)(x_2+x_3-x_4-x_5)^2 + \cdots + (1/4)(x_{n+m-3}+x_{n+m-2}-x_{n+m-1}-x_{n+m})^2];$$

and so forth. Note that Hill's TTLQV method avoids the restriction of block-sizes being equal to powers of 2 by calculating exhaustively variances for all block-sizes up to the maximum possible at (n+m)/2.

2.2 *Paired-Quadrat Variance Methods*. For the RPQV method, the variance at a spacing of one is given by

$$\mathrm{Var}(X)_1 = (1/p) \sum_{j=1}^{p} [(1/2)(x_j-x_{j+1})^2],$$

where j is a series of p random numbers between 1 and n+m. The quadrat pairs j and j+1 are taken without replacement. At a spacing of two

$$\mathrm{Var}(X)_2 = (1/q) \sum_{k=1}^{q} [(1/2)(x_k-x_{k+2})^2],$$

where k is another series of q random numbers ($k \neq j$) between 1 and n+m. The quadrat pairs k and k+2 are also taken without replacement. Note that by random sampling without replacement. Note that by random sampling without replacement, $\mathrm{Var}(X)_1$ and $\mathrm{Var}(X)_2$ are independent estimates for two different spacings. Thus standard tests of significance (e.g., variance ratio) can validly be applied to test hypotheses about any difference in variance at these two spacings.

For the PQV method, the variance at a spacing of one is given by

$$\mathrm{Var}(X)_1 = [1/(n+m-1)][(1/2)(x_1-x_2)^2 + (1/2)(x_2-x_3)^2 + \cdots + (1/2)(x_{n+m-1}-x_{n+m})^2];$$

at a spacing of two,

$$\mathrm{Var}(X)_2 = [1/(n+m-2)][(1/2)(x_1-x_3)^2 + (1/2)(x_2-x_4)^2 + \cdots + (1/2)(x_{n+m-2}-x_{n+m})^2];$$

and so forth. In general,

$$\mathrm{Var}(X)_k = [1/(n+m-k)][\sum_{i=k+1}^{m+n} (1/2)(x_i-x_{i-k})^2],$$

for spacing $k$ or lag $k$ in time-series. Note that at a spacing of one, the PQV calculation is the same as in the TTLQV method, since both are exhaustive pairings. However, at a spacing of two, the PQV calculation is based on paired-quadrats whereas the TTLQV calculation involves blocked-quadrats. Pairing does not confound in the variance estimate the effect of change in quadrat size, shape, and distance between centers as does blocking. Pairing gives a variance estimate between quadrats of a fixed size and shape (the original quadrats). Note also that variance estimates up to a spacing of $(n+m-1)$ are possible with the PQV method.

## 3. METHODS OF COMPARISON

Usher (1975) used two sets of artificial quadrat data (160 quadrats each) and a large set of field data (two transects of 600 quadrats each) to compare his SBQV method with BQV and Hill's TTLQV method (and with SA). Usher's artificial data consisted of a basic abundance pattern with two scales of heterogeneity at 4 and 30 (the quadrat abundance score changed after every 4th and 30th quadrat) and a basic amplitude pattern also with two scales of heterogeneity at 4 and 30 (the mean abundance remained constant throughout the transect but the abundance difference or amplitude changed after every 4th and 30th quadrat). Using two sets of field data and Usher's abundance pattern (with a stochastic element added), Ludwig and Goodall (1978) compared their PQV and RPQV methods with the BQV, SBQV, and TTLQV methods. In addition to the random variation applied to Usher's abundance pattern scores, Ludwig and Goodall randomly varied patch width and distribution about 4 and 30, rather than having the patches uniform in size and regular in distribution. However, their comparisons did not include scales of pattern other than at 4 and 30.

In this paper, the PQV, RPQV, BQV, SBQV, and TTLQV methods are compared using scales of pattern varying in both complexity and position. Patch width and position is varied as a random Poisson variate with scales of pattern examined at 4, at 6 & 32, and at 6 & 25. Artificial abundance scores for 1000 quadrats were computer generated, with each score having random noise added (Table 1). FORTRAN computer programs for this artificial pattern generation and for the PQV, RPQV, BQV, SBQV, and TTLQV methods (as applied to belt transects) are available upon request.

*TABLE 1: An example of artificial abundance pattern with randomly varying patch sizes or widths and patch positions. The mean abundance between patches at the small-scale of pattern (mean width and inter-patch distance of* 6*) is* 1 *and within a patch is* 4. *The mean abundance between patches at the large-scale pattern (mean width and inter-patch distance of* 32*) is* 3 *and within a patch is* 6. *These abundances are means since random noise has been added. Quadrats are contiguous across rows with scores for* 15 *quadrats per row and a total of* 100 *quadrats shown in this example. Small-scale patches are underlined.*

| | | | | | | | | | | | | | | |
|---|---|---|---|---|---|---|---|---|---|---|---|---|---|---|
| 0.6 | 0.6 | 1.6 | 0.8 | 0.4 | 1.6 | 1.1 | 0.6 | 3.7 | 4.4 | 4.1 | 4.4 | 3.2 | 4.3 | 4.5 |
| 1.3 | 1.3 | 1.0 | 1.6 | 1.1 | 0.7 | 1.4 | 3.1 | 3.3 | 4.3 | 3.8 | 4.3 | 1.1 | 1.2 | 1.5 |
| 1.2 | 1.1 | 4.0 | 6.1 | 6.0 | 5.2 | 4.3 | 5.5 | 5.7 | 5.7 | 3.3 | 2.7 | 3.2 | 2.4 | 2.2 |
| 3.2 | 5.6 | 6.6 | 5.6 | 6.3 | 2.8 | 3.1 | 3.5 | 2.6 | 2.6 | 3.2 | 3.5 | 2.9 | 6.4 | 6.0 |
| 5.6 | 4.5 | 3.9 | 3.9 | 3.8 | 4.3 | 1.2 | 1.0 | 0.7 | 1.3 | 0.7 | 3.8 | 3.3 | 4.8 | 4.3 |
| 3.4 | 3.3 | 4.8 | 4.2 | 0.6 | 0.3 | 1.1 | 0.3 | 1.0 | 0.4 | 4.3 | 4.1 | 3.9 | 3.8 | 4.6 |
| 3.8 | 3.8 | 0.5 | 0.9 | 2.6 | 2.5 | 3.3 | 5.9 | 5.8 | 6.1 | ........ | up to 1000 | | | |

## 4. RESULTS

*4.1 Simple Pattern.* Plots of variance against block-size or spacing for the BQV, SBQV, TTLQV, PQV, and the RPQV methods when there is a single scale of pattern at 4, i.e., expected mean size and distance between patches is 4 units, are shown in Fig. 1. The BQV and SBQV methods both show a strong single peak at a block-size of 4 and thus both accurately describe the known pattern. However, the BQV method also shows a rise in

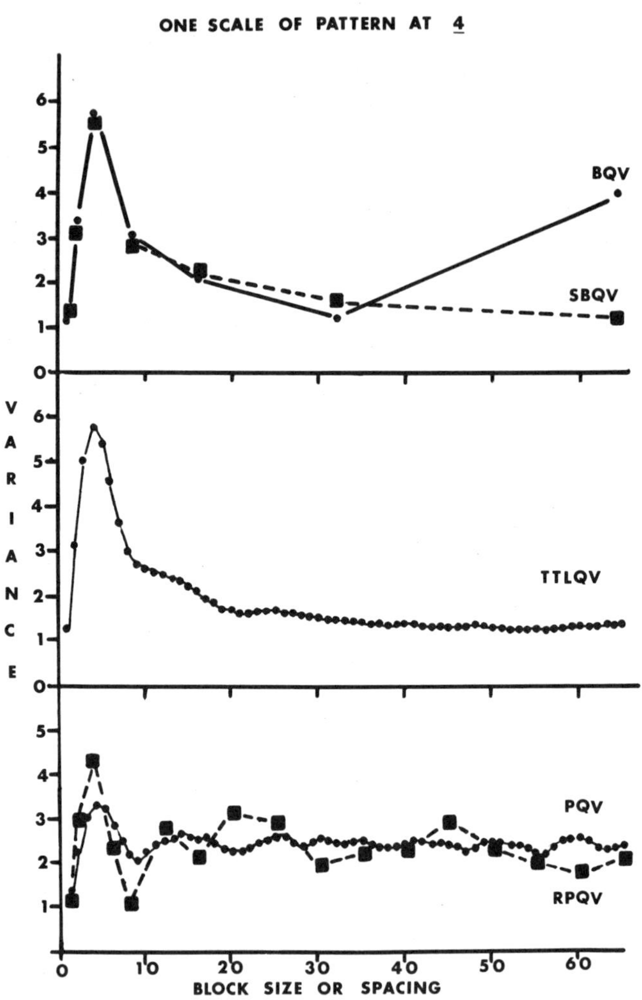


*FIG. 1: Spatial pattern analysis by the* BQV, SBQV, TTLQV, PQV, *and* RPQV *methods on one scale of artificial pattern at four.*

variance at the largest block-size (64), which reflects some effect of the starting position of blocking. When the starting position is stepped 50 times and the variance averaged over the 51 BQV analyses (the SBQV results), the rise in variance is not evident, which indicates that the SBQV method successfully eliminates the effect of starting position. The TTLQV method also shows a strong peak variance at a block-size of 4, thus accurately depicting the known pattern.

The PQV and RPQV methods also show peak variances at a spacing of 4. The PQV method displays a rather steady oscillation at a periodicity of 8. When the RPQV method was used to estimate variances at 17 different spacings (between 27 and 30 degrees of freedom for each spacing), the performance of the variance estimator was rather variable (when compared with the expected variances from the PQV method). This result substantiates the conclusion of Ludwig and Goodall (1978) that the RPQV method is not well suited to the exploration of pattern. It is well suited to the testing of variance patterns since independent variance estimates are required and since the variance estimates will be more precise when only a few spacings are being examined (df will be greater).

*4.2 Complex Patterns.* When two scales of pattern were generated at 6 and 32, the different methods exhibited noticeably different results (Fig. 2). Since the BQV method has been shown to be sensitive to starting position and since the RPQV method has been shown not to be well suited to the exploration of pattern (the purpose here), the results for these methods will not be shown. The SBQV method cannot show a peak variance at 6 since it is limited to block-sizes in powers of two and only estimates variances at 4 and 8, which in this case are about equal. This could lead to the false conclusion that either there is no small-scale pattern or that there is a weak pattern at 4. The SBQV method does accurately show the large-scale pattern at 32 since it does estimate a variance at that block-size.

The TTLQV method accurately depicts both scales of pattern by producing variance peaks at 6 and 30, with the variance at the large-scale pattern of 30 greater than that at the small-scale pattern of 6. The PQV method strongly shows the small-scale pattern at 6 and then, with oscillation, depicts the large-scale pattern at 31. The small-scale pattern is more strongly emphasized by having a variance peak about equal to that at 31 and by having clear oscillations with a periodicity of about 12.

Further comparative aspects of these methods are shown when two scales of pattern were generated at 6 and 25 (Fig. 3).

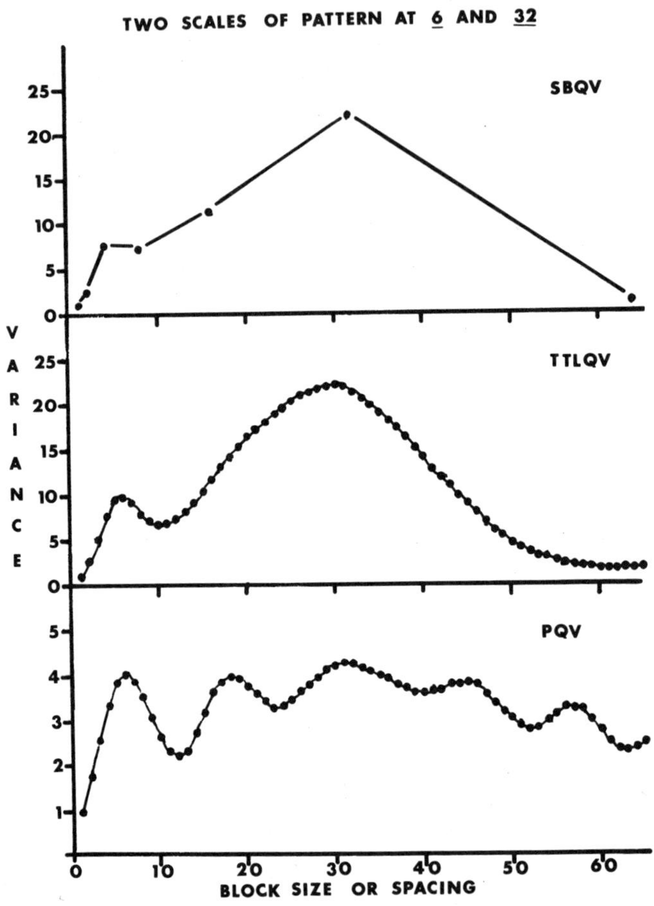


*FIG. 2: Spatial pattern analysis by the* SBQV, TTLQV, *and* PQV *methods on two scales of artificial pattern at six and thirty-two.*

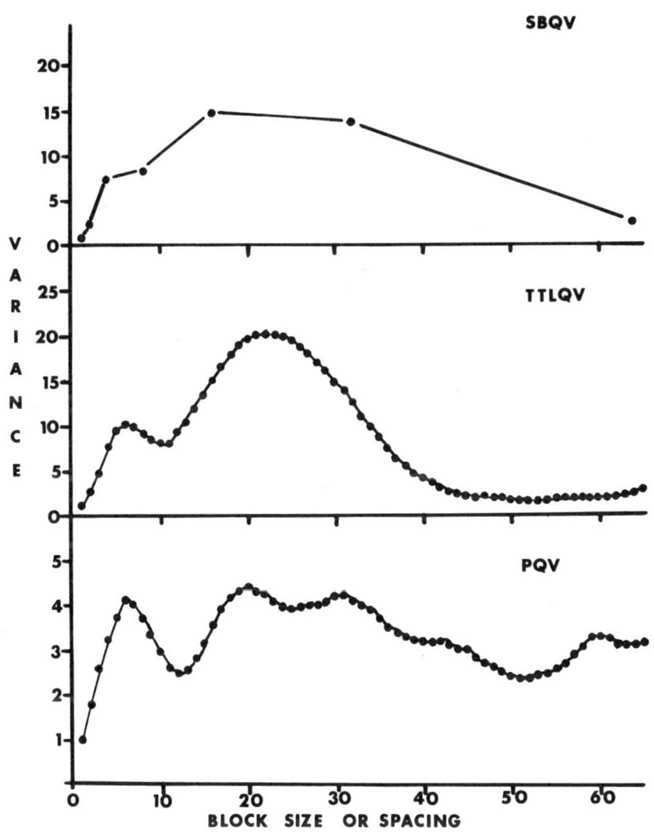


*FIG. 3: Spatial pattern analysis by the* SBQV, TTLQV, *and* PQV *methods on two scales of artificial pattern at six and twenty-five.*

The SBQV method now cannot show either the small- or large-scale pattern since the true scales of pattern are both intermediate to block-sizes at which this method estimates variances. The result is a rather rounded curve with no distinct variance peaks. The TTLQV method clearly shows the small-scale pattern at 6 and, although underestimating the actual peak, strongly indicates the large-scale pattern near 25. The PQV method clearly shows the small-scale pattern, but since the periodicity of this pattern is out of phase with a spacing of 25, the large-scale pattern is not shown.

## 5. DISCUSSION

This comparison of quadrat variance methods for the analysis of spatial pattern using artificial abundance patterns has emphasized some differences in these methods related to their abilities to discover known scales of pattern. The BQV and SBQV methods, being limited to variance estimates at powers of two, cannot detect scales of pattern intermediate between these points or can only show the pattern spread over the two closest points. The BQV method, being sensitive to starting position of blocking when regular pattern exists (Goodall, 1974), may produce variance peaks where none exist (Fig. 1). The SBQV method, being an average of many BQV or Stepped BQV analyses, is not sensitive to this problem. When a regular pattern exists, the results of the SBQV method are certainly superior to accepting the results of a single BQV analysis at the first quadrat position. However, this averaging of variances may involve the statistical problem of summing non-homogeneous variances (Usher, 1975). The SBQV method does require the sampling of extra (beyond some power of two) quadrats.

The TTLQV method characteristically produced a smooth variance curve and consistently showed variance peaks at or near the known scales of pattern (similar results were obtained for other artificial scales of pattern not reported here). It tends to emphasize the larger-scale of pattern when two exist by producing a higher variance peak. If the two scales of pattern were close together, the first or smallest-scale of pattern would only be indicated as a variance curve with a shoulder at that scale. The principal advantage of the TTLQV method over the other blocked-quadrat variance methods (BQV and SBQV) is that it allows estimates of variance at all possible block-sizes up to half the total number of quadrats; thus for the exploration of pattern it is of considerable advantage.

The PQV method characteristically produced an oscillating variance curve with a periodicity related to the small-scale pattern. When two scales of pattern exist, the large-scale pattern could either be emphasized (Fig. 2) or obscured (Fig. 3) depending on its position relative to the harmonics of the small-scale oscillation. The PQV method tends to be more sensitive to and more clearly shows the small-scale pattern. The RPQV method potentially should show the same variance pattern as the PQV method, but being based on random pairing of quadrats without replacement, it may have limited accuracy (even with 1000 quadrats and 17 spacings as shown in Fig. 1). The procedure of sampling without replacement also introduces a type of non-independence from one spacing to another. This type of dependence has been investigated by using replicate random sampling without replacement on the same data set; it can be important if the data set is small or if the degrees of freedom are low, such as at higher spacings. Both pairing of quadrat methods produce very sharp oscillations if the pattern is perfectly regular (Ludwig and Goodall, 1978). The randomly varying patch widths and positions used here are more realistic than such regular patterns, but only field data are truly real. The principal advantage of the artificial patterns used here is that the methods can be tested against known scales of pattern.

A main difference between the blocked-quadrat methods (BQV, SBQV, and TTLQV) and the paired-quadrat methods (PQV and RPQV) is that blocking methods estimate variance from quadrat differences where the quadrats change in both size and inter-quadrat distance; thus the variance is a confounding of these two effects (Ludwig and Goodall, 1978). With the pairing methods, the only effect on variance is inter-quadrat distance or spacing since the original quadrats are used and no blocking or pooling of these quadrats is performed. Also with the blocking methods a rigid sampling (contiguous quadrats) and analysis procedure (adjacent quadrats blocked or pooled) must be followed. With the pairing methods, quadrats need not be contiguous as long as inter-quadrat distance is constant or known, and quadrats may be missing or intentionally skipped. In fact, quadrats along the transect may be randomly partitioned into two subsets and one set used to explore for spatial pattern (and form hypotheses) by the PQV method and then the other set used to test any hypotheses by the RPQV method, as recommended by Ludwig and Goodall (1978). This recommended procedure would thus provide for a complete analysis of spatial pattern on a single transect of quadrat data. The transect data can also be split in two and the consistency of results compared. However, Ludwig and Goodall (1978) found little consistency in variances when Hill's TTLQV method was applied to transect subsets. This procedure would be most practical if a very large number of contiguous quadrats were available.

## 6. CONCLUSIONS

For the analysis of spatial pattern using quadrat data, Hill's (1973) TTLQV method clearly provides the most accurate evaluation of the scales and intensity of existing pattern. When two-scales of pattern exist, it tends to emphasize the larger-scale pattern. However, like the other methods based on blocking of quadrats, the TTLQV method confounds quadrat size and spacing in the variance estimate, and a rather rigid sampling and analysis design is required.

The PQV method accurately evaluates existing small-scale pattern. However, when a large-scale pattern also exists, the PQV method may obscure its presence if the small-scale variance oscillations are out of phase with its spacing. Being based on pairing of quadrats, it does not confound quadrat size and spacing effects in the variance estimate.

## REFERENCES

Bartlett, M. S. (1964). The spectral analysis of two-dimensional processes. *Biometrika*, 51, 299-311.

Burden, P. C. (1970). *Field and computer studies in the pattern of some calcicolous plants*. B.A. Thesis, University of York, York.

Chatfield, C. (1975). *The Analysis of Time-Series: Theory and Practice*. Chapman and Hall, London.

Errington, J. C. (1973). The effect of regular and random distributions on the analysis of pattern. *Journal of Ecology*, 61, 99-105.

Goodall, D. W. (1954). Minimal area: a new approach. *VIIIth International Botanical Congress, Ecology*, 19-21.

Goodall, D. W. (1974). A new method for the analysis of spatial pattern by random pairing of quadrats. *Vegetatio*, 29, 135-146.

Greig-Smith, P. (1952). The use of random and contiguous quadrats in the study of the structure of plant communities. *Annals of Botany*, 16, 293-316.

Greig-Smith, P. and Chadwick, M. J. (1965). Data on pattern within plant communities. III. Acacia-Capparis semi-desert scrub in the Sudan. *Journal of Ecology*, 53, 465-474.

Hill, M. O. (1973). The intensity of spatial pattern in plant communities. *Journal of Ecology*, 61, 225-235.

Kershaw, K. A. (1957). The use of cover and frequency in the detection of pattern in plant communities. *Ecology*, 38, 291-299.

Ludwig, J. A. and Goodall, D. W. (1978). A comparison of paired-with blocked-quadrat variance methods for the analysis of spatial pattern. *Vegetatio*, 38, 49-59.

Mead, R. (1974). A test for spatial pattern at several scales using data from a grid of contiguous quadrats. *Biometrics*, 30, 295-307.

Pielou, E. C. (1977). *Mathematical Ecology*. Wiley, New York.

Thompson, H. R. (1958). The statistical study of plant distribution patterns using a grid of quadrats. *Australian Journal of Botany*, 6, 322-342.

Usher, M. B. (1969). The relation between mean square and block size in the analysis of similar patterns. *Journal of Ecology*, 57, 505-514.

Usher, M. B. (1975). Analysis of pattern in real and artificial plant populations. *Journal of Ecology*, 63, 569-586.

[*Received: April* 1977. *Revised: January* 1979]

R. M. Cormack and J. K. Ord, (eds).,
*Spatial and Temporal Analysis in Ecology*, pp. 305-332. 

# THE DESCRIPTION OF POINT PATTERNS

S. A. L. M. KOOIJMAN

Department of Biology
Division of Technology for Society TNO
Organization for Industrial Research
P. O. Box 217, 2600 AE Delft, The Netherlands

SUMMARY. Statistical methods for analysis of point patterns are discussed. Special attention is paid to their application in ecological research. New procedures are presented, that seem to be better suited to the needs of the ecologist. These procedures concern the estimation of the local intensity, the covariance curve, and the variance curve from several sampling designs. A statistic which describes the total amount of interaction is proposed, and its relationship with the covariance curve is pointed out. The use of these estimators is illustrated by applying them to the dispersal patterns of barnacles, anemones, and glassworts, and their performance is studied by computer simulation.

KEY WORDS. distance method, grids, intensity, covariance curve, variance curve, interaction, estimation.

## 1. INTRODUCTION

The need to study point patterns arises at two points in the empirical cycle of ecological research:

- At the very first stage of the research, when ideas have to be formulated about the nature of the environmental variables and mutual relationships between organisms which may possibly play a significant role.

- In the verification stage, when it has to be shown that the postulated density fluctuations and mutual relationships are present.

The requirements that are imposed on the mathematical tools differ in these two stages. In the first, the purpose is to describe in as much detail as possible any deviation from a merely accidental dispersion of the individuals in a plane. When a statistical test is used to distinguish between accidental and reproducible phenomena, the test should be free of specific assumptions about the nature of the observational data. In fact, nothing meaningful can be asserted about them at this stage of the research. The test should be sensitive for a large class of alternative hypotheses, for in the main nothing meaningful can be said about these hypotheses either. This differs in the verification stage, in which some detailed ideas about the factors determining the location and numbers of organisms exist, and in this case, tests designed against well specified alternative hypotheses are required.

The aim of this paper is twofold. Firstly, it gives the ecologist algorithms for statistics which describe point patterns. Secondly, it is meant as a contribution to the discussion among statisticians about the methodology of adequately describing such patterns and choosing a sampling strategy. I am aware of the fact that this paper ignores some potentially useful developments, and for a review of other aspects of the subject, see Cormack (1979) and Diggle (1979).

## 2. FIRST ORDER PROPERTIES OF POINT PATTERNS

The most pronounced feature of a point pattern is its mean (local) intensity, i.e., the expected number of points or individuals per unit area. The intensity $\lambda(\underset{\sim}{z})$ at a site $\underset{\sim}{z}$ can be defined more exactly by $\lambda(\underset{\sim}{z}) = \lim_{A \to \underset{\sim}{z}} EN(A)/\{\text{area}(A)\}$, where the limit is taken as the surface area of a plot $A$, called area$(A)$, shrinks down to the site $\underset{\sim}{z}$. Here $EN(A)$ denotes the expected number of points in $A$. All properties of dispersal patterns referring to intensity only are called *first order properties*. Therefore, these properties have nothing to do with the simultaneous occurrence of individuals at different places.

There are two reasons motivating the study of first order properties in the context of the analysis of dispersal patterns. The first, and possibly the most important, is that correlations found between the intensity of the points and environmental variables can serve as a reasonable basis for an experimental study concerning a causal relationship. The second reason is that, in principle, the first order properties have to be known before higher order properties can be studied. For homogeneous as well as inhomogeneous Poisson patterns, the study of the first order properties is exhaustive. Therefore this section concentrates primarily on such patterns.

It is my personal feeling that, when the estimation of the (mean) intensity is of primary interest, distance sampling cannot be recommended over random plot sampling, if one has the choice. In order to give arguments for this, the estimation of intensity from distance samples will be discussed first in Section 2.1. In Section 2.2, the estimation of the intensity will be compared under both sampling designs. For special applications it makes sense to estimate the intensity as a function of coordinates in the plane, rather than to estimate the mean intensity. Section 2.3 will deal with this problem.

*2.1 Estimation of the Intensity from Plotless Samples.* There is a considerable literature dealing with the problem of how to estimate the intensity from plotless samples. Although many kinds of plotless sampling have been proposed (e.g., Diggle, *et al.*, 1976), this section deals only with samples consisting of repeated measurements of the distance between a randomly chosen sampling point (called *site*) and the nearest point (i.e., individual).

Unlike other estimators, the following one can be applied to any orderly point pattern; it makes use of the behavior of the probability density function (PDF), $f(r)$, near $r=0$, where the random variable $R$ is the distance between a randomly selected site and its nearest point. The first derivative of $f(r)$ with respect to $r$ is $2\pi\lambda$, for $r=0$, while second and higher derivatives vanish for $r=0$. Thus, near $r=0$, the distribution function is $F(r) = \pi r^2\lambda$ and the estimator $\hat{\lambda}_1$ is found by minimizing $\int_0^{r_m} w(r)\{\hat{F}(r)-\pi r^2\lambda\}^2 dr$ with respect to $\lambda$. The distribution function $F$ of $R$ is estimated by $\hat{F}(r) = (k-1)/n$ if $r_{k-1}<r<r_k$, with $r_0=0$ and $r_1, r_2, \cdots, r_n$ the ordered observations on $R$. For the choice $w(r) = 1/r$, this leads to

$$\hat{\lambda}_1 = \frac{2}{\pi m y_m^2}(m y_m - \textstyle\sum_{k=1}^m y_k),$$

where $y_i = r_i^2$ $(i=1,\cdots,n)$ and $m\leq n$.

When we know more about the point pattern, it is possible to improve the estimation of mean intensity. This will be shown for a class of patterns, called *local Poisson patterns*. Such patterns can be constructed formally in the following way. A given area is partitioned into cells, in which the points are placed according to Poisson processes whose intensities are themselves independent trials from an (unknown) PDF. Now $\Lambda$ is a random variable, whose moments can be estimated. It is assumed that the random site and its nearest point lie in the same cell with probability close to unity. We can then make use of the relationship $\ln G_Y(y) = \Psi(-\pi y)$, where $\Psi$ is the cumulant generating function of the intensity $\Lambda$ and $G_Y(y) = 1 - F_Y(y)$ the survivor function of $Y$. So, we see that we have to estimate the derivatives of the log survivor function of $Y$ at $0$ in order to obtain the moments of $\Lambda$. That is, near $y=0$ we have the relation

$$\ln G_Y(y) \simeq \sum_{j=1}^{\infty} (j!)^{-1}(-\pi y)^j \kappa_j .$$

This suggests the estimator $\hat{\lambda}_2$, found by minimizing

$$\int_0^{y_m} w(y)\,[\ln \hat{G}_Y(y) - \sum_{j=1}^{p} (j!)^{-1}(-\pi y)^j \kappa_j]^2 dy$$

as a function of the cumulants $\{\kappa_j\}_1^p$ of $\Lambda$ for some choice of $m$ and $p$. The function $\ln G_Y$ is estimated by $\ln \hat{G}_Y(y) = \ln(n-k+1) - \ln(n)$ if $y_{k-1} \leq y < y_k$.

For the choice $w(y) = 1/y$, it is possible to obtain an explicit formula for the estimators of the cumulants. This is accomplished by equating to zero the derivatives with respect to the cumulants of the function mentioned above. This leads to the matrix equation

$$\begin{bmatrix} \frac{(-\pi)^1}{1!}\hat{\kappa}_1 \\ \vdots \\ \frac{(-\pi)^p}{p!}\hat{\kappa}_p \end{bmatrix} = \begin{bmatrix} \frac{y_m^{1+1}}{1+1} & \cdots & \frac{y_m^{1+p}}{1+p} \\ \vdots & & \vdots \\ \frac{y_m^{p+1}}{p+1} & \cdots & \frac{y_m^{p+p}}{p+p} \end{bmatrix}^{-1}$$

$$\times \begin{bmatrix} \sum_{k=1}^{m} \frac{y_k^1 - y_{k-1}^1}{1} \ln[(n-k+1)/n] \\ \vdots \\ \sum_{k=1}^{m} \frac{y_k^p - y_{k-1}^p}{p} \ln[(n-k+1)/n] \end{bmatrix} .$$

Simulation studies indicate that p should be chosen to be as small as possible. Thus, when we want to estimate the mean intensity $\kappa_1$ only, the best choice is p=1. If an estimate of the variance $\kappa_2$ is also required, the best choice is p=2, etc. Some sampling properties of this estimator, called $\hat{\lambda}_2$, are given in Table 1, together with those of $\hat{\lambda}_1$, as well as the maximum likelihood estimators in the cases of a Poisson pattern and a gamma local Poisson pattern. A gamma local Poisson pattern is a local Poisson pattern in which the intensities in each cell are independent trials from a gamma distribution, i.e., $\Lambda$ has PDF $\lambda^{\alpha-1}[\beta^{\alpha}\Gamma(\alpha)]^{-1}\exp(-\lambda\beta^{-1})$. The choice of $\alpha=2$ and $\beta=1/2$ has been made in the preparation of Table 1. We can conclude that it has been possible to improve the estimation of the intensity by making use of knowledge about the structure of the point pattern. In the next section we will compare these results with those from random plot samples.

*TABLE 1: The estimated first and second moments of intensity estimates based on distance samples, compared with the moments based on a random plot. The moments based on distance samples are estimated from* 100 *intensity estimates, each based on* 100 *observations, obtained by computer simulation. Only the first* m *of the ordered observations are used to evaluate the estimators. The moments based on* 100 *random plots are calculated for a plot size equal to* $\pi$ *times the expected squared nearest point distance. Both patterns have intensity equal to* 1.0.

| Estimator | Poisson pattern: Mean | Poisson pattern: Mean square error | Gamma local Poisson pattern: Mean | Gamma local Poisson pattern: Mean square error |
|---|---|---|---|---|
| *Distance methods* | | | | |
| $\hat{\lambda}_1$, m = 20 | .918 | .061 | .908 | .055 |
| $\hat{\lambda}_2$, m = 20 | 1.033 | .064 | .975 | .056 |
| $\hat{\lambda}_2$, m = 100 | .999 | .011 | | |
| ML | 1.010 | .009 | .999 | .024 |
| *Random plot* | 1. | .01 | 1. | .02 |

*2.2 Comparison of Intensity Estimation from Plotless Samples and from Plot Samples.* As the intensity is a description of the counting process, we can expect to obtain better results by estimation from, for example, random quadrats. Indeed, when we use a quadrat of size $A$, the unbiased estimate for the intensity is given by $x_+/(nA)$, where $x_+$ is the sum of the counts in $n$ quadrats. If the pattern is Poisson, its variance is $\lambda/(nA)$. In order to compare estimation from plot samples and plotless samples, we choose $A$ to be $\pi$ times the expected squared nearest point distance (i.e., $\pi$ times $1/(\pi\lambda)$, which makes the expected total surface area examined for the presence of points equal in both cases. The unbiased estimate for the intensity from the random quadrats has a variance equal to the biased ML estimate of the distance method. For the variance of the latter, see e.g., Kendall and Moran (1963).

If $\lambda$ is compounded with a gamma distribution, the number of points $X$ in a random quadrat of size $A$ is given by the negative binomial distribution:

$$\Pr(X=x) = \frac{\Gamma(\alpha+x)}{\Gamma(\alpha)x!}\left(\frac{\beta A}{1+\beta A}\right)^x \left(1 - \frac{\beta A}{1+\beta A}\right)^\alpha .$$

This distribution has mean $\alpha\beta A$ and variance $\alpha\beta A(1+\beta A)$. If $\alpha=2$, $\beta=1/2$, and $A$ is equal to $\pi$ times the expected squared nearest point distance (i.e., $\pi$ times $1/[\pi\beta(\alpha-1)]$), the variance is .02 for $n=100$. Contrary to the case of Poisson patterns, this variance is smaller than that of the corresponding ML estimate of the distance method (see Table 1). It should be emphasized that, in comparing the sample properties of the ML intensity estimates, the distance method uses complete knowledge of the probabilistic properties of the pattern, while the random quadrat method does not.

The moments of the intensity can easily be found from the moments of the counts by means of the relationship $\phi_\Lambda(s) = \phi_X[\ln(1+s)]$, where $\phi$ is the moment generating function. More explicitly ${}_\Lambda\mu'_k = {}_X\mu_{(k)}$, where ${}_X\mu_{(k)}$ is the *kth* factorial moment of the counts $X$. The corresponding relationship between the cumulants ${}_\Lambda\kappa_k = {}_X\kappa_{(k)}$ is perhaps of more practical interest.

*2.3 Estimation of the Intensity as a Function of Coordinates in the Plane.* When samples from a point pattern are available in the form of counts in a grid or as a list of point-coordinates in a plot, it is possible to estimate the local intensity as a function of the coordinates in the plane. Let us call this function $f(\underset{\sim}{z})$, where $\underset{\sim}{z}$ is in the plot or in the grid frame. Note that $f$ has to be non-negative for the selected $\underset{\sim}{z}$ values. A suitable function would be some two-dimensional polynomial in the exponent. The parameters of this polynomial can be estimated, for instance, by the ML method, but the resulting numerical complications make this approach unattractive. Ordinary two-dimensional polynomials lead to much simpler numerical procedures, but have the disadvantage of not necessarily being positive for every point on the grid. Nevertheless, in many practical applications, the condition of being non-negative on the grid is fulfilled. For the one-dimensional problem see Gustavsson and Svensson (1976). The point pattern is assumed to be a realization of an inhomogeneous Poisson process. There are a total of $N = x_{++}$ points in the grid. The probability $q$ that a point falls in cell $c_{ij}$ is

$$q_{ij} = \int_{\underset{\sim}{z} \varepsilon c_{ij}} f(\underset{\sim}{z}) d\underset{\sim}{z} \left( \int_{\underset{\sim}{z} \varepsilon \bigcup_{ij} c_{ij}} f(\underset{\sim}{z}) d\underset{\sim}{z} \right)^{-1} .$$

The estimation of the parameters will be illustrated for the function $f(\underset{\sim}{z}) = C_0' + C_1' z_1 + C_2' z_2$, where $\underset{\sim}{z} = (z_1, z_2)$. The coordinate axes are chosen in such a way that the corners of the grid are given by (0,0), (0,1), (1,1), and (1,0).

*Estimation from grid counts.* For an $n{\times}m$ grid, the log likelihood function is given by $\Sigma_{ij} x_{ij} \ln q_{ij}$, where the $x_{ij}$ are the counts and

$$q_{ij} = [2+C_1(2i-1)m^{-1}+C_2(2j-1)n^{-1}]/[mn(2+C_1+C_2)]$$

with $C_1 = C_1'/C_0'$ and $C_2 = C_2'/C_0'$. The ML estimates for $C_1$ and $C_2$ are given by the solutions of

$$\frac{1}{m} \Sigma_{ij} x_{ij} \frac{i - \frac{1}{2}}{1+\hat{C}_1(2i-1)(2m)^{-1}+\hat{C}_2(2j-1)(2n)^{-1}} = \frac{x_{++}}{2+\hat{C}_1+\hat{C}_2} ,$$

$$\frac{1}{n} \Sigma_{ij} x_{ij} \frac{j - \frac{1}{2}}{1+\hat{C}_1(2i-1)(2m)^{-1}+\hat{C}_2(2j-1)(2n)^{-1}} = \frac{x_{++}}{2+\hat{C}_1+\hat{C}_2} ,$$

where $x_{++}$ indicates the sum of the counts and $\Sigma_{ij}$ denotes summation over all $i$ and $j$.

*Estimation from point-coordinates.* For a plot, the log likelihood function is given by

$$\Sigma_i \ln(1+C_1p_{1i}+C_2p_{2i}) - N\ln(1+\tfrac{1}{2}C_1+\tfrac{1}{2}C_2),$$

where $(p_{1i}, p_{2i})$ $(i=1,\cdots,N)$ are the $N$ point coordinates. Again $C_1 = C_1'/C_0'$ and $C_2 = C_2'/C_0'$. The ML estimates for $C_1$ and $C_2$ are the solutions of

$$\Sigma_i p_{ji}(1+\hat{C}_1p_{1i}+\hat{C}_2p_{2i})^{-1} = N(2+\hat{C}_1+\hat{C}_2)^{-1} \quad \text{for} \quad j=1,2.$$

*Examples of tests for homogeneity.* One advantage of using maximum likelihood estimators is that it is possible to test the null hypothesis $C_1 = C_2 = 0$ against the alternative that $C_1$ and/or $C_2$ is different from zero by means of a likelihood ratio (LR) test. The LR statistic is given by $2\Sigma_{ij}x_{ij}\ln \hat{q}_{ij} - 2x_{++}\ln(mn)$ in the case of grid counts, and by $2\Sigma_i\ln(1+\hat{C}_1p_{1i}+\hat{C}_2p_{2i}) - 2N\ln(1+\frac{1}{2}\hat{C}_1+\frac{1}{2}\hat{C}_2)$ in the case of point coordinates. Under the null hypothesis, the LR statistic has asymptotically a chi-square distribution with 2 degrees of freedom.

The results of applying the LR test to the patterns in the Appendix are given in Table 2. The direction of the steepest increase from $(0,0)$ to $(\hat{C}_1, \hat{C}_2)$, which was significant in the patterns of both barnacles and glasswort, is indicated on the figures in the Appendix with an arrow.

Pielou (1974) presents a procedure for estimating the intensity, which starts by counting the number of points in a circle which is then moved over the plot. The approach presented here has the advantage of being independent of the arbitrarily chosen circle size, and allows the possibility of performing tests on the intensity variation. In the absence of *a priori* knowledge about the functional form of the intensity, the above procedure of fitting polynomials is recommended only where a test is desired.

*TABLE 2: The* LR *test on a linear trend in the intensity applied to the patterns given in the Appendix. The first test is based on grid counts, the other two are based on point coordinates.*

| Data | Barnacles | Anemones | Glasswort |
|---|---|---|---|
| $\hat{C}_1$ | .959 | .405 | -.234 |
| $\hat{C}_2$ | .155 | .697 | 1.076 |
| LR statistic | 6.636 | 4.9 | 43.96 |
| Significance | + | - | + |

## 3. SECOND ORDER PROPERTIES OF POINT PATTERNS

We shall consider the simultaneous occurrence of individuals (of one or more species) in different plots. It is described by a simultaneous PDF. The properties of this simultaneous PDF, which depend on only its first and second moments and not on higher ones, are called *second order properties.*

Non-zero correlations in the occurrences of individuals in different places (conditional on the intensity) correspond with individual interactions; this is the main reason for using second order properties in the analysis of dispersal patterns. For the estimation of these correlations it is necessary to know the intensity, and in this section we shall assume that the intensity is constant over the entire plot. When the scale of the real intensity variation is much larger than the scale on which interactions take place, the falsity of the assumption of homogeneous intensity will have no serious consequences for the estimates of the covariance and variance curves for small arguments. However, when both take place on the same scale, it is not possible to estimate the covariance curve and the intensity variation simultaneously. In my opinion, this problem greatly reduces the applicability of this type of pattern description for ecological purposes.

The usual way of measuring covariances in grid samples will first be reviewed briefly in Section 3.1. The theory has been generalized to cope with counts on several species. In Section 3.2 the definitions of the autocovariance curve and the cross-covariance curve are given, in order to explain the exact relation between the usual covariance measures and the covariance curve. It will be shown how a test can be performed on the whole covariance curve at once. As this is one of the aims of spectral analysis, it can be an alternative in some sense, which is simpler both in interpretation and computation. The estimation of the covariance curve will also be discussed in this section, as it is closely related to this topic. Finally, in Section 3.2, it will be shown how the usual analysis of patterns in ecology can be reformulated in such a way that nonparametric tests are possible, without making use of computerized randomization procedures. However, it is my personal feeling that, except in special circumstances, such a procedure is not to be preferred to the covariance approach in ecological work. Interactions between species cannot be studied in this way, and the variance curve is difficult to interpret.

*3.1 Covariance Measures in a Grid.* If we have a sample in the form of a grid, the covariance is usually studied by the Moran statistic. For details and applications see, for instance, Cliff and Ord (1969, 1970, 1971, 1973) or Ord (1975). For species that may differ from each other, the Moran statistic can be defined as

$$I(k,\ell;\underset{\sim}{W}) = (m_{k2}m_{\ell 2})^{-\frac{1}{2}} \underset{\sim}{x}_k' \underset{\sim}{W} \underset{\sim}{x}_\ell (\underset{\sim}{1}' \underset{\sim}{W} \underset{\sim}{1})^{-1} ,$$

where $\underset{\sim}{x}_k$ represents the vector of centered cell contents of species $k$ present in the grid. The cells are read along the rows in this section. The mutual position of the cells is preserved in the matrix of specified weight coefficients $\underset{\sim}{W}$. The cell contents are centered in such a way that $E\underset{\sim}{x}_k = 0$. A vector of 'ones' is represented by $\underset{\sim}{1}$, having such a length that the matrix multiplication is defined. Also, $m_{k2}$ denotes the variance of the cell contents of species $k$.

A PDF of $I$ can be constructed on the basis of one of the following assumptions: either every permutation of the cell contents, carried out for each species separately, has the same probability (i.e., the cell contents are randomly mingled), or the cell contents are independently and identically distributed (IID) for each species separately.

The observations $\underset{\sim}{x}_k$ and $\underset{\sim}{x}_\ell$ are said to be spatially dependent or correlated with respect to a shift, specified by the matrix $\underset{\sim}{W}$, if the upper tail probability that $I$ exceeds its observed value is smaller than the type one error. Under the two assumptions of random mingling and IID, the first moment of $I$ is given by

$$E[I(k,\ell;\underset{\sim}{W})] = \delta_{k\ell}[N\Sigma_i w_{ii} - w_{++}]/[(N-1)w_{++}],$$

where $N$ is the number of cells and $\delta_{k\ell} = 1$ if $k=\ell$ and 0 otherwise. The second moment under random mingling is given by

$$E[I^2(k,\ell;\underset{\sim}{W})] = (m_{k2}m_{\ell 2}w_{++}^2)^{-1}\, E(\underset{\sim}{x}_k'\underset{\sim}{W}\underset{\sim}{x}_\ell)^2,$$

where $\underset{\sim}{x}_k'\underset{\sim}{1} = 0$ and the moments $m_{ki}$ are by definition $m_{ki} = \frac{1}{N}\Sigma_j x_{kj}^i$. The second factor can be found in Table 3. If the cell contents are independently normally distributed for each species separately, the second moment of $I$ can be written in the same form. In this case, $m_{ki}$ represents the *ith* (reduced) moment of the cell contents of species $k$. $m_{k2}$ will be close to its estimated value, given by $\frac{1}{N}\Sigma_i x_{ki}^2$. Note that $m_{k4}$ is fixed, once $m_{k2}$ is known. If the $x_{ki}$'s are centered Poisson distributed variables, and the mean of the counts is not too small, the formula for the second moment can be conceived as an approximation. Note that in this case even $m_{k2}$ is fixed when the mean of the counts is known. If, however, the moments of the counts are unknown, it is not obvious what the effect will be on the conclusions of the test, when the moments are replaced by their estimates.

For $k=\ell$, the diagonal elements of $\underset{\sim}{W}$ are often chosen to be zero, and this simplifies the expressions considerably. When we deal with the variance curve, the same table will be used for $k=\ell$ with non-zero diagonal elements.

*3.2 The Isotropic Covariance Curve.* This is defined (see, e.g., Bartlett, 1966) by

$$\gamma_{k\ell}(r) = \lambda_\ell(2\pi r)^{-1}\frac{\delta}{\delta r}\, EN[k;C(\underset{\sim}{p}_{\ell x},r)] - \lambda_k\lambda_\ell$$

$$= \lambda_k(2\pi r)^{-1}\frac{\delta}{\delta r}\, EN[\ell;C(\underset{\sim}{p}_{kx},r)] - \lambda_k\lambda_\ell, \quad \text{where}$$

$\lambda_k$ is the intensity of the points of type $k$, $p_{kx}$ $(x=1,2,\cdots)$ are the coordinates of the points of type $k$, $C(\underset{\sim}{z},r)$ denotes a circle with its center on $\underset{\sim}{z}$ and radius $r$, and $EN[k;C(\underset{\sim}{p}_{\ell x},r)]$

*TABLE 3: The second moment of* $\Sigma_{ij} w_{ij} x_{ki} \bar{x}_{\ell i}$. *The moment is obtained as* $\mu_2 = \underset{\sim}{K}' \underset{\sim}{C} \underset{\sim}{R}'$, *where* $\underset{\sim}{K}$ *is either the second or third column,* $\underset{\sim}{C}$ *is the matrix of coefficients, and* $\underset{\sim}{R}$ *is the heading row. The column* $\underset{\sim}{K}$ *consists of the values taken on by the first column under the indicated assumption, where* $E_{k,abcd} = E[x_{ka} x_{kb} x_{kc} x_{kd}]$ *and* $E_{k,ab} = E[x_{ka} x_{kb}]$. *Further,* $m_{kj}$ *is the jth moment of* $x_{ki}$, *for arbitrary* i, *where* $m_{k1} = 0$. *The numbers in parentheses indicate descending factorial powers, for instance,* $(N-1)^{(2)} = (N-1)(N-2)$.

| $k = \ell$ | Random mingling | IID | $\Sigma_i w_{ii}^2$ | $\Sigma_i w_{ii}^2 + \Sigma_{ij} w_{ij}^2 + \Sigma_{ij} w_{ij} w_{ji}$ | $2\Sigma_i w_{ii}(w_{i+} + w_{+i})$ | $\Sigma_i (w_{i+} + w_{+i})^2 + 2\Sigma_i w_{ii} \Sigma_{ij} w_{ij}$ | $w_{++}^2$ |
|---|---|---|---|---|---|---|---|
| $E_{k,abcd}$ | $\frac{3Nm_{k2}^2 - 6m_{k4}}{(N-1)^{(3)}}$ | 0 | -6 | 1 | 2 | -1 | 1 |
| $E_{k,aabc}$ | $\frac{2m_{k4} - Nm_{k2}^2}{(N-1)^{(2)}}$ | 0 | 12 | -2 | -3 | 1 | |
| $E_{k,aaab}$ | $\frac{-m_{k4}}{N-1}$ | 0 | -4 | | 1 | | |
| $E_{k,aabb}$ | $\frac{Nm_{k2}^2 - m_{k4}}{N-1}$ | $m_{k2}^2$ | -3 | 1 | | | |
| $E_{k,aaaa}$ | $m_{k4}$ | $m_{k4}$ | 1 | | | | |

| $k \neq \ell$ | Random mingling | IID | $\Sigma_{ij} w_{ij}^2$ | $\Sigma_i w_{+i}^2$ | $\Sigma_i w_{i+}^2$ | $w_{++}^2$ |
|---|---|---|---|---|---|---|
| $E_{k,ab} E_{\ell,ab}$ | $\frac{m_{k2} m_{\ell 2}}{(N-1)(N-1)}$ | 0 | 1 | -1 | -1 | 1 |
| $E_{k,aa} E_{\ell,ab}$ | $\frac{-m_{k2} m_{\ell 2}}{N-1}$ | 0 | -1 | | 1 | |
| $E_{k,ab} E_{\ell,aa}$ | $\frac{-m_{k2} m_{\ell 2}}{N-1}$ | 0 | -1 | 1 | | |
| $E_{k,aa} E_{\ell,aa}$ | $m_{k2} m_{\ell 2}$ | $m_{k2} m_{\ell 2}$ | 1 | | | |

is the expected number of points of type $k$ within a distance $r$ from a randomly chosen point of type $\ell$. $\gamma_{k\ell}(r)$ has dimension $(\text{length})^{-4}$. Note that $\lim_{r\downarrow 0} EN[k;C(\underset{\sim}{p}_{\ell x},r)] = \delta_{k\ell}$. It is assumed that the intensity is independent of the coordinates, and the covariance of the numbers in two small plots will be a function of the distance between these plots only. When $k=\ell$, the curve is called the *autocovariance curve*.

*The total amount of interaction in a pattern.* If we scale the grid counts in such a way that $x_{k+} = 0$ and $\Sigma_i x_{ki}^2 = N$ for all species $k$, the Moran statistic takes the form $I(k,\ell;\underset{\sim}{W}) = \underset{\sim}{x}_k'\underset{\sim}{W}\underset{\sim}{x}_\ell(\underset{\sim}{1}'\underset{\sim}{W}\underset{\sim}{1})^{-1}$. We will show that the problem of assigning appropriate weight values can be avoided by studying the statistic

$$I^2(k,\ell) = \max_{\underset{\sim}{W}} I^2(k,\ell;\underset{\sim}{W})$$

under certain constraints on $\underset{\sim}{W}$. We shall see that $I^2(k,\ell)$ can be interpreted as a statistic on the total amount of interaction, and we shall elucidate the connection between the covariance curve and maximizing the weight coefficients.

In order to define a maximum for $I^2(k,\ell;\underset{\sim}{W})$ we have to fix $\Sigma_{ij}w_{ij}$ and $\Sigma_{ij}w_{ij}^2$ as, say, $\Sigma_{ij}w_{ij} = \alpha$ and $\Sigma_{ij}w_{ij}^2 = \beta$. Since we have $N$ counts in the case of $k=\ell$, or $2N$ counts in the case of $k\neq\ell$, and $N^2$ weight coefficients, we need the further constraint of $w_{ij} = f_s$ if the pair of cells $(i,j)$ is grouped into class $c_s$. For example, if the distance between the centers of cells $i$ and $j$ is $d_{ij}$, we may define $c_s$ as $c_s = \{i, j$ such that $m_{s-1} < d_{ij} \leq m_s\}$, where $m_{s-1}$ and $m_s$ are pre-selected distances defining the distinct classes. Using this approach, it can be shown that

$$I^2(k,\ell;\alpha,\beta) = \frac{N^2\beta-\alpha^2}{N^2-1}\, I^2(k,\ell;1,1),$$

so that $\alpha$ and $\beta$ simply appear as scaling constants. From now on, we choose $\alpha=\beta=1$ and write $I(k,\ell)$ rather than $I(k,\ell;1,1)$. If we set $a_s = \Sigma x_{ki}x_{\ell j}$ where the summation is over the $n_s$

pairs of cells that fall into class $c_s$, then

$$I^2(k,\ell) = N^{-2}(N^2-1)\Sigma_s\ a_s^2/n_s,$$

$$f_s = N^{-2} + N^{-2}I^{-1}(N^2-1)\ a_s/n_s \qquad (s=1,\cdots,K).$$

For the calculation of the moments of $I(k,k)$ under random mingling, see Kooijman (1976).

The interpretation of $I(k,\ell)$ becomes clear when we examine the continuous situation. This is done in two steps. First we decrease the cell size by putting more and more cells within the grid-frame. Then we remove the border effects by increasing the 'grid' circumference. For orderly patterns for which $\gamma(r)$ is $o(1)$ for increasing $r$, $I^2(k,\ell)$, divided by the surface area for which it is calculated, converges in probability to $J(k,\ell)$, say, where $J(k,\ell) = \frac{\pi}{\lambda_k\lambda_\ell}\sum_{s=1}^{K-1} \Sigma(m_s^2-m_{s-1}^2)\Gamma_{k\ell}^2(m_{s-1},m_s)$, in which $\Gamma_{k,\ell}(a,b)$ is the mean covariance for distances between $a$ and $b$, or $\Gamma_{k,\ell}(a,b) = \frac{2}{b^2-a^2}\int_a^b r\gamma_{k\ell}(r)dr$, and $\lambda_k$ denotes the intensity of points of type $k$. The upper class limit, $m_K$, is set greater than the largest of all $d_{ij}$. For the special choice in which $m_s = sds$ with $ds\to 0$ and $K\to\infty$ so that $m_{K-1} = T$ remains fixed, the formula becomes $J(k,\ell) = \int_0^T \frac{2\pi}{\lambda_k\lambda_\ell} r\gamma^2{}_{k\ell}(r)dr$.

Patterns of ecological interest will always have the property that the interaction between individuals is only local, i.e., the interaction dies out rapidly with increasing distance. This corresponds with the strong mixing condition (see Brillinger, 1976). For such patterns, the integral in the expression for $J$ will also converge when $T$ approaches infinity. It is now obvious that $I(k,\ell)$ is a statistic on the total amount of interaction. The covariance curve is squared before the integration. So, the spacing out mechanisms also contribute to $I(k,\ell)$. Indeed, for a Poisson pattern where $\gamma_{kk}(r)=0$, $J(k,k)=0$, indicating no interaction. Another example is, for instance, $\gamma_{kk}(r) = \kappa rK_1(\kappa r)$, where $\kappa$ is a parameter and $K_1$ the modified Bessel function, which has been fitted to actual data, as in Whittle (1954) and Matérn (1960). In this case $J(k,k) = 2\pi(\kappa\lambda)^{-2}$, where $\lambda$ is the intensity. For a center-satellite model, (see, e.g., Neyman and Scott, 1957; Dacey, 1969; Warren, 1971), we have the following characterization. If $\lambda$ denotes the intensity of the Poisson parent process, and the number of daughter points per parent point has a mean $\mu$ and a second factorial moment $\mu_{(2)}$, the covariance curves are given by the symmetric matrix

$$\underset{\sim}{\gamma}(r) = \frac{1}{2\pi r}\begin{pmatrix} 0 & \mu\lambda f(r) \\ \cdot & \mu_{(2)}\lambda f_2(r) \end{pmatrix},$$

where $f(r)$ is the PDF of the distance between a particular parent point and one of its randomly chosen daughter points, and $f_2(r)$ is the PDF of the distance between two randomly chosen daughter points from the same parent point (see Bartlett, 1960, 1966). Now the values for $J(k,\ell)$ are given by

$$\frac{1}{2\pi}\begin{pmatrix} 0 & \mu\int_0^\infty r^{-1}f^2(r)dr \\ \cdot & \mu_{(2)}^2\mu^{-2}\int_0^\infty r^{-1}f_2^2(r)dr \end{pmatrix}.$$

For the special case in which $f(r) = r\sigma^{-2}\exp(-\frac{1}{2}r^2\sigma^{-2})$, we find that $f_2(r) = \frac{1}{2}r\sigma^{-2}\exp\{-\frac{1}{4}r^2\sigma^{-2}\}$, and that the values for $J(k,\ell)$ are given by

$$(8\pi\sigma^2)^{-1}\begin{pmatrix} 0 & 2\mu \\ \cdot & \mu_{(2)}^2\mu^{-2} \end{pmatrix}.$$

For many probability distributions for the number of daughter points per parent point, it may be shown that both the pattern of the daughter points and the pattern of the parent points will be nearly indistinguishable from the Poisson pattern for small $\mu$, while it is possible to trace the relationship between parent and daughter points by means of $J(1,2)$.

We first showed the relationship between the maximized Moran statistic and the covariance curve. Then, it was shown that this statistic can be used to characterize models for spatial point processes. It is possible to invert the argument: If we want to compare the 'goodness of fit' of two models, each of which specifies the covariance curve for a particular grid sample, we can calculate the Moran statistics with matrices of weight coefficients based on these two covariance curves and decide that the model with the larger Moran statistic best describes the data. In order to know whether this description is good or bad, we can compare it with the maximized Moran statistic.

*Estimation of the isotropic covariance curve.* For the estimation of the covariance curve from a list of $N_k$ and $N_\ell$ coordinates in a plot $A$, it seems reasonable to estimate $\gamma_{k\ell}(r_s)$ by $\hat{\Gamma}_{k\ell}(r_s - \frac{1}{2}\Delta, r_s + \frac{1}{2}\Delta)$ for a small number $\Delta$, where $\hat{\Gamma}_{k\ell}(a,b)$ is given by

$$\frac{\hat{\lambda}_\ell}{\pi(b^2-a^2)} \{\hat{E}N[\ell;C(\underset{\sim}{p}_{kx},b)] - \hat{E}N[\ell;C(\underset{\sim}{p}_{kx},a)]\} - \hat{\lambda}_k\hat{\lambda}_\ell .$$

Obviously, the intensity $\lambda_k$ is estimated by $\hat{\lambda}_k = N_k/\text{area}(A)$. Furthermore, $EN[\ell;C(\underset{\sim}{p}_{kx},r_s)]$ is estimated by

$$\sum_{x=1}^{N_k} \frac{2\pi r_s w_{xs}}{\text{cir}[C(\underset{\sim}{p}_{kx},r_s)]\cap A} N[\ell;C(\underset{\sim}{p}_{kx},r_s)\cap A],$$

where $\text{cir}[C(\underset{\sim}{p}_{kx},r_s)]\cap A$ is written for that part of the circle which has its center on coordinates $\underset{\sim}{p}_{kx}$, and has a radius $r_s$ that falls within plot A. Also, $N[\ell;C(\underset{\sim}{p}_{kx},r_s)\cap A]$ denotes the number of points of type $\ell$ in plot A, which are within distance $r_s$ from the point of type k, having coordinates $\underset{\sim}{p}_{kx}$ (see Figure 1). The $w_{xs}$ are weight coefficients, such that $w_{+s}=1$, and it is reasonable to choose smaller coefficients for points close to the border and larger ones for points near the center of the plot. A convenient choice is

$$w_{xs} = \frac{\text{cir}[C(\underset{\sim}{p}_{kx},r_s)]\cap A}{\Sigma_x \text{cir}[C(\underset{\sim}{p}_{kx},r_s)]\cap A} ,$$

in which case $\hat{\gamma}_{k\ell}(r_s) = N_k N_\ell \text{area}^{-2}(A)[-1 + \text{area}(A)S/T]$ where

$$S = \Sigma_x N\{\ell; C[\underset{\sim}{p}_{kx}, r_s + \tfrac{1}{2}\Delta]\cap A\} - \Sigma_x N\{\ell; C[\underset{\sim}{p}_{kx}, r_s - \tfrac{1}{2}\Delta]\cap A\},$$

$$T = \Delta N_\ell \Sigma_x \operatorname{cir}[C(\underset{\sim}{p}_{kx}, r_s)]\cap A.$$

A computer program, for this formula with $k=\ell$, is given in Appendix B.

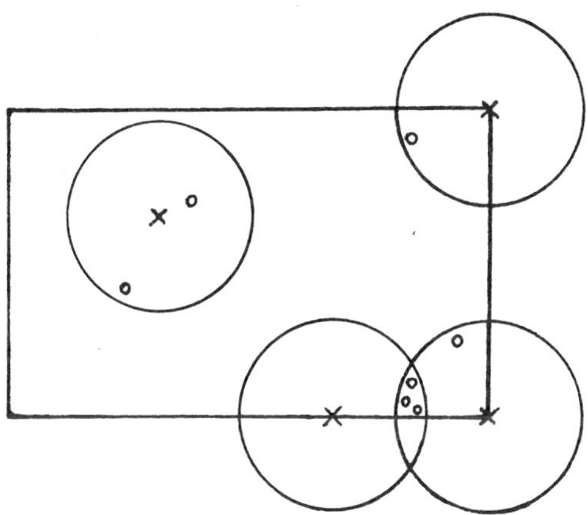

*FIG 1: Estimation of* $EN[\ell; C(\underset{\sim}{p}_{kx}, r_s)]$. *Points of type* $\ell$ *and* k *are indicated by small circles and crosses respectively. The numbers* $N[\ell; C(\underset{\sim}{p}_{kx}, r_s)\cap A]$ *are* 2, 3, 4, *and* 1, *if* $r_s$ *is the radius of the circles. The values for* $2\pi r_s\{\operatorname{cir}[C(\underset{\sim}{p}_{kx}, r_s)]\cap A\}^{-1}$ *are* 1, 2, 4, *and* 4 *respectively. The weight coefficients* $w_{xs}$ *are* $\frac{1}{2}$, $\frac{.5}{2}$, $\frac{.25}{2}$, *and* $\frac{.25}{2}$ *respectively. The estimate for* $EN[\ell; C(\underset{\sim}{p}_{kx}, r_s)]$ *is* $\frac{1}{2} \times 1 \times 2 + \frac{.5}{2} \times 2 \times 3 + \frac{.25}{2} \times 4 \times 4 + \frac{.25}{2} \times 4 \times 1 = 5$.

An example of such an estimate is given in Figure 2. The estimated autocovariance curve is plotted for the anemone-data of Appendix A. $\Delta$ is equal to an interval indicated on the abscissa. We can conclude that there is some spacing out within a short distance, and this may indicate that the animals tend to avoid touching each other with their tentacles when these are extended during high tide. The variance of the estimated covariance curve is estimated by

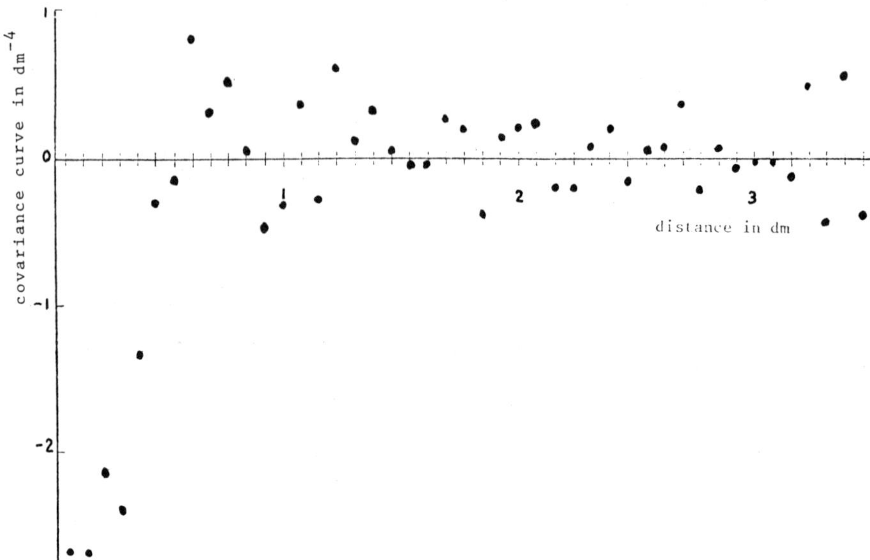


*FIG. 2: The estimated autocovariance curve from a* 14.5 × 9.75 *dm plot with sea-anemones.*

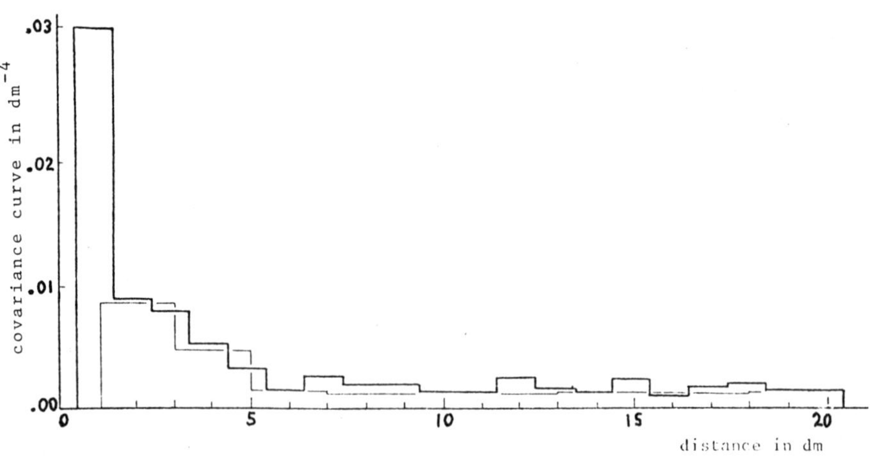


*FIG. 3: The estimated autocovariance curve from a* 70 × 70 *(thick line) and a* 35 × 35 *grid (thin line) covering the same* 100 × 75 *dm plot with glasswort plants.*

$$\hat{\text{Var}}(\hat{\gamma}_{k\ell}(r_s)) = N_k^3\{\Delta \text{ area}(A) \Sigma_x \text{cir}[C(\underset{\sim}{p}_{kx}, r_s)]\cap A\}^{-2}$$

$$\times \hat{\text{Var}}\{N[\ell;C(\underset{\sim}{p}_{kx}, r_s + \tfrac{1}{2}\Delta)\cap A] - N[\ell;C(\underset{\sim}{p}_{kx}, r_s - \tfrac{1}{2}\Delta)\cap A] .$$

When the covariance curve has to be estimated from $N$ grid counts, we have already seen that we only have to rescale the weight coefficients which maximize the Moran statistic, i.e.,

$$\hat{\gamma}_{k\ell}(r_s) = N^2\text{area}^{-2}(A)(a_s/n_s)\surd(m_{k2}m_{\ell 2}),$$

for $r_s = \frac{1}{2}m_{s-1} + \frac{1}{2}m_s$, and $m_{k2}$ is the estimated variance of the cell counts. An example of such an estimate is given in Figure 3. The estimated autocovariance curve is plotted for the glasswort data of Appendix A where the $m$'s are chosen as $m_s = \Delta s - \Delta/2$. The points were counted in a $35 \times 48$ grid and also in a $70 \times 96$ grid. We can conclude that there is some clustering within a short distance. Although there is a linear trend in the intensity, this will not have much influence on the covariance estimates, as the scale on which the interactions take place is small in comparison with the variation in intensity. The variance of the estimated covariance curve is estimated by

$$\hat{\text{Var}}(\hat{\gamma}_{k\ell}(r_s)) = n_s^{-2}N^4\text{area}^{-4}(A)m_{k2}m_{\ell 2}$$

$$\times \{\Sigma_{ij:d_{ij}\varepsilon c_s} x_{ki}^2 x_{\ell j}^2 - n_s^{-1}a_s^2\},$$

where, as before, the $x_{ki}$'s are the centered cell counts.

*3.3 The Variance Curve.* Traditionally, in plant ecology, patterns are studied by means of the Greig-Smith method (Greig-Smith, 1952). Originally, this was an analysis of grids of $2^m \times 2^n$ square cells. The method provided for the estimation of the variance of non-overlapping blocks of $2^{m-i} \times 2^{n-j}$ cells, where $i=1,\cdots,m$ and $j=1,\cdots,n$ were mostly chosen to increase simultaneously. The blocks were obtained by partitioning into increasingly larger blocks, keeping them as square as possible. Several authors have tried to reformulate the analysis, in order to make it possible to test randomness in the counts made in these blocks for a collection of choices of i and j (see Thompson, 1955; Mead, 1974; Zahl, 1974). For a further discussion see the papers of Diggle (1979) and Cormack (1979).

It is doubtful whether it would bother the ecologist very much that simultaneous inference is difficult although this is of theoretical interest. One can often avoid much work and difficulty by obtaining a new set of data by a nearby grid sample and performing a statistical test for some fixed block size, indicated by a previous hypothesis-generating analysis. The requirement that the blocks do not overlap in the Greig-Smith method, makes the resulting variance estimate highly dependent on the positioning of the grid. This holds especially for small grids. Small displacements can result in clusters falling within one block, instead of several. The requirement that the block size must have a power of 2 implies a waste of information stored in the pattern. Dropping both requirements, the variance for a square block of size $s$ is estimated by

$$V(s) = d^{-1} \sum_{i=1}^{1+m-s} \sum_{j=1}^{1+n-s} \left( \sum_{k,\ell=0}^{s-1} x_{i+k,j+\ell} \right)^2 - d^{-2} \left( \sum_{i=1}^{1+m-s} \sum_{j=1}^{1+n-s} \sum_{k,\ell=0}^{s-1} x_{i+k,j+\ell} \right)^2 ,$$

where $d = (1+m-s)(1+n-s)$ and $x_{ij}$ is the number of points in the cell of row $i$ and column $j$. There are $m$ rows and $n$ columns, $s=1,2,\cdots,\min(m,n)$.

We can express this statistic as the quadratic form, $V(s) = \Sigma_{i,j,k,\ell} w_{ijk\ell}(s) x_{ij} x_{k\ell}$, with $w_{ijk\ell}(s)$ given by

$$d^{-1} \times \max\{0, \min(1+m-s, 1+m-i, 1+m-k, s-i+k, s-k+i, i, k)\}$$
$$\times \max\{0, \min(1+n-s, 1+n-j, 1+n-\ell, s-j+\ell, s-\ell+j, j, \ell)\}$$
$$- d^{-2} \times \min(1+m-s, 1+m-i, s, i) \min(1+m-s, 1+m-k, s, k)$$
$$\min(1+n-s, 1+n-j, s, j) \min(1+n-s, 1+n-\ell, s, \ell) .$$

Thus $w_{ijk\ell} = w_{k\ell ij}$. The expectation of $V(s)$ in the case of randomly mingled cell contents is $EV(s) = mn(mn-1)^{-1} m_2 \Sigma_{ij} w_{ijij}$, with $m_2 = \frac{1}{mn} \Sigma_{ij} x_{ij}^2$, and in the case of IID cell contents is

$EV(s) = m_2 \Sigma_{ij} w_{ijij}$, where $m_2$ is the variance of the cell contents. The second moment of $V(s)$ is given in Table 3. As $V(s)$ does not change when a constant is added to the cell contents, $w_{ij++} = 0$. Therefore, only the first two columns of Table 5 contribute to the second moment. For large grids, the trace of $\underset{\sim}{W}$ can be approximated by $mns^2$. In this case, the first moment of $V(s)$ under both hypotheses, can be approximated by $s^2 \Sigma_{ij} x_{ij}^2$.

As an example, $V(s)$ is calculated for the barnacle data of Appendix A. The moments under the two null hypotheses differ very little in this case. For random mingled cell contents, they are given in Table 4.

If it is permissible to approximate the PDF of $V(s)$ under the null hypothesis by an appropriate normal distribution, we can conclude that the first three blocks show significant deviations from the null hypothesis of random mingling. For the autocovariance of these data, see Kooijman (1976). The interpretation of the variance curve deserves some care, as its relationship with the covariance curve is less obvious than in the one-dimensional

*TABLE 4: The variances* $V(s)$ *for block size* $s=2,\cdots,9$, *estimated from the counts of barnacles in the* $10 \times 10$ *grid in Figure 5 (see Appendix A). The mean and standard deviation are given in case of random mingled grid counts. The normalized values have zero mean and unit variance under random mingling.*

| Block size | Observed variance $V(s)$ | $E[V(s)]$ | $Var^{1/2}[V(s)]$ | $\frac{V(s)-E[V(s)]}{Var^{1/2}[V(s)]}$ |
|---|---|---|---|---|
| 2 | 34.19 | 16.93 | 3.05 | 5.66 |
| 3 | 82.61 | 35.43 | 10.94 | 4.31 |
| 4 | 119.74 | 55.25 | 23.68 | 2.72 |
| 5 | 108.81 | 69.37 | 35.82 | 1.10 |
| 6 | 65.39 | 73.69 | 42.01 | -.09 |
| 7 | 38.61 | 70.58 | 44.59 | -.20 |
| 8 | 63.78 | 59.49 | 43.21 | -.10 |
| 9 | 81.69 | 38.75 | 37.27 | 1.15 |

situation, i.e., for some plot A, $V(A) = \lambda\ \text{area}(A) + \int_0^\infty \gamma(r)\,d\eta(r)$ with $\eta(r) = \int_{\underset{\sim}{z} \in A} \text{area}[C(\underset{\sim}{z}, r) \cap A]\,dz$, whereas for the one-dimensional situation we have the relation $V(A) = \lambda R + 2\int_0^R (R-r)\gamma(r)dr$ for a time period A from 0 till R ( cf Cox and Lewis, 1966).

## ACKNOWLEDGEMENTS

The research for this paper has been done at the Institute of Theoretical Biology of the University of Leiden. For critically reading the manuscript I am obliged to Prof. Dr. K. Bakker, Prof. Dr. J. Fabius, Mrs. N. Garnier, Mrs. C. Hengeveld, Prof. Dr. M. Jeuken, Mr. W. van Laar, Mr. E. Meelis, and Mr. J. A. J. Metz. I thank Mr. D. C. Brandt for the assistance in sampling the glasswort data, Mr. A. Prins for performing the conversion of the glasswort coordinates on separate sheets into the global coordinates, and Mr. I. Zorge for his help in locating suitable plots of glasswort and sea-anemones. This study has been made possible by a grant from the Netherlands Organization for the Advancement of Pure Research.

## REFERENCES

Bartlett, M. S. (1960). *Statistical Population Models in Ecology and Epidemiology*. Methuen, London.

Bartlett, M. S. (1966). *An Introduction to Stochastic Processes*. *2nd* ed. Cambridge University Press, Cambridge.

Brillinger, D. R. (1976). Estimation of the second-order intensities of a bivariate stationary point process. *Journal of the Royal Statistical Society, Series B*, 38, 60-66.

Cliff, A. D. and Ord, J. K. (1969). The problem of spatial autocorrelation. In *Studies in Regional Science*, A. J. Scott, ed. Pion, London. 25-55.

Cliff, A. D. and Ord, J. K. (1970). Evaluating the percentage points of a spatial autocorrelation coefficient. *Geographical Analysis*, 3, 51-62.

Cliff, A. D. and Ord, J. K. (1971). A regression approach to univariate spatial forecasting. In *Regional Forecasting*, M. D. I. Chisholm, *et al.*, eds. Butterworth, London.

Cliff, A. D. and Ord, J. K. (1973). *Spatial Autocorrelation*. Pion, London.

Cormack, R. M. (1979). Spatial aspects of competition between individuals. In *Spatial and Temporal Analysis in Ecology*,

R. M. Cormack and J. K. Ords, eds. Satellite Program in Statistical Ecology, International Co-operative Publishing House, Fairland, Maryland.

Cox, D. R. and Lewis, P. A. W. (1966). *The Statistical Analysis of Series of Events.* Methuen, London.

Dacey, M. F. (1969). Some properties of a cluster point process. *The Canadian Geographer,* 13, 128-140.

Diggle, P. J. (1979). Statistical methods for spatial point pattern in ecology. In *Spatial and Temporal Analysis in Ecology,* R. M. Cormack and J. K. Ord, eds. Satellite Program in Statistical Ecology, International Co-operative Publishing House, Fairland, Maryland.

Diggle, P. J., Besag, J., and Gleaves, J. T. (1976). Statistical analysis of spatial point patterns by means of distance methods. *Biometrics,* 32, 659-667.

Eberhardt, L. L. (1967). Some developments in distance sampling. *Biometrics,* 23, 207-216.

Greig-Smith, P. (1952). The use of random and contiguous quadrats in the study of the structure of plant communities. *Annals of Botany, London N.S.,* 16, 293-316.

Gustavsson, J. and Svensson, A. (1976). A Poisson regression model applied to classes of road accidents with small frequencies. *Scandinavian Journal of Statistics,* 3, 49-60.

Kendall, M. G. and Moran, P. A. P. (1963). *Geometrical Probability.* Griffin, London.

Kooijman, S. A. L. M. (1976). Some remarks on the statistical analysis of grids especially with respect to ecology. *Annals of Systems Research,* 5, 113-132.

Matérn, B. (1960). Spatial variation. *Meddelanden från Statens Skogsforskningsinstitut,* 49, 5.

Mead, R. (1974). A test for spatial pattern at several scales using data from a grid of contiguous quadrats. *Biometrics,* 30, 295-307.

Neyman, J. and Scott, E. L. (1957). On a mathematical theory of populations conceived as conglomerations of clusters. In *Cold Spring Harbor Symposium on Quantitative Biology, Vol. XXII,* Population Studies: Animal Ecology and Demography. The Biology Laboratory, Cold Spring Harbor, Long Island, New York.

Ord, J. K. (1975). Estimation methods for models of spatial interaction. *Journal of the American Statistical Association*, 70, 120-126.

Pielou, E. C. (1962). The use of plant-to-neighbor distances for the detection of competition. *Journal of Ecology*, 50, 357-367.

Pielou, E. C. (1974). *Population and Community Ecology: Principles and Methods*. Gordon and Breach, New York.

Thompson, H. R. (1955). Spatial point processes with application to ecology. *Biometrika*, 42, 102-115.

Warren, W. G. (1971). The center-satellite concept as a basis for ecological sampling. In *Statistical Ecology, Vol. 2*, G. P. Patil, E. C. Pielou, and W. E. Waters, eds. The Pennsylvania State University Press, University Park. 87-118.

Whittle, P. (1954). On stationary processes in the plane. *Biometrika*, 41, 434-449.

Zahl, S. (1974). Application of the S-method to the analysis of spatial pattern. *Biometrics*, 30, 513-524.

## APPENDIX A

### A1. Glasswort, *Salicornia europaea*

In February 1974, the coordinates of 1,768 Glasswort plants were fixed in a $9.6 \times 7$ meter plot at 'De Kwade Hoek.' See Figure 4. This is a salt marsh west of Stellendam (Zuid-Holland, The Netherlands). So there is one plant per 3.8 $dm^2$. The plot was fairly water level. Glasswort was the only cormophyte in the plot. The coordinates have been fixed by means of a panthograph on $9 \times 12$ separate sheets. The sheets have been fixed relative to each other by means of 4 reference points at the corners of each sheet. Subsequently, the sum of the squared distances between the points corresponding to the same reference point has been minimized to ensure as accurate a representation as is possible of the plant locations.

### A2. Beadlet anemone, *Actinia equina*

In May 1976, the coordinates of 231 anemones were fixed in a $9.75 \times 14.5$ dm plot at Quiberon (Bretagne). See Figure 5. The plot is a part of the north face of a boulder well above

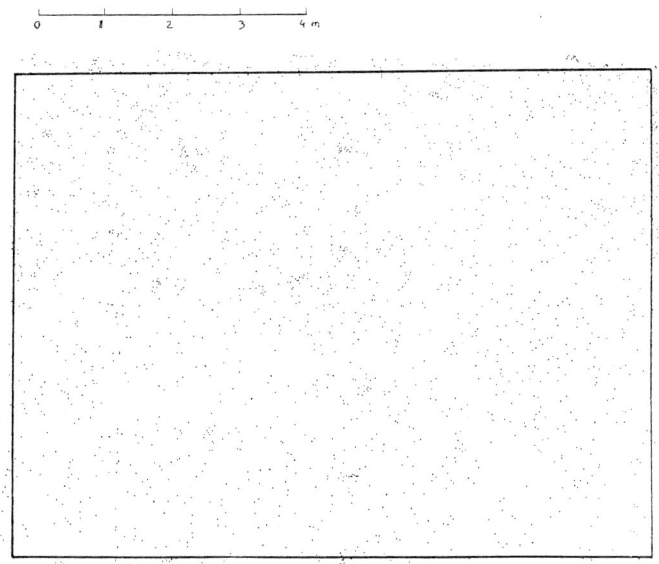


*FIG. 4: Glasswort specimens* (Salicornia europaea) *in a plot.*

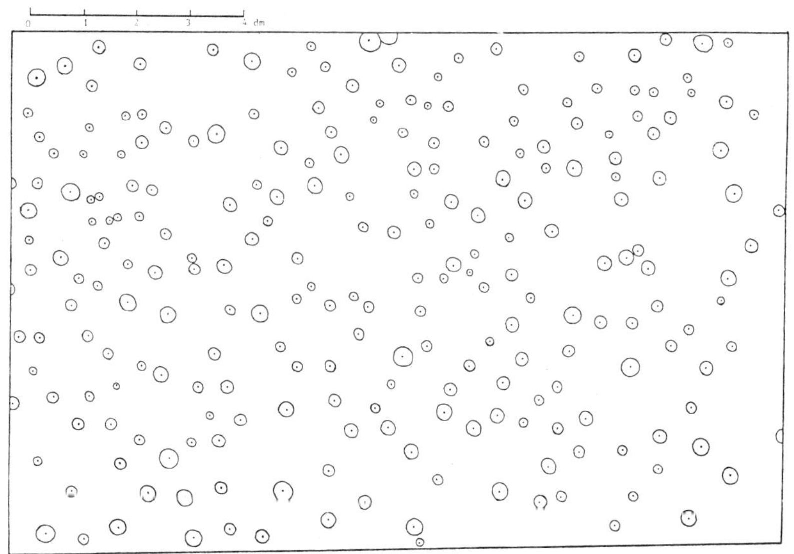


*FIG. 5: Beadlet anemones* (Actinia equina) *in a plot.*

low-tide level. Beside the anemones, star barnacles (*Chthamalus stellatus*) were abundant.

A3. *Acorn barnacle,* Balanus balanoides

In December 1974, the 166 animals were counted in a grid of 10 × 10 cells on a cutter, put in dock at Scheveningen (Zuid-Holland, The Netherlands). See Figure 6. The gridded plot is part of a strip of .75 meter width, normally just below water level, abundantly occupied by barnacles.

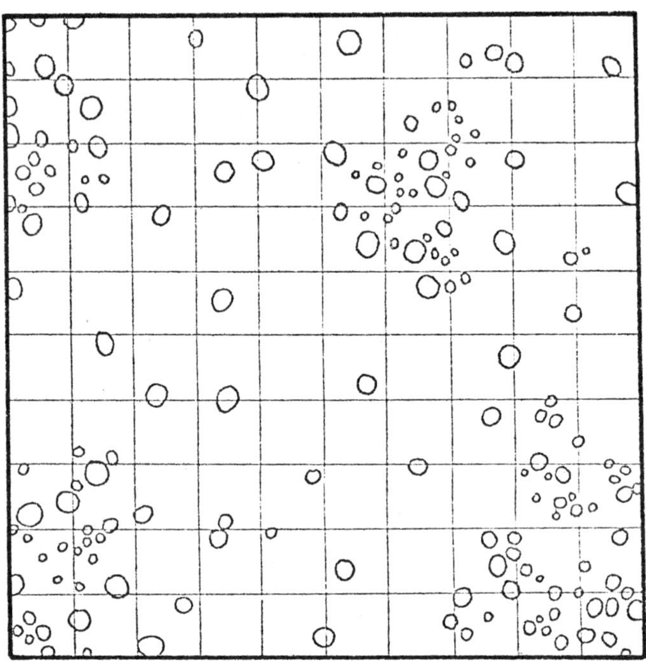

*FIG. 6: Acorn barnacles* (Balanus balanoides) *in a plot.*

## APPENDIX B
## COMPUTER PROGRAM IN APL FOR THE ESTIMATION OF THE COVARIANCE CURVE FROM A LIST OF COORDINATES IN A PLOT

```
∇ISOCORF[⎕]∇

    ∇ C←P ISOCORF X;N;I;N;R;GR;Z;K;B

[1]   C←H←0×GR←(÷B←2×K÷⌊/P)×¯0.5+R←⍳K←⌊0.25×N←(I←1)↑⍴X
[2]  L:C←C++⌿(⌈B×(+/(X-(N,2)⍴X[I;])*2)*0.5)∘.=R
```

```
[3]  H←H-(○0.5×2=Z)+(1+3=Z←+⌿Z=1)×+/¯2○Z←1⌊(X[I;],P-X[I;])∘.÷CR
[4]  →(N≥I←I+1)/L
[5]  C←GR̄,[1.5](Z÷○2×GR)×(B×C×○2÷H+○2×N)-GR×○2×Z←N÷×/P

     ∇
```

*Input*

Left argument: a vector of length 2 representing the coordinates of the right upper point of the plot. The lower left point is assumed to be on (0,0).

Right argument: A N × 2 matrix with the point coordinates.

*Output*

Matrix with first column being the abscissa values and second column being the estimates of the isotropic covariance curve.

[*Received: July* 1978. *Revised: January* 1979]

R. M. Cormack and J. K. Ord, (eds).,
*Spatial and Temporal Analysis in Ecology,* pp. 333-346. 

# THE ANALYSIS OF SPATIAL PATTERNS OF SOME GROUND BEETLES (COL. CARABIDAE)

R. HENGEVELD

Department of Geobotany
Catholic University
Toernooiveld
Nijmegen, The Netherlands

SUMMARY. A non-linear model was applied to data from a meadow in The Netherlands in order to describe the spatial distribution of Carabid beetles in causal terms. An optimum relationship between the catches of the beetles and an environmental variable has been modelled by a Gaussian curve. The response curves fitted for nine species are given for the catches of two monthly collecting periods. Furthermore, attempts are made to identify the possible causative variable by comparison of the trend-surfaces deduced from the factor-estimates of the sampling points, and also from a linear fit of those factor-estimates with the height of the sampling points. The paper concentrates on the methodological aspects of spatial analysis.

KEY WORDS. spatial distribution, non-linear relationship, Gaussian curve, trend-surface analysis, methodology, Carabid beetles.

## 1. INTRODUCTION

Geographers often refer to Greig-Smith's (1957) book to indicate that in botany the development of spatial analysis had a relatively early start. Yet, with the publication of that early book this development did not really get into full swing, except for the advances made by relatively few workers such as Pielou (1969, 1977).

In their books, both Greig-Smith and Pielou concentrate on the description and analysis of spatial patterns and the association between species. However, this approach is only one of several which it is possible to adopt.

For example, Curtis' Wisconsin School (cf. Curtis, 1959) tried to explain the numerical abundance of plants using the value of an environmental variable which varies monotonically along a line transect, an idea developed further by Whittaker (1967) who called the method Gradient Analysis. In this, the ordination of sampling plots is still done subjectively using previous knowledge, both of the nature of the variable and of the existence of a gradient. In a further development of this analysis, Gauch, Chase, and Whittaker (1974) fitted a Gaussian curve to the points resulting from their ordered observations, but I feel that this procedure is not statistically sound.

Much earlier, Gause (1930) and others (e.g., Braun-Blanquet, 1928), also stated that a Gaussian curve could be applied as a model for describing reactions of a species to its environment. The adoption of this model for a non-linear reaction of species to environmental variables in the form of an optimum curve (cf. Allee, *et al.*, 1949) was especially important for Gause, since it enabled him to measure the species' *plasticity*, i.e., tolerance, and also to estimate the different optimal environmental conditions for two or more species from the numerical responses. This latter idea, that of the estimation of the optima for the different species, seems later to have become lost in Gradient Analysis. However, in other respects Gause's approach is similar to that of Curtis and Whittaker.

Still another approach has been adopted by Gittins (1968), who calculated higher-order spatial polynomial trends by applying a trend-surface analysis on factor-scores from a Principal Components Analysis on plant data by Goodall (1954). This is less subjective than Gradient Analysis, but the species responses to environmental variables are still described by a linear model. The consequence of this is usually a poor fit to the data. Also it is not possible to estimate the tolerance of the species to the hypothetical causative environmental variables.

In the preceding paragraphs I have covered only some of the important ideas occurring in the literature, rather than painstakingly making a complete review of the literature on spatial analysis in ecology. To illustrate these ideas and their drawbacks, I shall apply them to the analysis of the distribution of a ground beetle species in a meadow, sampled at 252 points in a rectangular $12 \times 21$ grid system (Figure 1). Table 1 shows the numbers of this species, *Dyschirius globosus* (Hbst.), caught at different points in this field totalled over ten fortnightly collections during 1976.

To describe this distribution pattern we can either draw contour lines for a number of density classes over the plan of collecting points in the field, or plot the cumulative percentages of the 21 sampling sites for each of the 12 columns on normal probability paper (Figure 2). The plan with the contour lines only gives a picture of the actual situation, but there is no attempt to do this in terms of a causative variable. Conversely, a representation of the cumulative percentages on probability paper does suggest a numerical response of this species to some variable. If the intensities of the environmental variable are assumed to show a linear gradient in that part of the field, then Figure 2 suggests that the response is non-linear.

## 2. DESCRIPTION OF THE STUDY

In pilot studies such as those mentioned, it would be more appropriate to make suggestions about causative variables as a result of preliminary investigations, rather than unfounded statements about them beforehand. However, the assumption of a non-linear numerical response with respect to a causative environmental variable, like that of Gause, seems quite reasonable. In this way the aim of our research was to give a description of spatial pattern in terms of the hypothetical environmental variables.

To begin with, as is the case in any analysis, we distinguish between the descriptive and causal aspects. The descriptive aspect concentrates on questions of whether or not spatial patterns are homogeneous, and on the estimation of the spatio-temporal scale of such patterns. The choice of scale at which sampling should be carried out is, initially, a matter of subjective guess-work. However, this scale can be adjusted properly after examination of the results of pilot experiments. The causal aspects concentrate on the explanation of the observed patterns by those variables which vary on a comparable spatial scale in the area under study.

With all this in mind, a program was set up to investigate the distribution of some ground beetle species in 40 hectares of meadowland in one of the three newly reclaimed IJsselmeer polders in The Netherlands (Figure 1). Approximately one-third of the field was drained, and a ditch runs down the length of the field. The field was sampled at 252 sites with 40 meters between sites in each direction. Five pitfall traps were placed at each site with five meters between traps. The nature of the pitfall traps and their spatial pattern at a site are illustrated in Figure 1. The distance between any two sampling sites in this field was assumed to be sufficient to consider the samples as independent. It is even possible that the distance between the

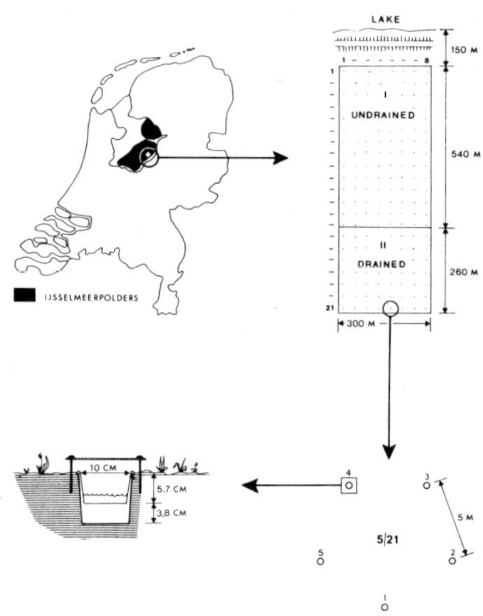


*FIG. 1: Location of the sampling area in The Netherlands, and diagram of the grid system of* 8 × 21 *sampling sites (in some cases extended to* 12 × 21*) each consisting of* 5 *sampling devices. The area has an undrained* (I) *and a drained* (II) *part, and is separated from a large inland lake by a dike and a hedgerow.*

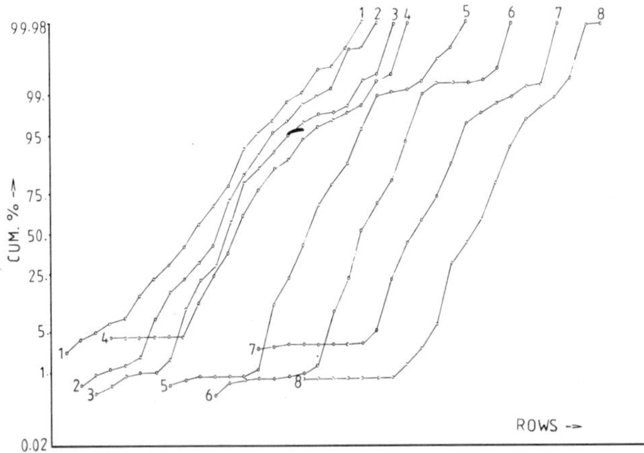


*FIG. 2: Cumulative percentages of the numbers of individuals of* **Dyschirius globosus** *collected at each sampling site in the Ecological Reserve in* 1976*, represented on probability paper. The* 8 *curves represent the catches in columns* 1 *to* 8 *of Table 1, the* 21 *points representing the rows. Row* 1 *is represented by the lowest points of the curves.*

*TABLE 1: Total numbers of individuals of* Dyschirius globosus *caught at each sampling site in* 1976.

| | 1 | 2 | 3 | 4 | 5 | 6 | 7 | 8 | 9 | 10 | 11 | 12 |
|---|---|---|---|---|---|---|---|---|---|---|---|---|
| 1 | 16 | 5 | 4 | 39 | 6 | 4 | 20 | 2 | 10 | 3 | 0 | 1 |
| 2 | 10 | 3 | 2 | 3 | 2 | 3 | 3 | 0 | 0 | 0 | 0 | 0 |
| 3 | 7 | 2 | 3 | 1 | 1 | 1 | 1 | 0 | 0 | 1 | 0 | 0 |
| 4 | 14 | 1 | 1 | 1 | 0 | 0 | 0 | 0 | 0 | 0 | 0 | 0 |
| 5 | 5 | 6 | 0 | 0 | 0 | 0 | 0 | 0 | 0 | 0 | 0 | 0 |
| 6 | 52 | 51 | 10 | 1 | 0 | 1 | 0 | 0 | 0 | 0 | 0 | 0 |
| 7 | 58 | 78 | 92 | 70 | 4 | 1 | 0 | 0 | 0 | 1 | 0 | 0 |
| 8 | 53 | 58 | 137 | 154 | 125 | 5 | 1 | 2 | 0 | 1 | 0 | 0 |
| 9 | 69 | 76 | 75 | 90 | 142 | 87 | 16 | 4 | 2 | 0 | 0 | 0 |
| 10 | 106 | 101 | 321 | 228 | 235 | 131 | 138 | 10 | 0 | 0 | 0 | 0 |
| 11 | 91 | 262 | 255 | 140 | 334 | 309 | 176 | 70 | 1 | 0 | 0 | 0 |
| 12 | 82 | 108 | 68 | 99 | 136 | 195 | 120 | 37 | 20 | 0 | 0 | 0 |
| 13 | 87 | 65 | 56 | 32 | 110 | 119 | 116 | 44 | 0 | 0 | 1 | 0 |
| 14 | 26 | 43 | 43 | 45 | 111 | 142 | 114 | 62 | 12 | 0 | 0 | 0 |
| 15 | 10 | 12 | 16 | 19 | 37 | 49 | 64 | 33 | 12 | 0 | 0 | 0 |
| 16 | 10 | 15 | 11 | 8 | 1 | 5 | 8 | 12 | 12 | 4 | 0 | 0 |
| 17 | 3 | 3 | 2 | 5 | 2 | 0 | 6 | 3 | 5 | 1 | 2 | 0 |
| 18 | 5 | 4 | 5 | 7 | 4 | 0 | 2 | 2 | 2 | 0 | 2 | 6 |
| 19 | 0 | 6 | 12 | 7 | 5 | 1 | 3 | 2 | 0 | 0 | 0 | 0 |
| 20 | 1 | 0 | 1 | 1 | 1 | 2 | 0 | 1 | 5 | 0 | 1 | 0 |
| 21 | 1 | 1 | 5 | 4 | 1 | 3 | 5 | 0 | 6 | 4 | 1 | 5 |

*TABLE 2: Total numbers of individuals of* Pterostichus coerulescens *caught at each sampling site during the ten collecting periods from April to September* 1976. *The line between columns* 8 *and* 9 *represents the ditch which runs through the field.*

| | 1 | 2 | 3 | 4 | 5 | 6 | 7 | 8 | 9 | 10 | 11 | 12 |
|---|---|---|---|---|---|---|---|---|---|---|---|---|
| 1 | 173 | 275 | 137 | 74 | 109 | 81 | 83 | 19 | 61 | 55 | 35 | 8 |
| 2 | 169 | 104 | 88 | 75 | 64 | 53 | 54 | 21 | 38 | 29 | 32 | 16 |
| 3 | 185 | 207 | 153 | 59 | 43 | 33 | 36 | 8 | 22 | 25 | 23 | 4 |
| 4 | 242 | 224 | 81 | 68 | 18 | 26 | 10 | 12 | 39 | 11 | 8 | 2 |
| 5 | 401 | 354 | 53 | 43 | 17 | 13 | 10 | 8 | 4 | 3 | 5 | 3 |
| 6 | 307 | 269 | 291 | 80 | 16 | 10 | 12 | 11 | 3 | 1 | 1 | 1 |
| 7 | 363 | 362 | 332 | 465 | 100 | 32 | 22 | 20 | 6 | 15 | 9 | 6 |
| 8 | 390 | 596 | 345 | 264 | 297 | 104 | 25 | 49 | 22 | 23 | 38 | 13 |
| 9 | 352 | 615 | 313 | 225 | 220 | 216 | 74 | 36 | 12 | 17 | 6 | 11 |
| 10 | 267 | 423 | 277 | 301 | 187 | 269 | 213 | 131 | 23 | 9 | 7 | 1 |
| 11 | 367 | 380 | 273 | 272 | 286 | 321 | 270 | 167 | 51 | 3 | 1 | 0 |
| 12 | 307 | 349 | 326 | 296 | 195 | 352 | 342 | 163 | 191 | 1 | 5 | 0 |
| 13 | 699 | 458 | 448 | 265 | 240 | 429 | 395 | 310 | 181 | 17 | 1 | 0 |
| 14 | 369 | 499 | 508 | 592 | 356 | 332 | 409 | 348 | 184 | 9 | 0 | 0 |
| 15 | 435 | 662 | 487 | 566 | 499 | 441 | 369 | 405 | 307 | 59 | 1 | 0 |
| 16 | 843 | 753 | 717 | 672 | 412 | 443 | 520 | 375 | 242 | 147 | 16 | 3 |
| 17 | 974 | 1262 | 800 | 990 | 416 | 601 | 471 | 545 | 493 | 256 | 142 | 28 |
| 18 | 866 | 604 | 799 | 837 | 454 | 490 | 643 | 460 | 281 | 153 | 215 | 181 |
| 19 | 908 | 710 | 668 | 793 | 508 | 466 | 528 | 400 | 299 | 270 | 219 | 217 |
| 20 | 629 | 651 | 765 | 641 | 474 | 776 | 516 | 720 | 322 | 359 | 273 | 152 |
| 21 | 660 | 963 | 1414 | 887 | 970 | 664 | 801 | 650 | 297 | 363 | 345 | 140 |

traps within a sampling site was great enough to allow consideration of the individual trap-counts as independent. In the first year, 1974, the beetles were collected twice, each collection taking place after a catching period of one month, at the end of August and of September, respectively. In these initial experiments, the non-homogeneity of the patterns and the numbers of individuals caught were such that in the main experiments, which were carried out in the next two years, it seemed feasible to have catching periods of two weeks' duration and to put samples from the five traps per site together. Pooling the catches of the five traps should reduce the sampling variability.

Ground beetles were chosen for this study because they have a number of properties which are useful for spatial analysis. Firstly, these beetles are easy to identify because they are quite large insects, and as a family their systematics have been well worked out. It is also useful that these beetles are mainly surface dwellers, which reduces the sampling space to two dimensions, making ecological analysis relatively simple. It also makes sampling with pitfall traps possible, reducing to zero the possible contamination of the observations through human interference at the time of trapping. This delivers data of particular ecological interest, for they can be interpreted as a measure of the beetles' chance of meeting something else (cf. Heydemann, 1953; Thiele, 1977), be this an individual of the other sex, a predator, food, a more favorable site, etc. Therefore, sampling Carabid beetles with pitfall traps results in a better population description than, for example, the mere abundances of animal populations can offer. Another property which makes ground beetles useful for spatial analysis is their great mobility. This makes successive samples independent, so that the analysis of the dynamics of spatial pattern is relatively easy. Furthermore, their great mobility reduces the time delay between a change in the environment and the response to the change reflected in the catches. Thus, it is feasible to make assumptions about the probability distribution of the samples, about the spatial patterns (in terms of the first or second moments of a point process), as well as about the kind of response, linear or nonlinear, to environmental variables. If plants or animals of low mobility are used, the delay between the change in the intensity of the environmental variable and the response of a species to it will highly complicate the analysis, especially when differences in the spatial distributions between succeeding generations also come into play. The formulation of a model describing a spatial pattern in causal terms will therefore be exceedingly difficult in such cases.

## 3. DESCRIPTION OF THE DATA

Taking the results for an individual species from these initial experiments, Table 1 gives a picture of the distribution of *Dyschirius globosus*. This species shows a clear preference for a particular area in the center of the field. Its numbers are highest in the center of this local range of distribution and taper regularly towards the margin. The location of this species as well as its spatial dispersion seem to be the same both within and between the years.

The distribution of *Pterostichus coerulescens* (L.) lies mainly in one corner of the field and varies during the year. In this area a high concentration of beetles is found in spring (Table 2) and this is both preceded and followed by a more dispersed distribution over the field. In autumn the callow, or young, beetles also occur in this part of the field. This temporary concentration possibly indicates that this is a reproduction area into which old imagines migrate in early spring and from which they migrate after reproduction. The evident drop in the numbers of callow beetles just over the ditch which runs through the field, also indicates a movement away from this possible reproduction area (Table 3).

*TABLE 3: Total numbers of individuals of* Pterostichus coerulescens *caught at each sampling site during three collecting periods in September and October* 1976.

| | 1 | 2 | 3 | 4 | 5 | 6 | 7 | 8 | 9 | 10 | 11 | 12 |
|---|---|---|---|---|---|---|---|---|---|---|---|---|
| 1 | 112 | 132 | 32 | 42 | 66 | 43 | 25 | 28 | 8 | 12 | 16 | 9 |
| 2 | 79 | 58 | 36 | 31 | 31 | 30 | 12 | 37 | 9 | 6 | 6 | 5 |
| 3 | 144 | 182 | 49 | 110 | 46 | 20 | 14 | 18 | 2 | 3 | 13 | 4 |
| 4 | 223 | 319 | 59 | 34 | 46 | 27 | 18 | 17 | 2 | 1 | 2 | 1 |
| 5 | 430 | 414 | 267 | 104 | 67 | 25 | 13 | 20 | 1 | 0 | 0 | 0 |
| 6 | 442 | 201 | 449 | 156 | 90 | 27 | 18 | 40 | 0 | 0 | 0 | 1 |
| 7 | 330 | 427 | 316 | 259 | 155 | 102 | 74 | 90 | 0 | 1 | 0 | 0 |
| 8 | 651 | 440 | 501 | 348 | 201 | 106 | 103 | 73 | 1 | 3 | 3 | 1 |
| 9 | 290 | 433 | 504 | 137 | 180 | 170 | 153 | 144 | 7 | 3 | 2 | 6 |
| 10 | 149 | 265 | 116 | 114 | 109 | 148 | 70 | 372 | 1 | 4 | 0 | 0 |
| 11 | 231 | 233 | 317 | 266 | 405 | 625 | 258 | 219 | 4 | 1 | 0 | 0 |
| 12 | 825 | 558 | 582 | 485 | 745 | 467 | 279 | 414 | 18 | 2 | 2 | 0 |
| 13 | 947 | 734 | 842 | 833 | 579 | 649 | 594 | 713 | 9 | 4 | 0 | 0 |
| 14 | 1063 | 897 | 1018 | 800 | 716 | 530 | 596 | 1092 | 27 | 11 | 0 | 0 |
| 15 | 1017 | 1130 | 1241 | 1385 | 1219 | 964 | 416 | 928 | 51 | 23 | 4 | 2 |
| 16 | 1356 | 1444 | 1157 | 1473 | 1272 | 1070 | 652 | 768 | 38 | 30 | 13 | 3 |
| 17 | 1653 | 1578 | 1637 | 1045 | 1121 | 1368 | 1489 | 784 | 55 | 83 | 68 | 22 |
| 18 | 1357 | 1809 | 1729 | 1609 | 1232 | 1318 | 973 | 665 | 114 | 59 | 210 | 86 |
| 19 | 1480 | 1031 | 919 | 1128 | 714 | 714 | 800 | 775 | 240 | 89 | 136 | 99 |
| 20 | 847 | 1468 | 1266 | 972 | 1664 | 1189 | 765 | 1354 | 135 | 249 | 207 | 140 |
| 21 | 1358 | 1847 | 2518 | 1706 | 1283 | 829 | 625 | 177 | 129 | 210 | 208 | 107 |

Other distribution patterns are displayed by the catches of other species. The spatial dispersion of congeneric species resemble each other more than do those of species of different genera. For most species these patterns are stable from year to year.

## 4. ANALYSIS OF THE DATA

For the causal approach, a hypothesis generating procedure was used which Kooijman (1977) developed for studies of this kind. It is assumed that arrivals at the sampling devices follow a Poisson distribution. The arrival rate is thought to be specific for each device and depends upon the hypothetical environmental variable or factor. Also, the rate for each species depends upon its commonness and upon the optimum conditions and the degree of tolerance with respect to the hypothetical variable. The postulated model for $X_{ij}$, the count for the *jth* species at the *ith* sampling site, is

$$X_{ij} = \mathrm{Poi}[a_j e^{-b_j(f_i-\mu_j)^2}], \text{ where}$$

$a_j$ is the commonness, or relative abundance, of the *jth* species;
$\mu_j$ denotes the optimum level of the hypothetical variable for the *jth* species;
$b_j$ is the inverse of the tolerance; and
$f_i$ describes the value of the hypothetical variable at the *ith* site.

Therefore, if there are $R$ sites and $K$ species, we have an $R \times K$ 'abundance' matrix of observations, and there are $R + 3K$ parameters to be estimated from the data. Computer programs have been written which provide maximum likelihood estimators (see Kooijman, 1977). These estimates are calculated according to the Newton-Raphson procedure, starting from initial guesses for values of a hypothetical variable, obtained from a variant on principal components analysis. It was possible to show that the values must be close to the first principal component, if it is applied to the doubly centered covariance matrix of the logarithm of the catches. This idea has been generalized to two hypothetical environmental variables (Kooijman, 1977).

Figure 3 shows the individual points and the curves which express the expected numerical response of *Dyschirius globosus* as a function of a hypothetical variable for the two collecting periods. The factor values are adjusted so that the optimum appears

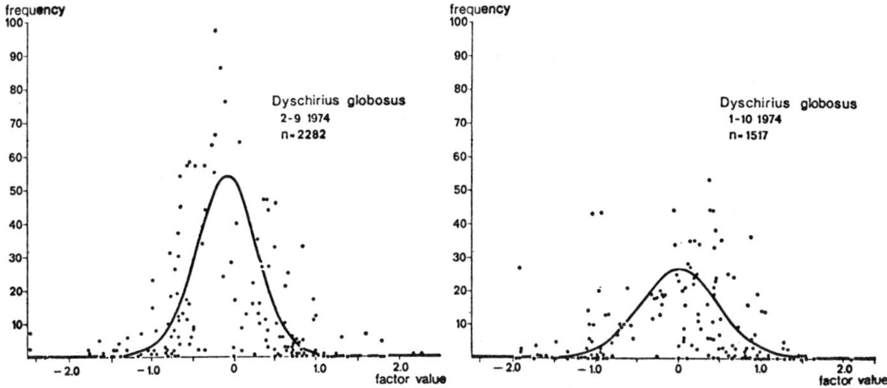


*FIG. 3: Graph of* Dyschirius globosus *in the two sampling periods, showing curves with a Gaussian distribution for the expected values of the number of a given species at each sampling point. The dots represent the actual number of individuals collected. x axis: estimated factor value of the sampling site; y axis: expected or actual number of individuals.*

at zero. Nine fairly abundant species were observed in all. The species and their abundances are given in Table 4. Response curves for the common species are given in Figure 4 for the same collecting periods. Also, the curves are given separately for each species and in more detail in Kooijman and Hengeveld (1979).

Figure 5 shows, for the two collecting periods respectively, the estimated values of the hypothetical causative variable calculated from the frequencies of the nine most common species. There is a clear pattern in these values, and this has been made easier to read by applying a trend-surface analysis to these estimates. It will be clear from these two figures that the gradient is not a simple monotonic function of distance, which is one of the assumptions made (unintentionally) in the subjective judgment of Gradient Analysis. Further, it is clear that the two patterns resemble each other, but that they are not identical. This dissimilarity indicates that the spatial pattern of the hypothetical causative environmental variable varies in time, although it may be linked to some constant factor. Figure 6 shows a good fit between the relative heights of the pitfalls as a constant factor and the estimated values of the variables. There is also a good fit between these values and the lutum content of the soil at each site (Kooijman and Hengeveld, 1979).

*TABLE 4: Total number of specimens per species collected during two one-month periods in* 1974 *in the Ellerslenk Ecological Reserve.* Period A: 1-8 *to* 2-9; Period B: 2-9 *to* 1-10.

| Species | Code letters | Period A | Period B |
|---|---|---|---|
| *Amara communis* | -- | 70 | 71 |
| *Amara familiaris* | -- | 141 | 32 |
| *Bembidion properans* | Bp | 100 | 253 |
| *Clivina fossor* | -- | 194 | 23 |
| *Clivina collaris* | -- | 70 | 21 |
| *Dyschirius globosus* | Dg | 2282 | 1517 |
| *Pterostichus coerulescens* | Pco | 6576 | 6832 |
| *Pterostichus cupreus* | Pcu | 649 | 1115 |
| *Trechus obtusus* | To | 1449 | 2020 |

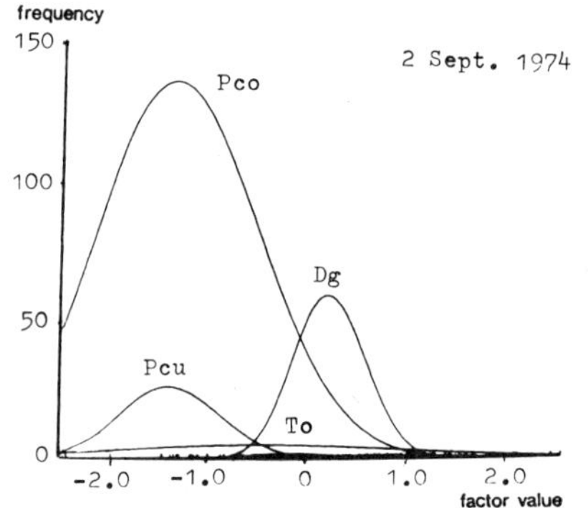


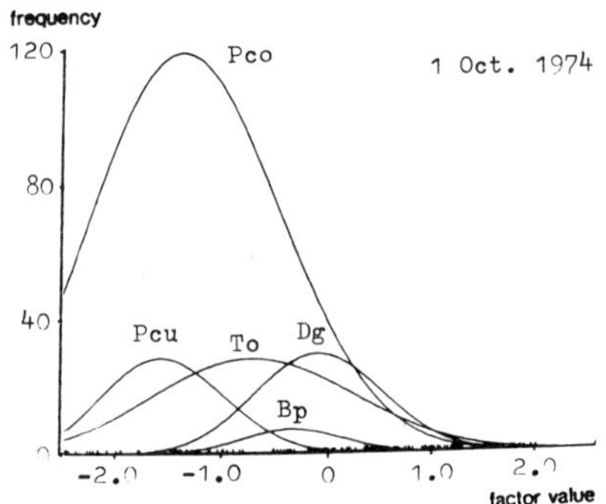


*FIG. 4: Graphs of* 4 *and* 5 *species from the two sampling periods respectively, showing curves with a Gaussian distribution for the expected values of the number of a given species at each sampling point. Of the nine species on which the calculations were based, the frequencies of the remaining* 5 *and* 4 *species respectively were too low to be represented on this scale. See Kooijman and Hengeveld* (1979) *for these curves and for the actual frequencies of the catches.*

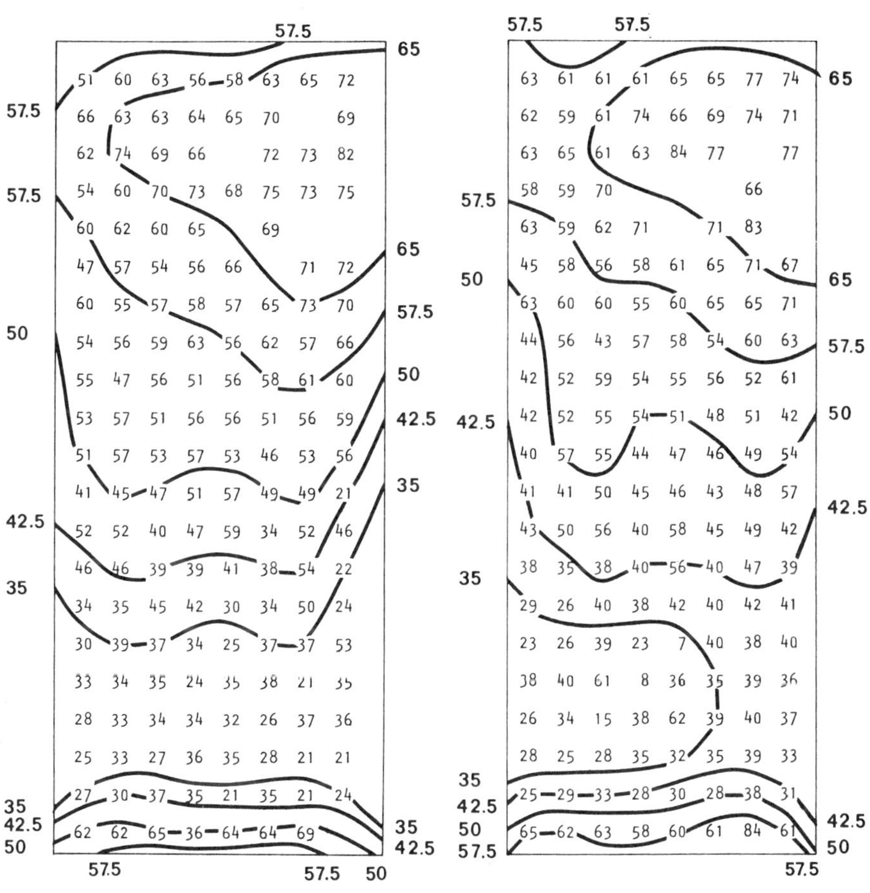


*FIG. 5: Factor values estimated per sampling point, with the two-dimensional 5th-degree polynomial fitted to these estimates. Seven points were omitted in these two cases because none of the species had been collected there. The sampling dates were: 2nd September, 1974 and 1st October 1974.*

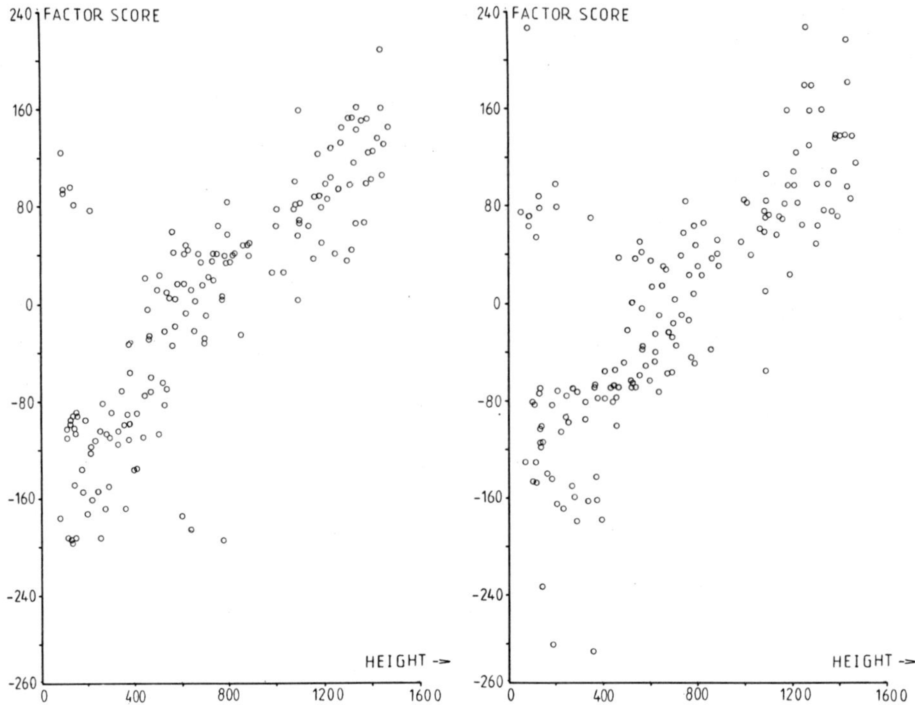


*FIG. 6: Regression of the factor values from Figure 5 plotted against the relative heights of the sampling sites.*

The variable factor which is related to both relative height and lutum content could be the moisture content of the top layer of the soil. Therefore this factor was estimated in 1976, together with the penetrability of this layer. These results have still to be evaluated.

Next autumn, laboratory experiments are planned to test the relationship between moisture content and the preference of these nine species, in order to close the empirical cycle. The research program will be continued by studying other phenomena shown by these species in this field. Among these phenomena are the spatial distribution of wing forms of *Trechus obtusus* Er., and the dynamics of the dispersion pattern in relation to reproduction as, for example, in *Pterostichus coerulescens*.

## 5. CONCLUSIONS

The result of this investigation is that we can describe spatial non-homogeneity for the different species. The general form of the non-homogeneous patterns is stable from year to year during the period of investigation, and probably it is determined by a variable factor such as the moisture content of the soil. These ground beetle species are assumed to react in a non-linear way to the different values of this factor which are estimated and described by means of a fifth order polynomial surface.

The optimum of these non-linear numerical responses with regard to the hypothetical causative factor, as well as their relative tolerances, can be calculated for the different species. Furthermore, the results of another research project show that not only the optima of the numerical responses vary over different parts of the species' geographical range, but that the same is especially true for the tolerances of the species (Hengeveld and Haeck, 1979).

The spread of the observations around the estimated Gaussian curve still shows a considerable scatter, as is quite usual when more complicated models are applied to actual observations. Therefore, not too much weight should be given to goodness of fit tests. The only function of models of this kind is to provide suggestions for further experiments.

## ACKNOWLEDGMENTS

I wish to thank Dr. S. A. L. M. Kooijman for making many improvements in the text, and my wife Claire for correcting my English.

## REFERENCES

Allee, W. C., Emerson, A. E., Park, O., Park, T., and Schmidt, K. P. (1949). *Principles of Animal Ecology*. Saunders, Philadelphia.

Braun-Blanquet, J. (1928). *Pflanzensoziologie*. Springer, Berlin.

Curtis, J. T. (1959). *The Vegetation of Wisconsin*. University of Wisconsin Press, Madison.

Gauch, H. C., Chase, C. B., and Whittaker, R. H. (1974). Ordination of vegetation samples by Gaussian distributions. *Ecology*, 55, 1382-1390.

Gause, G. F. (1930). Studies on the ecology of Orthoptera. *Ecology*, 11, 307-325.

Greig-Smith, P. (1957). *Quantitative Plant Ecology*. Butterworth, London.

Gittins, R. (1968). Trend-surface analysis of ecological data. *Journal of Ecology*, 56, 845-869.

Goodall, D. W. (1954). Objective methods for the classification of vegetation: III. An essay on the use of factor analysis. *Australian Journal of Botany*, 2, 304-324.

Hengeveld, R. and Haeck, J. (1979). The distribution of abundance, a biogeographical and ecological study (in preparation).

Heydemann, B. (1953). *Agraroekologische Problematik*. Thesis, University of Kiel.

Kooijman, S. A. L. M. (1977). Species abundance with optimum relations to environmental variables. *Annals of Systems Research*, 6, 123-138.

Kooijman, S. A. L. M. and Hengeveld, R. (1979). The description of a non-linear relationship between some carabid beetles and environmental factors (in preparation).

Pielou, E. C. (1969). *An Introduction to Mathematical Ecology*. Wiley, New York.

Pielou, E. C. (1977). *Mathematical Ecology*. Wiley, New York.

Thiele, H. U. (1977). *Carabid Beetles In Their Environment*. Springer-Verlag, Berlin.

Whittaker, R. H. (1967). Gradient analysis of vegetation. *Biological Reviews*, 42, 207-264.

[*Received:* *October* 1978. *Revised:* *April* 1979]

AUTHOR INDEX

# SUBJECT INDEX

## INTERNATIONAL STATISTICAL ECOLOGY PROGRAM

The International Statistical Ecology Program (ISEP) consists of the activities of the Statistical Ecology Section of the International Association for Ecology and of the Liaison Committee on Statistical Ecology of the International Statistical Institute, the Biometric Society, and the International Association for Ecology. The ISEP is a non-profit program formulated to serve the needs of interdisciplinary research and training in the newly emerging fields of Statistical Ecology and Ecological Statistics.

## SATELLITE PROGRAM IN STATISTICAL ECOLOGY

The Second International Congress of Ecology was held in Jerusalem during September 1978. In this connection, ISEP organized a Satellite Program in Statistical Ecology during 1977 and 1978. The emphasis was on research, review, and exposition concerned with the interface between quantitative ecology and relevant quantitative methods. Both theory and application of ecology and ecometrics received attention. The Satellite Program consisted of instructional coursework, seminar series, thematic research conferences, and collaborative research workshops.

Research papers and research-review-expositions were specially prepared for the program by concerned experts and expositors. These materials have been refereed and revised, and are now available in a series of ten edited volumes listed on page ii of this volume.

The Satellite Program takes as its theme the better melding of fundamental ecological concepts with rigorous empirical quantification. The overall result should be progress toward a stronger body of general ecologic and ecometric theory and practice.

## FUTURE DIRECTIONS

The satellite-like-programs help create and sustain enthusiasm, inward strength, and working efficiency of those who desire to meet a contemporary social need in the form of some interdisciplinary work. It should be only proper and rewarding for everyone involved that such programs are planned from time to time.

Plans are being made for a satellite program in conjunction with the next Biennial Conference of the International Statistical Institute and the next International Congress of Ecology. Care should be exercised that the next program not become a mere replica of the present one, however successful it has been. Instead, the next program should be organized so that it helps further the evolution of statistical ecology as a productive field.

The next program is being discussed in terms of subject area groups. Each subject group is to have a coordinator assisted by small committees, such as a program committee, a research committee, an annual review committee, a journal committee, and an education committee. This approach is expected to respond to the need for a journal on statistical ecology, and also to the need of bringing out well planned annual review volumes. The education committee would formulate plans for timely modules and monographs. Interested readers may feel free to communicate their ideas and interests to those involved in planning the next program. The mailing address is: International Statistical Ecology Program, P. O. Box 218, State College, PA 16801, USA.